Frontiers in Clinical Drug Research: Anti-Infectives
(Volume 1)

Editor

Atta-ur-Rahman, FRS

Kings College
University of Cambridge
Cambridge
UK

CONTENTS

CHAPTERS

PREFACE

The treatment of infectious diseases has come a long way since the pioneering days of Anton von Leeuwenhoek. Significant milestones were made by the discoveries of Pasteur and Kock, followed much later by the discovery of penicillin by Fleming and the development of vaccines by Jenner, Salk, and Sabin. The 20th century saw several progressive strides being taken to fight other infectious organisms, particularly with the development of several pharmaceutical agents to combat them. Anti-infective drug research has, therefore, evolved into a multidisciplinary field which involves the collaboration of medical professionals, microbiologists, pharmaceutical scientists and epidemiologists worldwide.

The first volume of **Frontiers in Clinical Drug Research – Anti Infectives** brings updated reviews to readers interested in recent advances in the development of pharmaceutical agents for the treatment of infectious diseases. The chapters in this volume of the eBook series are written by leading experts in their respective fields. There are several topics relevant to this field in this introductory volume covering antibiotics, vaccines, drug resistance and more.

In the first chapter of this volume, Brandenburg *et al.* present their findings on a new class of polypeptide antibiotics against bacterial and viral infections. These researchers have discovered a set of polypeptide compounds which exhibit broad spectrum activity against bacterial pathogenic factors such as lipopolysaccharide and endotoxin. This effort is important in the discovery of novel compounds to target bacterial infections. These infections already pose a serious threat to human health, which is further increased by the incidence of bacterial resistance to existing drugs.

Next, in chapter 2, Kelly and Khardori provide a brief review on antimicrobial agents and highlight new mechanisms of actions that have been discovered in recent years. This chapter briefly covers the history of anti-microbial drug discovery and highlights new drugs including streptogramins, lipoglycopeptides, and fluoroquinolones.

Luo *et al.* discuss antisense oligonucleotide (ASO) based drugs against bacteria and viruses in chapter 3. This is a comprehensive review which provides a wealth of information on ASO technology, targeted organisms and host factors, and mechanisms of ASO action. The authors have also cited new drugs which are undergoing clinical trials.

In chapter 4, Vennison and Sankar have reviewed the major vaccines currently available against flaviviruses and new approaches utilized for designing novel vaccine candidates and antiviral agents targeting various pathways in the flavivirus life-cycle. The article also lists molecules that inhibit flavivirus functions and includes several useful illustrations for readers.

In chapter 5, Altindis and coworkers present a summary of the progress made on plants used for edible vaccine research. Edible vaccines can prove to be a novel and economic therapeutic solution to treat infectious diseases.

In Chapter 6, Saini *et al.,* review the recentprogress made in the field of antibiotics against gram negative bacteria. This chapter provides information on antibiotic agents used to fight infections caused by specific bacterial species such as *K. pneumoniae, E. coli* and *H. influenza.* They address concerns about MRSA and VRSA strains of *S. aureus* as well as the requirements of future R&D efforts to develop new drugs.

Chapter 7 is a summary of the current developments against enteric pathogens. Haque and colleagues bring together information on the properties of these pathogens, the antimicrobials currently in use and the mechanisms of drug resistance shown by these bacteria.

Finally, Huang *et al.* provide an interesting review of the use of bacteriophages against gram positive bacteria over the last 20 years. The authors of this chapter hint at new mechanisms to tackle clinically significant microorganisms including multi-drug resistant strains. They demonstrate the bacteriophage therapy is a novel way to address the growing problem of hospital acquired infections.

I hope that readers will find value in this collection of reviews and draw inspiration for conducting further drug discovery research in the field of anti-infective drug research.

I am grateful for the timely efforts made by the editorial personnel, especially Mr. Mahmood Alam (Director Publications), Ms. Maria Baig (Manager Publications) and Mr. Taimur at Bentham Science Publishers for their assistance.

Atta-ur-Rahman, FRS

Kings College
University of Cambridge
Cambridge
UK

List of Contributors

Aamir Ali

Human Enteric Pathogens Group, Health Biotechnology Division, National Institute for Biotechnology and Genetic Engineering, Faisalabad, Pakistan

Abdul Haque

Human Enteric Pathogens Group, Health Biotechnology Division, National Institute for Biotechnology and Genetic Engineering, Faisalabad, Pakistan

Aline Dupont

Medizinische Hochschule Hannover, Carl-Neuberg-Str. 1, D-30625 Hannover Germany

Ashok Rattan

Department of Laboratory Medicine, Medanta-The Medicity, Sector-38, Gurgaon, Haryana-122 001, India

Asma Haque

Department of Bioinformatics and Biotechnology, GC University, Faisalabad, Pakistan

Aynur Gurel

Ege University, Faculty of Engineering, Department of Bioengineering, 35100 Bornova, Izmir, Turkey

Beate Schittl

Technische Universität München / Helmholtz Zentrum München, Institut für Virologie, Trogerstrasse 30, 81675 München, Germany

David B. Huang

ContraFect Corporation, 28 Wells Avenue, Third Floor, Yonkers, New York, USA

Emrah Altindis

Harvard Medical School, Microbiology and Immunobiology Department, 200 Longwood Avenue, Boston, 02115 MA, USA

Eva Krause

Heinrich-Pette-Institut, Leibniz-Institut für Experimentelle Virologie, Martinistr. 52, 20251 Hamburg, Germany

Guillermo M. de Tejada

Universidad de Navarra, Department of Microbiology, Irunlarrea 1, E-31008 Pamplona, Spain

Hoonmo L. Koo

Baylor College of Medicine, USA

Hui Bai

Department of Biotechnology, Institute of Radiation Medicine, Academy of Military, Medical Sciences, Beijing, 100850, China, No. 451 Hospital of Chinese People's Liberation Army, Xi'an, 710054, China

Ismet Deliloglu Gurhan

Ege University, Faculty of Engineering, Department of Bioengineering, 35100 Bornova, Izmir, Turkey

Joachim Hauber

Heinrich-Pette-Institut, Leibniz-Institut für Experimentelle Virologie, Martinistr. 52, 20251 Hamburg, Germany

Julie Lucifora

Technische Universität München / Helmholtz Zentrum München, Institut für Virologie, Trogerstrasse 30, 81675 München, Germany

Kulvinder S. Saini

Research & Development, Eternal University, Baru Sahib-173 101, Himachal Pradesh, India; Department of Biological Sciences, Faculty of Science, King Abdulaziz University, Jeddah 21589, Saudi Arabia

Klaus Brandenburg

Forschungszentrum Borstel, Division of Biophysics, Parkallee 10, D-23845 Borstel, Germany

Lena Heinbockel

Forschungszentrum Borstel, Division of Biophysics, Parkallee 10, D-23845 Borstel, Germany

Marcel Krepstakies

Heinrich-Pette-Institut, Leibniz-Institut für Experimentelle Virologie, Martinistr. 52, 20251 Hamburg, Germany

Mathias Hornef

Medizinische Hochschule Hannover, Carl-Neuberg-Str. 1, D-30625 Hannover, Germany

Mehmet Ozgun Ozen

Ege University, Faculty of Engineering, Department of Bioengineering, 35100 Bornova, Izmir, Turkey

Nancy Khardori

Division of Infectious Disease, Department of Internal Medicine, Eastern Virginia, Medical School, Norfolk, Virginia, USA

Patrick Garidel

Martin-Luther- Universität Halle-Wittenberg, von-Danckelmann-Platz 4, D-06120 Halle, Germany

Pinar Nartop

Ege University, Faculty of Engineering, Department of Bioengineering, 35100 Bornova, Izmir, Turkey

Raymond Schuch

ContraFect Corporation, 28 Wells Avenue, Third Floor, Yonkers, New York, USA

S. Gowri Sankar

Department of Biotechnology, Anna University-BIT Campus, Tiruchirappalli-620 024, Tamil Nadu, India

S. John Vennison

Department of Biotechnology, Anna University-BIT Campus, Tiruchirappalli-620 024, Tamil Nadu, India

Sabine Dömming

Universitätsklinikum Aachen, Department of Intensive Care, Pauwelsstr. 30, D-52074 Aachen Germany

Shengqi Wang

Beijing Institute of Radiation Medicine, the Academy of Military Medical Sciences, Beijing, China

Stefanie Schmeiser

Institut für Virologie, Stiftung Tierärztliche Hochschule Hannover, Buenteweg 17, 30559, Hannover, Germany

Sultan Gulce Iz

Ege University, Faculty of Engineering, Department of Bioengineering, 35100 Bornova, Izmir, Turkey

Susana Sánchez-Gómez

Universidad de Navarra, Department of Microbiology, Irunlarrea 1, E-31008 Pamplona, Spain

Thomas Gutsmann

Forschungszentrum Borstel, Division of Biophysics, Parkallee 10, D-23845 Borstel, Germany

Tobias Schürholz

Universitätsklinikum Aachen, Department of Intensive Care, Pauwelsstr. 30, D-52074 Aachen, Germany

Ulrike Protzer

Technische Universität München / Helmholtz Zentrum München, Institut für Virologie, Trogerstrasse 30, 81675 München, Germany

Ursula Kelly

Division of Infectious Disease, Department of Internal Medicine, Eastern Virginia, Medical School, Norfolk, Virginia, USA

V. Samuel Raj

Department of Microbiology, Daiichi-Sankyo India Pharma Pvt. Ltd. (DSIN), Sector-18, Udyog Vihar Industrial Area, Gurgaon 122 015, Haryana, India

Vincent A. Fischetti

Laboratory of Bacterial Pathogenesis, The Rockefeller University, USA

Volker Moennig

Institut für Virologie, Stiftung Tierärztliche Hochschule Hannover, Buenteweg 17, 30559 Hannover, Germany

Wolfram Brune

Heinrich-Pette-Institut, Leibniz-Institut für Experimentelle Virologie, Martinistr. 52, 20251 Hamburg, Germany

Xiaochen Bo

Beijing Institute of Radiation Medicine, the Academy of Military Medical Sciences, Beijing, China

Xiaoxing Luo

Department of Pharmacology, School of Pharmacy, Fourth Military Medical University, Xi'an, China

Yani Kaconis

Forschungszentrum Borstel, Division of Biophysics, Parkallee 10, D-23845 Borstel, Germany

Yasra Sarwar

Human Enteric Pathogens Group, Health Biotechnology Division, National Institute for Biotechnology and Genetic Engineering, Faisalabad, Pakistan

Yu You

The Military General Hospital of Beijing, Beijing, 100700, China, The 309th Hospital of Chinese People's Liberation Army, Beijing, 100091, China

2

Send Orders for Reprints to reprints@benthamscience.net

CHAPTER 1

Anti-Infective Polypeptides for Combating Bacterial and Viral Infections

Lena Heinbockel[1], Guillermo M. de Tejada[2], Susana Sánchez-Gómez[2], Yani Kaconis[1], Eva Krause[3], Wolfram Brune[3], Stefanie Schmeiser[4], Volker Moennig[4], Tobias Schürholz[5], Sabine Dömming[5], Mathias Hornef[6], Aline Dupont[6], Joachim Hauber[3], Patrick Garidel[7], Beate Schittl[8], Julie Lucifora[8], Thomas Gutsmann[1], Marcel Krepstakies[3], Ulrike Protzer[8] and Klaus Brandenburg[1,*]

[1]*Forschungszentrum Borstel, Div. of Biophysics, Parkallee 10, D-23845 Borstel, Germany;* [2]*Universidad de Navarra, Department of Microbiology, Irunlarrea 1, E-31008 Pamplona, Spain;* [3]*Heinrich-Pette-Institut, Leibniz-Institut für Experimentelle Virologie, Martinistr. 52, 20251 Hamburg, Germany;* [4]*Institut für Virologie, Stiftung Tierärztliche Hochschule Hannover, Buenteweg 17, 30559 Hannover, Germany;* [5]*Universitätsklinikum Aachen, Department of Intensive Care, Pauwelsstr. 30, D-52074 Aachen, Germany;* [6]*Medizinische Hochschule Hannover, Carl-Neuberg-Str. 1, D-30625 Hannover, Germany;* [7]*Martin-Luther-Universität Halle-Wittenberg, von-Danckelmann-Platz 4, D-06120 Halle, Germany and* [8]*Technische Universität München / Helmholtz Zentrum München, Institut für Virologie, Trogerstrasse 30, 81675 München, Germany*

Abstract: Bacterial and viral infections are still a major threat of human health. The increasing resistance of bacterial isolates to common antibiotics and the lack of new compounds reaching the clinic leading to serious problems in health care. The variability of different virus families with individual entry pathways and replication strategies make the development of suitable therapeutics with cross-species activity complicated. Furthermore, the infections often cause each other, so that an initial virus infection is frequently accompanied by a bacterial 'superinfection' with severe consequences.

We developed a new class of compounds based on polypeptides, which exhibit broad-spectrum antiviral activity with simultaneous inhibition of important bacterial

*Address correspondence to Klaus Brandenburg: Forschungszentrum Borstel, Leibniz-Zentrum for Medicine and Biosciences, Parkallee 10, D-23845 Borstel, Germany; Tel: +49-4537-1882350; Fax: +49-4537-1886320; E-mail: kbrandenburg@fz-borstel.de

pathogenicity factors such as lipopolysaccharide (LPS, endotoxin) and lipoproteins. Here, we summarise recent results and discuss them in the context of the progress made in the field of polypeptides as novel anti-infective therapeutic agents.

Keywords: Antimicrobial peptides, bacterial and viral infection, classical swine fever, clinical trials, endotoxins, heparan sulfate, hepatitis B, HIV, inflammation, isothermal titration calorimetry, lipopolysaccharides, lipoproteins, MRSA, polypeptides, sepsis, shock, small-angle X-ray scattering, superinfection, tumor-necrosis-factor-alpha.

INTRODUCTION

Infectious diseases remain a major concern for human health care. In many countries millions of people are still dying from severe infections, making them the first cause of death. In industrialised countries the increasing resistance to multiple standard antibiotics, the lack of new antimicrobial agents and the demographic development lead to a high emergence of systemic infections such as bacterial sepsis [1, 2]. In Europe 25,000 deaths directly related to microbial resistance were reported in 2007. However, recent data indicate that this number may be much higher. It was reported that in Germany more than 100,000 people suffer from MRSA infections with 10-14,000 death cases annually (Deutsche Sepsis-Gesellschaft, http://www.sepsis-gesellschaft.de).

Since the 1990's a great deal of work was done to develop new antibacterial drugs based on naturally occurring defence proteins, such as lactoferrin, bactericidal permeability-increasing protein (BPI), lipopolysaccharide-binding protein (LBP), Limulus-anti-LPS factor (LALF, in recombinant form endotoxin-neutralising protein ENP), granulysin, and NK-lysin (overview in [3]). Also, antimicrobial peptides such as cathelicidins and defensins as well as derivatives of them were investigated [4]. The motives for using peptides as anti-microbial and anti-infective drugs are manifold. They range from broad-spectrum activity (bacteria, fungi and viruses) over rapid microbial killing to low levels of resistances. However, the disadvantages of using such compounds are high manufacturing costs (depending on the chemical structure) and low return on investment, systemic and local toxicity, a reduced activity of cationic peptides in the presence of specific cations (Mg^{2+}, Ca^{2+}) and a limited stability of the peptides under physiological conditions.

Several of the examined compounds revealed a promising antimicrobial activity, some of them also a reasonable lipopolysaccharide (LPS)-neutralising activity, but a breakthrough in the development of a new anti-infectious drug is still lacking [5, 6]. A multi-center study on the use of recombinant lactoferrin (talactoferrin® from Agennixin) for the treatment of severe sepsis failed. The clinical phase II/III study was stopped, since the Data Safety Monitoring Boards, analysing intermediate results of the study, found that the 28-days-mortality in the talactoferrin-arm of the study was higher than in the placebo-arm (www.agennix.com).

The treatment of infections caused by viruses is also challenging. Considering for example the situation for hepatitis B virus (HBV) where a reliable prophylactic vaccine exists. There is still a high number of infected individuals, with more than 350 million chronically infected, due to inadequate coverage of vaccination in many countries [7]. For many other virus infections no therapeutics are available, because suitable therapeutic targets are lacking or development costs of antivirals are regarded as too expensive. Host defense peptides and derivatives of them are investigated for their antiviral functions. Some are described as to be able to inhibit the entry of viruses into mammalian cells, making them to promising antiviral candidates. As an example, retrocyclin, a θ-defensin, inhibits the entry of human deficiency virus HIV-1 into peripheral blood mononuclear cells (PBMC) [8].

USE OF AMP AGAINST BACTERIAL INFECTIONS

Antimicrobial peptides (AMPs) are an attractive alternative to common antibiotics for combating infections also from drug-resistant pathogens. Many of these agents exhibit a broad anti-bacterial, anti-viral and anti-fungal spectrum and induce rapid microbial killing. The mechanism of action from AMPs often involves binding to structurally essential molecules on the microbial surface, averting the development of drug resistance by the targeted microorganism. Indeed, AMPs are becoming occasionally the last efficient therapeutic, since they often maintain their activity even in the presence of resistance to multiple classes of antimicrobials. Colistin (polymyxin E) for example, a cyclic cationic lipopeptide developed 60 years ago, is still successfully used in the treatment of infections due to multi-resistant *Pseudomonas aeruginosa* and other Gram-negative bacteria. Colistin is available in two forms, colistin sulphate and colistimethate sodium (or colistin methanesulfonate), a pro-drug, which is hydrolysed *in vivo*. The latter compound is less potent than colistin sulphate but it is also less toxic than the parent molecule [2]. Another polymyxin, polymyxin B (PMB) is used in Japan in an extra corporal device for the filtration of blood (a recent review about the use of PMB in sepsis therapy [9]).

Furthermore some AMPs display potent immunomodulatory and anti-inflammatory activities [10, 11] and act additive with conventional antibiotics [12-14]. These characteristics have prompted an intense research on AMP-based therapies for different applications, including both, peptides as the sole antimicrobial component or as supplement in combination with conventional antibiotics.

The applicability of the AMPs as anti-infectious drugs is restricted by some drawbacks, including a relevant level of toxicity, reduced activity in serum or in the presence of physiological concentrations of salts, susceptibility to proteolysis and high manufacturing costs. However, several strategies have been optimized in the last years to overcome those difficulties and as a consequence the list of AMPs reaching clinical trials is growing markedly.

Concerning toxicity, numerous studies have examined the structural basis governing the interaction of AMPs with prokaryotic and eukaryotic cell membranes at the molecular level. This knowledge is being successfully applied to enhance selectivity of some AMPs *via* modification of their hydrophobicity or amphipathicity [15]. However, more investigations are needed to better understand the toxicity and further improve AMP selectivity *in vivo*.

To extend the half-life of systemically administered AMPs, strategies to avoid a proteolytic degradation were developed. Classical examples of this approach involve the substitution of standard amino acids with non-natural amino acids (*e.g.,* D-amino acids) [16, 17], peptide cyclisation or terminal modification *via* amidation, alkylation or acetylation [16, 18]. Another option to increase the bioavailability of AMPs is the replacement of tryptophan or histidine residues with the bulky amino acids β-naphthylalanine and β-(4,4′-biphenyl) alanine which reduce the salt sensitivity of AMPs [19]. However, the most promising strategy is probably the synthesis of peptoids, compounds whose side chain is connected to the nitrogen of the peptide backbone, instead of the α-carbon [17, 20, 21]. Peptoids display an enormous structural diversity and are completely resistant to enzymatic proteolysis. Additionally, immobilisation of AMPs onto surfaces represents an interesting way to avoid peptide degradation while providing local activity. This approach is currently pursued to prevent the microbial attachment and colonisation of medical devices and implants [22]. Finally, AMPs can be encapsulated in delivery systems such as liposomes or nanoparticles to increase their *in vivo* stability [23]. This modification also lowers their toxicity but increases the high costs of AMP production. A possible solution for the expensive manufacturing is the recombinant synthesis of AMPs in genetically manipulated microorganisms. This method may facilitate the commercial scale production of the compounds to an affordable limit [24].

Over the past two decades a large number of AMPs have been characterised (reviewed in [25-27] and an Antimicrobial Peptide Database (APD) has been established based on an extensive literature search. To date, this database contains detailed information about 2221 peptides [28].

Although a vast number of peptides have undergone preclinical studies, only a few reached the clinical evaluation [29-31] mainly due to a failing in the phase II trial. However, success rates for many other types of drugs are also rather modest. A decrease from 28% (2006-2007) to 18% (2008-2009) was recorded for the success of phase II projects in total, obviously depending on the field of therapy [32]. The current likelihood of a drug successfully progressing from phase III to launch is 50% [33].

A peptide-based drug approved for clinical use was daptomycin, an acidic lipopeptide isolated from cultures of *Streptomyces roseosporus*. This antibiotic received approval by the Food and Drug Administration (FDA) in 2003 for the treatment of skin and soft tissue infections caused by Gram-positive bacterial pathogens. Three years later the FDA extended the therapeutic indication of this drug to the treatment of bacteremia and right-sided endocarditis caused by *S. aureus* and methycillin-resistant *S. aureus* (MRSA). The proposed mechanisms of action of the cyclic peptide involves its insertion into the bacterial membrane, followed by membrane depolarization and additionally the inhibition of DNA, RNA and protein synthesis [34].

The therapeutic options for the treatment of infections caused by Gram-positive bacteria are in expansion and some of the new drugs have novel mechanisms of action (*i.e.,* linezolid, daptomycin, tigecycline). Also Gram-negative pathogens such as *Pseudomonas aeruginosa, Acinetobacter baumannii, Stenotrophomonas maltophilia,* and several members of the Enterobacteriaceae, are rapidly evolving resistance to most of the currently available antibiotic agents. Thus, efforts including AMP development are currently focused on the treatment of Gram-negative pathogens or agents with broad-spectrum activity.

There are several candidate peptides in clinical trials, but only four of these compounds have been approved for Phase III clinical trials (Table **1**). The therapeutic indications for these agents include oral mucositis (iseganan), diabetic foot ulcers (pexiganan), catheter infections (omiganan) and sepsis (opebacan, Neuprex®, rBPI21). Only the latter drug is intended for systemic use, whereas the rest of them are made for topical treatment.

Table 1: Antimicrobial peptides involved in clinical trials.

Name	Company	Origin	Spectrum/Mode of Action	Trial Phase (*via*)
CZEN-002	Zengen	Peptide derived from alpha-Melanocyte-Stimulating Hormone (a-MSH)	GPB, GNP, Candida. Yeast regulatory mechanisms, interference by cAMP induction, anti-inflamatory [61-63].	I/II: Vulvovaginal candidiasis (topical)
Daptomycin	Cubicin	Cyclic lipopeptide derived from Streptomyces roseosporus	GPB. Depolarisation of membrane potential, inhibition of protein, DNA and RNA synthesis [34].	In Market (Parenteral)
EA-230	Exponential Biotherapies	Short peptide derived from β-HCG (human chorionic gonadotropin)	Anti-inflammatory [64]. Sepsis and renal failure protection.	I/II: www.expobio.com/clinical-trials/index.php (intravenous)

Table 1: contd....

Friulimicin B	MerLion Pharmaceuticals GmBH	Cyclic lipopeptide produced by *Actinoplanes friuliensis*	GPB. Cell wall synthesis inhibition [65, 66].	I; healthy volunteers NCT00492271 (intravenous)
hLF1-11	AM-Pharma	Cationic peptide fragment comprising the first eleven residues of human lactoferricin	GPB, GNB, fungi. Membrane disruption, DNA binding, immunomodulatory [67-70].	I/II: Infectious complications among haematopoietic stem cell transplant recipients and infections with Candida NCT00509938 [71] (intravenous)
IMX942	Inimex	Synthetic cationic peptide derived from IDR1 (Innate defense regulator)	Broad-spectrum. Immunomodulatory, no direct antimicrobial activity. (www.inimexpharma.com)	I/II: severe acute bacterial skin and skin structure infections (intravenous)
Iseganan (IB-367)	Ardea Biosciences	Synthetic peptide derived from protegrin 1 from pig	GPB, GNB, fungi. Membrane disruption [72].	III: Preventing oral mucositis in patients undergoing radiation therapy for head and neck cancer NCT00022373 [73] (mouth rinse solution)
Locilex™ (pexiganan acetate 1% cream)	MacroChem corporation/ Gaenera Corporation	Pexiganan (MSI-78) analog of the magainin peptides isolated from frog skin	GPB, GNB. Membrane disruption, anti-endotoxic [36, 74].	III: Treatment of diabetic foot ulcers NCT00563394 NCT00563433 [35] (topical)
Lytixar ® (LTX-109)	Lytix Biopharma	Synthetic antimicrobial peptide	GPB, GNB. Membrane disruption [75-77].	II: Gram-positive, skin infections mild eczema/ dermatosis NCT01223222 I/II: nasal carrier MRSA NCT01158235 (topical)
Omiganan (MX-226/ CPI-226/ CLS001)	Migenix/ Cadence Pharmaceutical/Cutanea Life Science	Synthetic cationic peptide derived from indolicidin	GPB, GNB, fungi. Membrane disruption [78-80].	III: Skin Antiseptic NCT00608959 III: and prevention of catheter infection/ colonization NCT00231153 (topical)
OP-145	OctoPlus	Synthetic peptide derived from the human cathelicidin, LL-37	GPB, GNB. Anti-endotoxic. (www.octoplus.nl)	II: Chronic otitis media (ear drops)

Table 1: contd....

Opebacan (NEUPREX®/rBP I21)	XOMA	A modified recombinant fragment of human bactericidal/permeability-increasing protein	GNB. Anti-endotoxic [81].	II: burn (intravenous) NCT00462904 III Meningoccemia [39] (intravenous)
PAC-113	Pacgen Biopharma-ceuticals	Synthetic peptide derived from histatin	Candida spp. Membrane disruption, mitochondria interaction. (www.pacgenbiopharm.com)	IIb: Oral candidiasis in HIV seropositive patients (mouth rinse solution)
PMX-30063	PolyMedix	Arylamide oligomer mimetic of a defensin	GPB, GNB. Membrane disruption (www.polymedix.com)	II; patients with acute bacterial skin and skin structure infections (topical)
Polymyxins (B, E or colistin)	Generic drug	Lipopeptide derived from *Bacillus polymyxa*	GNB. Membrane disruption, anti-endotoxic [2]	In Market (Otic, ophtalmical, parenteral, topical)
RDP58	Genzyme	Semisynthetic D-amino acid decapeptide derived from HLA class I	Anti-inflammatory [82, 83].	II: Ulcerative colitis (topical)
XOMA 629	Xoma	Synthetic peptide compound derived from bactericidal/permeability-increasing protein	Broad spectrum specially GPB. Not driven by pore formation. (www.xoma.com)	IIa: Impetigo/mild to moderate acne (topical)

Informations based on previous reports [29-31] and updated according to data from company press releases and the Clinical Trials Database of NIH (www.clinicaltrials.gov).
GNB; Gram-negative bacteria, GPB; Gram-positive bacteria.

Iseganan (Ardea Bioscience Inc; Table **1**) is a synthetic protegrin derived from pigs. In a phase III clinical trial, patients under radiation therapy for head and neck cancer received the peptide as mouth rinse to prevent ulcerative oral mucositis. However, the treatment did not demonstrate benefit in comparison with placebo. More recently, another phase II/III trial was conducted with the compound to test its ability to prevent ventilator-associated pneumonia, but it was ended due to safety concerns. After these setbacks, the company decided to drop the further development of iseganan.

Pexiganan (MSI-78) is a synthetic analogue of magainin, an AMP isolated from the skin of the African Clawed Frog (Table **1**). The efficacy of a cream containing 1% of pexiganan acetate (Locitex TM) was compared in a phase III trial with the oral administration of the chinolone ofloxacin to treat diabetic foot ulcer. The study revealed that both treatments had similar efficacy, ofloxacin-resistant bacteria were isolated from some patients receiving ofloxacin, whereas no significant resistance to pexiganan was detected in patients treated with this drug [35]. It displays a broad-spectrum antimicrobial activity, whereas the applicability is restricted to topical formulations due to its sensitivity to proteolytic degradation. However, a number of current studies [36] focus on enhancing the serum stability of pexiganan, making the peptide a promising candidate for the treatment of systemic infections. A biopharmaceutical company, Dipexium Pharmaceuticals LLC, is planning a new phase III trial with pexiganan.

Omiganan (MX-226) is an analogue of indolicidin, a natural AMP purified from bovine neutrophils (Table **1**). It was originally developed by Migenix and licensed to Cadence Pharmaceuticals for catheter-related infections (MX-226/MBI-226) and Cutanea Life Sciences for dermatological diseases (CLS001). The companies declared the compound as a topical skin antiseptic drug and conducted phase III clinical trials to evaluate its efficacy for the prevention of catheter colonization and catheter-related infections in patients with central venous catheters. As a mode of action the interaction with Gram-positive cell walls was described [37]. Although secondary end points (*i.e.,* reduction of systemic bacterial infection) were achieved, the treatment failed to reach the primary objective, namely the prevention of local catheter site infection. Omiganan passed another recently finished Phase III trial as a topical skin antiseptic drug in healthy adults [38].

Opebacan is a recombinant fragment of human bactericidal/permeability increasing protein (BPI) with anti-endotoxic properties (Table **1**). Data from a phase III trial to analyse the ability of this drug to control severe menigococcemia in children revealed a substantial reduction in morbidity. Although differences in mortality between the treated and untreated groups were not detected, there was a trend towards an enhanced survival in the Opebacan-treated group [39].

Additional peptides are in early clinical trials (I/II; Table **1**) to evaluate a potential toxicity and their pharmacokinetic profile. Most of these clinical trials involve the non-parenteral administration (*i.e.,* topical, oral or in-ear) of the compound. Topical application provides an attractive strategy for the control of skin infections because it allows the direct delivery of the drug to the infection site and therefore high local drug levels. This is also the route of choice for compounds exhibiting potential toxicity after systemic administration. Oral administration remains uncommon for peptide due to a lack of absorption in the intestine and high enzymatic degradation [40].

Peptides investigated in clinical trials include Xoma-629 for the treatment of impetigo, PAC-113 for oral Candidiasis in HIV seropositive individuals, CZEN-002 for vulvovaginal candidiasis, Lytixar™ (LTX-109) for Gram-positive skin infection, OP-145 for chronic middle ear infection, RDP-58 for ulcerative colitis and PMX-30063 for acute bacterial skin and skin-structure infection (ABSSSI). Although the latter compound is being administered topically, PolyMedix plans to seek FDA approval for an intravenous formulation and conduct additional clinical trials in patients with bloodstream infections, lung infections and oral mucositis. In addition to PMX-30063 further peptide-based drugs are being examined by the parenteral route including hLF-11 for patients with bacteremia due to *S. epidermidis*, opebacan (see above) for the treatment of burn injuries, IMX-942 for skin infections and EA-230 in patients with risk of developing renal failure.

Although most of the ongoing clinical trials target skin infections and concern topical administration, the intense research devoted to overcome limitations of peptide substances might result in a new generation of AMPs suitable also for systemic application.

Use of AMPs Against Sepsis

As described above the majority of therapeutic peptides, which is currently under investigation, is established not for the use as systemically administered reagents. Considering four features of AMPs, however, make them beneficial for the treatment of severe bacterial infections in the future. These are first of all peptides described to have the ability to kill bacteria efficiently (low MIC values), secondly the triggering of the immune system for improving its fight against infections, thirdly there is rare induction of resistance and fourthly there are peptides able to neutralise bacterial pathogenicity factors such as endotoxins and thus directly decrease inflammatory reactions [5, 41].

Table 2: Amino acid sequences of some variants and their molecular weights of the SALP- series [14, 49, 52, 84, 85] and unpublished.

Do-D	GIGKFSKKGAAARRRKVSLKAL	2387
Pep19-2.5	GCKKYRRFRWKFKGKFWFWG	2711
Pep19-2.5gek	GCKKYRRFRWKFKGK	1988
Pep19-2.5KO	KFGKWRFGKYRFCWKFRGWK	2711
Pep19-2.5LF	GCKKYRRFRWKFKGKLFLFG	2665
Pep19-2.8	GCKKYRRFRWKFKGKFFWWG	2814
Pep19-4	GKKYRRFRWKFKGKWFWFG	2750
Pep19-4gek	GKKYRRFRWKFKGK	1884
Pep19-8	GRRYKKFRWKFKGRWFWFG	2636
Pep19-8.3Acyl	SRRYKKFRWKFKGRWFWFGHexanCONH$_2$	2779
Pep19-12.1Zykl	GCKKFRRFRRWKYKGKFWFWCFG	3118
Pep19-12.1S	GCKKFRRFRRWKYKGKFWFWSFG	3102
Pep19-12.1SS	GSKKFRRFRRWKYKGKFWFWSFG	3086

To develop an efficient strategy for the treatment of systemic infections, leading to severe inflammation states such as sepsis, point four is highly relevant. Various approaches to design AMPs with the primary aim to neutralise bacterial endotoxins (lipopolysaccharides, LPS) have been performed. These investigations were based on the use of the LPS-binding domain of natural defense proteins for detoxification, such as lactoferrin, lipopolysaccharide-binding protein (LBP), bactericidal-permeability increasing protein (BPI), granulysin and Limulus-anti-LPS-factor (LALF). There are reports about recent progress made in this area [42, 43]. Both, the high peptide concentrations needed to efficiently neutralise LPS associated with significant cytotoxicity to human cells and the lack of translational efforts to progress in clinical phases have prevented a breakthrough [44]. For example, the nephrotoxicity and neurotoxicity of the cationic polypeptides polymyxin B and polymyxin E (colistin) restrict their application in patients. However, the observed toxic effects in previous clinical studies might have resulted from incorrect dosing based on a limited understanding of the pharmacodynamics and toxicodynamics of the reagents [9, 45]. Immobilised polymyxin B was used in a more recent study of hem operfused cartridges in abdominal septic shock (EUPHAS) [46]. The patients improved in mean arterial pressure and organ function. Furthermore, the 28-day mortality was lower in the polymyxin group compared to conventional therapy alone

(32% *vs.* 53%). Consequently, the ethics committee stopped the study, because it was unjustifiable with respect to the placebo group. Two additional studies are currently performed to confirm these results and to address further aspects of the treatment revealed in the EUPHAS study (EUPHRATES: clinicaltrials.gov-identifier NCT01046669 and EUPHAS 2 [47].

Comprehensively we examined in the last years LPS-neutralising polypeptides with special regard on peptides based on the LPS-binding domain of the LALF protein. Modifications in the amino acid (AA) sequence (see Table **2**) were introduced to enhance the binding affinity to LPS, in particular to the lipid A moiety of LPS and thus facilitate efficient cross-reactivity to the numerous LPS variants occurring in Gram-negative bacteria. These peptides were named synthetic anti-LPS peptides (SALP). We could show that the induction of tumor-necrosis-factor-α by LPS in human mononuclear cells is efficiently inhibited even at low Pep19-2.5: LPS ratios (see Fig. **1**, middle). Also, we demonstrated that the exact sequence is decisive *e.g.,* with the results obtained using the scrambled Pep19-2.5KO variant (Fig. **1**, right). As illustrated in Fig. **2**, the efficient neutralisation from the peptide correlates with a strong, saturable interaction with the bacteria, as determined by isothermal titration calorimetry [49].

Regarding the mechanism of the molecular interaction, it was found that the main effect of endotoxin neutralisation results from the conversion of the aggregate structure of LPS from a unilamellar/cubic to a multilamellar organization [48-51]. This finding is illustrated in Fig. **3** (upper panel) for LPS in the presence of Pep19-2.5, with the occurrence of two main reflections at 8.85 and 4.55 nm, corresponding to the 1st and 2nd order of a multilamellar structure. The conversion of LPS into multilamellar aggregates is less pronounced or does not take place for the scrambled Pep19-2.5KO version and an additional peptide, Pep19-8. For the latter the original non-lamellar aggregation of LPS is observed even in the presence of the peptide (Fig. **3**, bottom). The reduced immunostimulatory activity of LPS organised in multilamellar aggregates might result from a steric hindrance of the binding by proteins involved in LPS inflammatory signal cascade activation such as LBP, CD14 and TLR4 [41, 48, 52].

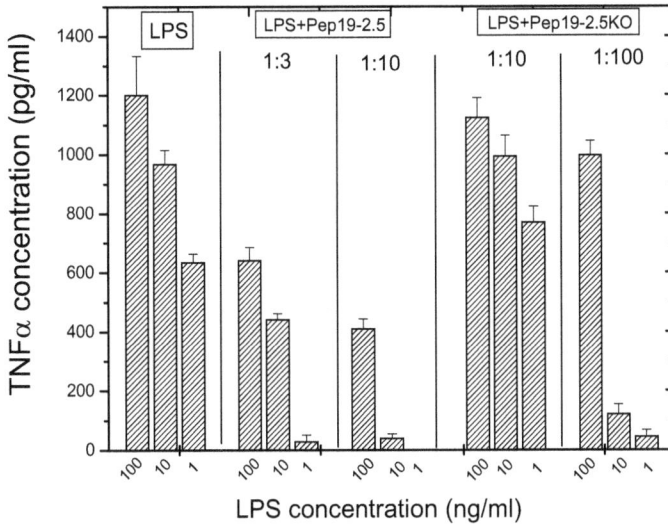

Figure 1: LPS-induced stimulation of tumor-necrosis-factor-α in human mononuclear cells in the absence and presence of the peptides Pep19-2.5 and Pep19-2.5KO at different concentrations [49].

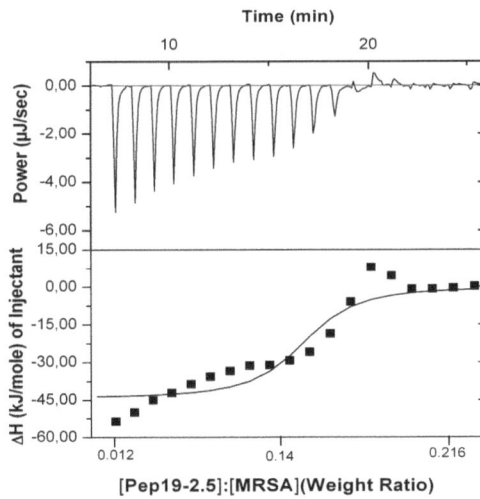

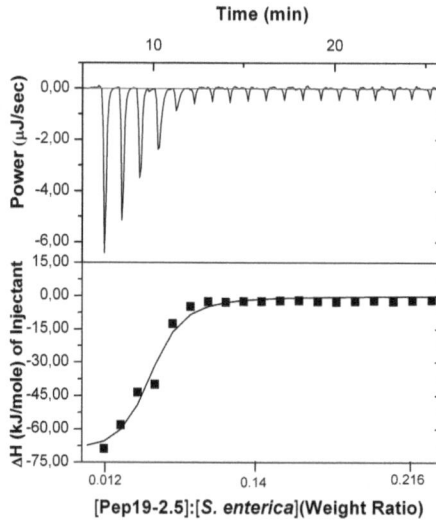

Figure 2: Calorimetric determination of the interaction of Pep19-2.5 with heat-killed Gram-negative (*S. enterica*) and Gram-positive (*S. aureus*) bacteria. To the bacteria dispersion (2 mg/ml) 1,5 µl of the peptide solution (2 mg/ml) were titrated and the calorimetric signal was recorded. The measurements were performed at 37 °C [14].

Recently it was discovered that Pep19-2.5 reveals also considerable activity against the inflammatory properties of cell wall components from Gram-positive bacteria. The versatile effectiveness represents a key point in the treatment of diseases like sepsis caused by a variety of different microorganisms [14].

USE OF AMPs AS ANTIVIRAL AGENTS

The property of many SALPs to be small amphiphilic polycationic compounds enables them to an exceptional cross-reactivity. Binding could be demonstrated both to viral envelopes with negative charges and to human cell surface receptors, mediating infection by viruses. These receptors include proteoglycans such as heparan sulfate, keratin sulfate and chondroitin sulfate. In the following chapter we present data for the inhibition of viral infections by blocking these receptors.

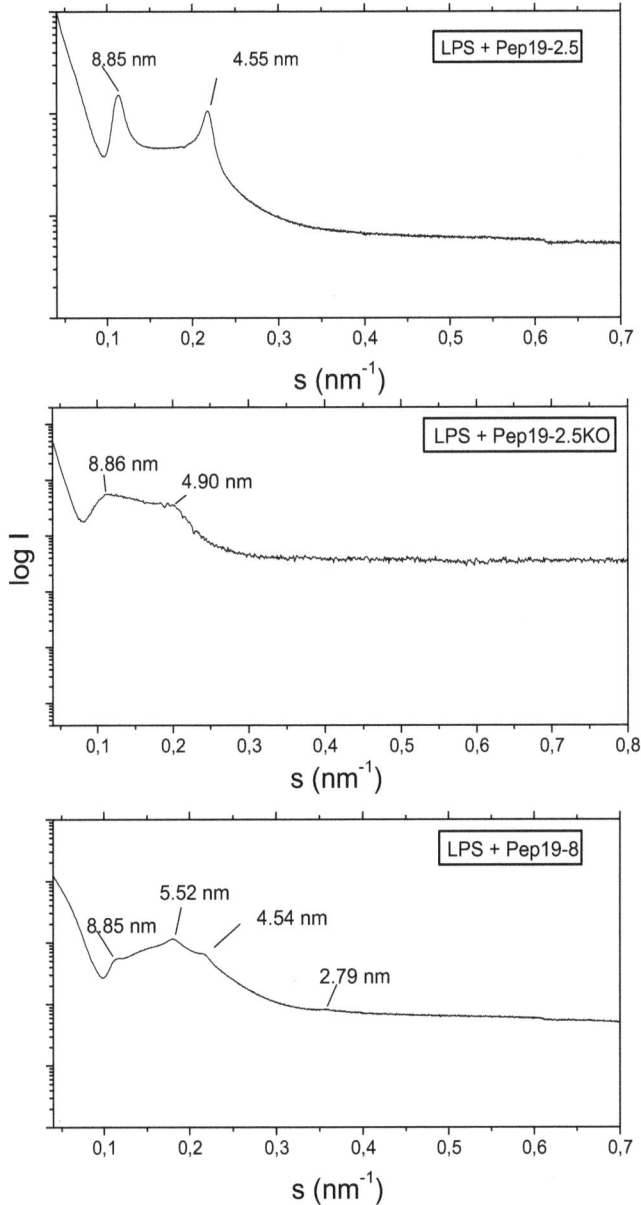

Figure 3: Small-angle X-ray scattering (SAXS) patterns of LPS R60 in the presence of Pep19-2.5 (top), Pep19-2.5KO (middle), and Pep19-8 (bottom). The concentration ratios were [LPS]:[Peptide] 3:1 weight%. The logarithm of the scattering intensity log I is plotted *vs.* the scattering vector s (= 1/d, d = lattice spacings) [49].

Interaction with Viruses Infecting *via* Binding to Heparan Sulfate

The interaction of SALPs with various animal or human cells involved in viral infection was examined [53]. It was demonstrated that some SALPs are able to inhibit the entry of human immunodeficiency virus type 1 (HIV-1), herpes simplex virus (HSV) 1 and 2, and hepatitis B and C to their respective host cells. The data are summarised in Table **3**, in which the 50% inhibitory concentrations IC50 are given for the viruses and their respective host cells. An efficient inhibition was exhibited in particular for HSV-1 (Table **3**) and hepatitis B, for which the IC50 concentrations were below 1 µM. Changes in the peptides AA revealed that the viral infection (in this case HIV) critically depends on the detailed SALP sequence. For example, the longer variants Pep12-1S and Pep12-1.SS (see Table **1**) and also Pep19-2.5LF, with the sequence LFLF at the C-terminal end instead of FWFW, revealed a lower activity than the Pep19-2.5 (Fig. **4**). Of particular importance is the observation that the entire inhibition is abolished using the shortened compound Pep19-2.5gek (Fig. **4**) in accordance with a strongly reduced anti-LPS activity (see above). Accordingly, it is important to notice that the Coulomb attraction of the positive charges of the peptides with the negative surface charges alone is not sufficient. Additionally the hydrophobic interaction of the C-terminal end with the target cell is required, probably *via* a direct membrane intercalation or a formation of a secondary structure that is favourable for an interaction.

Two possible mechanisms for the inhibition of the viral cell entry by peptides were examined. Firstly the peptides interact with the virus envelope and secondly the peptides interact with a common cell surface receptor such as heparan sulphate to block the penetration into the cell.

Table 3: Half Maximal Inhibitory Concentrations (IC50) for peptide mediated inhibition of study viruses

	Pep19-2.5		Pep19-4	
Virus	**IC50 (µg/mL)**	**IC50 (µM)**	**IC50 (µg/mL)**	**IC50 (µM)**
HIV-1BaL	8.0	2.9	22	8.0
HIV-1NL4-3	16	5.9	10	3.6
HSV-1	0.42	0.15	1.8	0.7
HCVpp	40	15	37	14
HBV	1.0	0.37	ND	ND

Abbreviations: HBV, hepatitis B virus; HCV, hepatitis C virus; HIV-1, human immunodeficiency virus type 1; HSV-1, herpes simplex virus 1; ND, not determined.

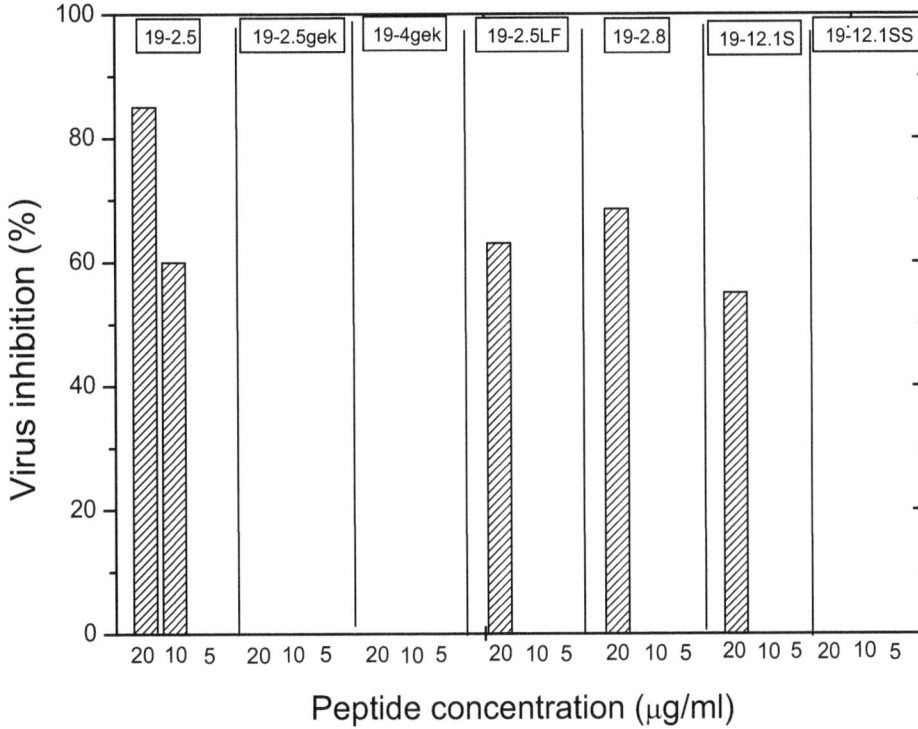

Figure 4: Inhibition of human immunodeficiency virus (HIV) infection at the level of attachment. PM1 cells were infected with R5-tropic HIV type 1 (HIV-1) BaL and cultivated in the presence of the indicated peptides at different Pep concentrations (see AA sequences in Table **1**).

The phospholipid matrix of HIV viruses was described to be strongly anionic, *i.e.,* to contain negatively charged phosphatidylinositol (PI), PI-diphosphate (PIP), and PI-triphosphate (PIP2) [54]. Since the peptides contain various cationic AA, the attractive Coulomb interaction might lead to viral inactivation. To study the interaction of Pep19-2.5 with the viral membrane, we have applied small-angle X-ray scattering (SAXS) with synchrotron radiation. The aggregate structures of PI, PIP and PIP2 in the absence and presence of Pep19-2.5 were elucidated. Data for PIP2 are presented in Fig. **5** and exhibit two scattering maxima at 4.81 and 2.53 nm for the phospholipid alone (top), and only slight changes in the presence of the peptide (bottom). These data are characteristic for the existence of a lipid bilayer, partially unilamellar, but without an aggregate structure change due to peptide addition. Similar results were found for the interaction of the peptide with PI and PIP (not shown). Thus, despite the strong anionic character of the viral envelope, there is only a slight change of the membrane architecture, making direct binding of the peptide to the virus unlikely.

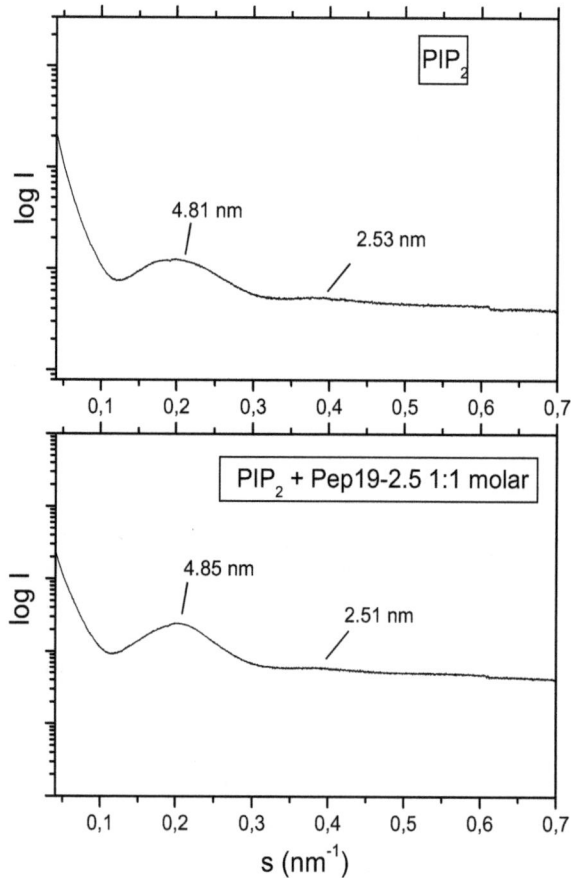

Figure 5: Small-angle X-ray scattering of phosphatidylinositol-4,5-bisphosphate (PIP2) in the absence and presence of Pep19-2.5 at 1:1 M/M. The logarithm of the scattering intensity log I is plotted *vs.* the scattering vector s (= 1/d, d = lattice spacings).

To study the second inhibitory possibility, isothermal titration calorimetry was applied for the determination of an interaction from heparan sulfate (HepS) and peptide. The titration of Pep19-2.5 to HepS exhibited an exothermic reaction with a S-shaped saturation at [Pep19-2.5]:[HepS] = 1:1 weight% [53]. The results indicate a peptide binding with high affinity to heparin. The proteoglycan is located on various cell surfaces and is used as an attachment target by many enveloped viruses including HIV, HSV, HBV and HCV, explaining the blockade of the viral infection due to a competitive binding mechanism or a steric hindrance of the viral entry mechanisms.

Furthermore, effects on cytomegalovirus could be demonstrated (Fig. **6**). The effect of Pep19-2.5 on human cytomegalovirus (HCMV) replication was investigated by a titer reduction assay. For this, an HCMV stock preparation was exposed to Pep19-2.5 or a control peptide. Treatment with Pep19-2.5 resulted in a reduction of virus titer compared to the control peptide, indicating an inhibition of HCMV replication due to Pep19-2.5 ((A) absolute, (B) relative values). The inhibitory effect of Pep19-2.5 was concentration-dependent and ranged between a 7-fold titer reduction at a concentration of 1 µg/ml and a 46-fold titer reduction at a concentration of 5 µg/ml. It was described that heparan sulfate functions as an attachment factor for HCMV and is strictly necessary for HCMV infection [56]. Thus, it can be presumed that Pep19-2.5 inhibits HCMV attachment to target cells, preventing the initial step of viral infection, by binding to heparin sulphate.

Overall, there is only limited knowledge about peptides in the treatment of viral infections. For HIV, the use of retrocyclin, a θ-defensin 18-residue peptide is described to inhibit the virus entry into cells [8]. Investigations revealed that retrocyclin forms patch-like aggregates on the surface of CD4^{+} cells, thus interfering with an early stage of viral infection.

Interaction with Animal Viruses

Classical Swine Fever is a highly contagious viral disease of pigs and as a major constraint to the pork production it is of huge economical concern. It is caused by Classical Swine Fever Virus (CSFV), a pestivirus from the family Flaviviridae, harbouring also important human viruses like Dengue virus, Yellow Fever virus and Hepatitis C virus. For CSFV the cellular receptor for virus entry is not known so far [57], whereas for another member of the genus Pestivirus, namely Bovine Viral Diarrhea Virus (BVDV) CD46 has been identified as major cellular receptor [57].

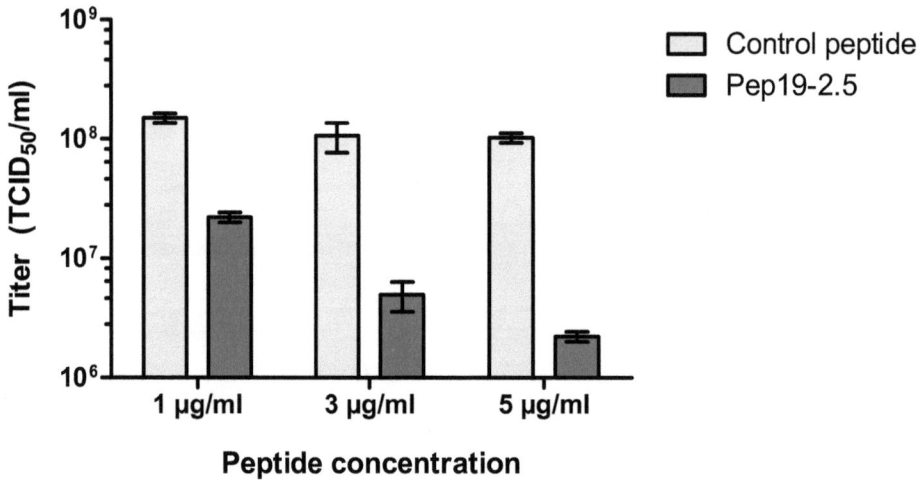

(A)

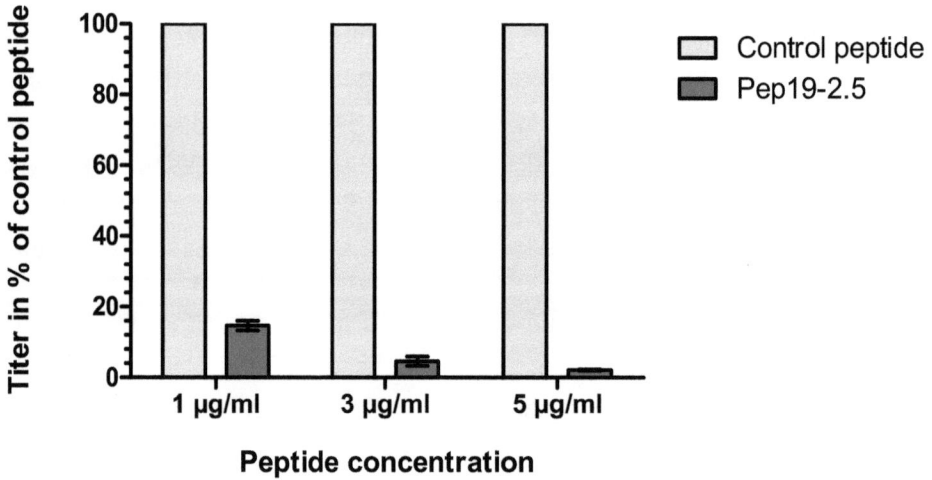

(B)

Figure 6: Inhibition of the infection from cytomegalovirus (HCMV AD169) by Pep19-2.5 at three different concentrations. A) Absolute values of the virus titers. B) Relative values of the virus titers in percentage. Error bars represent standard error of the mean.

Here, we present data from an *in vitro* study in which a test system was developed to find out if and in which concentration selected peptides were able to protect permissive porcine culture cells from an infection with CSFV. By comparison with a test system based on BVDV and bovine cells it was examined whether this protection was based on blocking of a CSFV specific cellular receptor.

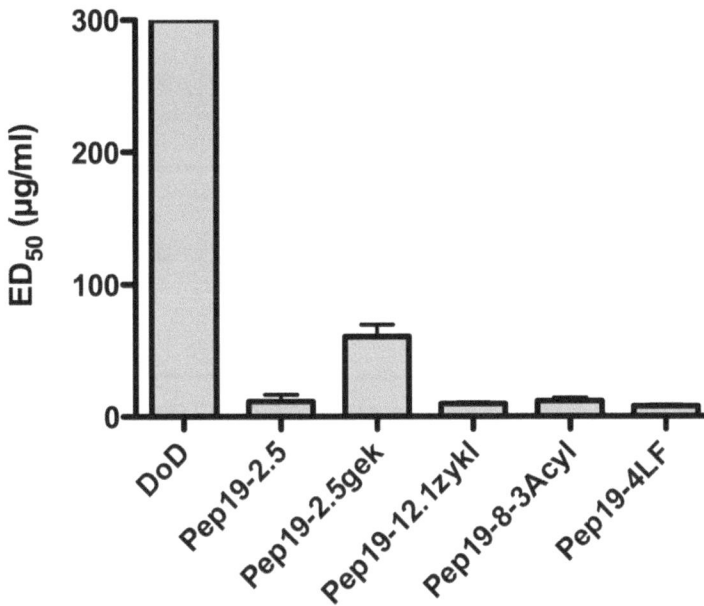

Figure 7: Inhibition of the infection of porcine kidney cells as ED_{50} value induced by classical swine fever virus (CSFV) in the presence of the indicated peptides.

Applying this test system to different peptide candidates a repeatable dose-dependent protection of cells was found. Fig. **7** depicts the results for different peptide candidates from three independent tests. Peptide DoD was included as negative control. It was found to have no protective effect in the highest concentration, therefore the calculated titer was given as >= 300 µg/ml ED50. For the other candidates an individual dose-dependent effect could be observed and it was found that the Pep19-2.5gek showed 6 times minor protective effects (mean value 60 µg/ml ED50) compared to Pep19-2.5 and other candidates (values around 10 µg/ml ED50) Therefore, the established test was found to be able to mirror the grade of inhibitory effects of the peptides on virus infection. In a similar test setup using BVDV and bovine cells such a dose-dependent protective effect could not be achieved (data not shown) which provides a hint that the protective effect is specific for CSFV infection and is linked to the relevant cellular receptor. The identification of this receptor will be subject of further studies and may be supported by the information gained from the peptide protection assay.

OUTLOOK

The data presented here are indicative that polypeptides with suitable physico-chemical characteristics may be excellent inhibitors of the bacteria- and LPS-induced inflammation and of viral infection. As shown, however, even small changes in the amino acid sequence may lead to drastic changes in the biological efficiency. For an understanding of this complex dependence sophisticated techniques and procedures such as crystallization experiments might be useful. Presently, we intend to apply also neutron small-scattering experiments by using specifically deuterated amino acids such as phenylalanine (F) for the determination of the exact location of the AA of the peptides during the interaction with the target structures.

Although it was demonstrated that the investigated peptides cause the inhibition of the entry from relevant viruses into human cells, this does not correspond –as for the anti-septic action – to a general antiviral property of the peptides. Furthermore, the concentrations necessary for the antiviral effects are noticeably higher then for the prevention of bacteria caused inflammation. No antiviral activity was detected against equine herpes viruses HZ1 (in collaboration with K. Osterrieder, Berlin), varicella zoster viruses (in collaboration with A. Sauerbrei, Jena) and Epstein Barr viruses (in collaboration with C. Muenz, Zürich). These viruses belong to the class of Herpesviridae and therefore the results are opposed to the findings described above for herpes simplex 1 and 2 viruses. Thus, the infection mechanisms might vary considerably already for the same family of viruses, not allowing to establish any general mechanism.

METHODS

X-ray Scattering

X-ray scattering (SAXS) measurements of LPS:peptide mixtures were performed at the European Molecular Biology Laboratory (EMBL) synchrotron radiation facility HASYLAB in Hamburg using the double-focusing monochromator-mirror camera X33 [58]. Diffraction patterns in the range of the scattering vector $0.01 < s < 0.0$ nm-1 ($s = 2 \sin \theta / \lambda$, 2θ scattering angle and λ the wavelength = 0.15 nm) were recorded at 20, 40 and 60 °C with exposure times of 1 min using an image plate detector with online readout (MAR345, MarResearch, Norderstedt/Germany). The s-axis was calibrated with Ag-Behenate, which has a periodicity of 5.84 nm. The diffraction patterns were evaluated as described previously [49] assigning the spacing ratios of the main scattering maxima to defined three-dimensional structures.

Inhibition of Cytomegalovirus

MRC5 cells were seeded on 96 well plates and treated with different concentrations of Pep19-2.5 or a control peptide for 1 h prior to infection. A virus stock preparation of HCMV strain AD169-GFP was serially diluted and added to the cells in the presence of the corresponding peptide concentration. Ten days post infection virus titers were determined using median tissue culture infective dose (TCID50) [59].

Inhibition of Classical Swine Fever Virus (CSFV)

For quantitative analysis of protective effects of the peptides, porcine kidney cells were pre-incubated with a dilution series of the relevant peptide candidates. In the following, CSFV was added in defined concentrations and after incubation times peptides and virus were removed from the cell layer. The cells were cultivated and after three days they were checked for CSFV infection. A test was found valid when the applied virus concentration was found to be within the calculated range and cell controls and virus controls were found as expected. Each test was repeated three times. The quantitative value for protective dose of peptides was calculated as endpoint titer in accordance to the formula of Spaermann and Kaerber [60] and was given with the unit ED50 (effective dose 50), as the peptide concentration which was capable to protect 50% of the culture of one dilution step.

ACKNOWLEDGEMENTS

The authors are indebted the German ministry (Ministerium für Bildung und Forschung) BMBF for financial help in the frame of a preclinical study 'Therapy of infectious diseases with special regards to bacterial sepsis ' (project: 01GU0824), and the Else-Kröner-Fresenius-Stiftung (EKFS) for support in the project 'Biophysikalische Charakterisierung der Interaktion von Endotoxinen mit nicht-steroiden antientzündlichen Wirkstoffen und der Modifikation durch antimikrobielle Peptide' (Project 2011 A_140).

We are grateful for the excellent technical assistance of Nina Hahlbrock and Christine Hamann. G.M.T. was funded by a grant from Ministerio de Sanidad y Consumo (FIS-PI050768), and from Proyectos de Investigación Universidad de Navarra (PIUNA-2008-11) Spain.

CONFLICT OF INTEREST

The author(s) has confirmed that there is no conflict of interest.

REFERENCES

[1]   Bone RC. The pathogenesis of sepsis. Ann Intern Med. 1991;115:457-69.
[2]   Landman D, Georgescu C, Martin DA, Quale J. Polymyxins revisited. Clinical Microbiol Rev. 2008;21(3):449-65. Epub 2008/07/16.
[3]   Andrä J, Gutsmann T, Garidel P, Brandenburg K. Mechanisms of endotoxin neutralization by synthetic cationic compounds. J Endotoxin Res. 2006;12(5):261-77.
[4]   Bensch KW, Raida M, M,,gert HJ, Schulz-Knappe P, Forssmann WG. hBD-1: a novel á-defensin from human plasma. FEBS Lett. 1995;368:331-5.
[5]   Munford RS. Detoxifying endotoxin: time, place and person. J Endotoxin Res. 2005;11(2):69-84. Epub 2005/06/14.
[6]   Nahra R, Dellinger RP. Targeting the lipopolysaccharides: still a matter of debate? Curr Opin Anaesthesiol. 2008;21(2):98-104. Epub 2008/04/30.
[7]   Williams R. Global challenges in liver disease. Hepatology. 2006;44(3):521-6. Epub 2006/08/31.
[8]   Munk C, Wei G, Yang OO, Waring AJ, Wang W, Hong T, *et al*. The theta-defensin, retrocyclin, inhibits HIV-1 entry. AIDS Res Human Retrovir. 2003;19(10):875-81. Epub 2003/10/31.
[9]   Garidel P, Brandenburg K. Current understanding of polymyxin B applications in bacteraemia/sepsis therapy prevention: Clinical, pharmaceutical, structural and mechanistic aspects. Anti-Infective Agents Med Chem. 2009;8(4):385.
[10]   Koczulla AR, Bals R. Antimicrobial peptides: current status and therapeutic potential. Drugs. 2003;63(4):389-406. Epub 2003/02/01.
[11]   Marr AK, Gooderham WJ, Hancock RE. Antibacterial peptides for therapeutic use: obstacles and realistic outlook. Curr Opin Pharmacol. 2006;6(5):468-72. Epub 2006/08/08.
[12]   Sanchez-Gomez S, Lamata M, Leiva J, Blondelle SE, Jerala R, Andrä J, *et al*. Comparative analysis of selected methods for the assessment of antimicrobial and membrane-permeabilizing activity: a case study for lactoferricin derived peptides. BMC Microbiol. 2008;8:196. Epub 2008/11/19.
[13]   Sanchez-Gomez S, Japelj B, Jerala R, Moriyon I, Fernandez Alonso M, Leiva J, *et al*. Structural features governing the activity of lactoferricin-derived peptides that act in

synergy with antibiotics against Pseudomonas aeruginosa *in vitro* and *in vivo*. Antimicrob Agents Chemother. 2011;55(1):218-28. Epub 2010/10/20.

[14]   Heinbockel L, Sanchez-Gomez S, Martinez de Tejada G, Dömming S, Brandenburg J, Kaconis Y, *et al*. Preclinical investigations reveal the broad-spectrum neutralizing activity of Peptide pep19-2.5 on bacterial pathogenicity factors. Antimicrob Agents Chemother. 2013;57(3):1480-7. Epub 2013/01/16.

[15]   Chen Y, Mant CT, Farmer SW, Hancock RE, Vasil ML, Hodges RS. Rational design of alpha-helical antimicrobial peptides with enhanced activities and specificity/therapeutic index. J Biol Chem. 2005;280(13):12316-29. Epub 2005/01/29.

[16]   Stromstedt AA, Pasupuleti M, Schmidtchen A, Malmsten M. Evaluation of strategies for improving proteolytic resistance of antimicrobial peptides by using variants of EFK17, an internal segment of LL-37. Antimicrob Agents Chemother. 2009;53(2):593-602. Epub 2008/11/26.

[17]   Miller SM SR, Ng S, Zuckermann RN, Kerr JM, Moss W. Proteolytic Studies of Homologous Peptide and N-Substituted Glycine Peptoid Oligomers. Bioorg Medic Chem Letrers. 1994;4(22):2662.

[18]   Nguyen LT, Chau JK, Perry NA, de Boer L, Zaat SA, Vogel HJ. Serum stabilities of short tryptophan- and arginine-rich antimicrobial peptide analogs. PloS One. 2010;5(9). Epub 2010/09/17.

[19]   Yu HY, Tu CH, Yip BS, Chen HL, Cheng HT, Huang KC, *et al*. Easy strategy to increase salt resistance of antimicrobial peptides. Antimicrob Agents Chemother. 2011;55(10):4918-21. Epub 2011/07/20.

[20]   Godballe T, Nilsson LL, Petersen PD, Jenssen H. Antimicrobial beta-peptides and alpha-peptoids. Chem Biol Drug Des. 2011;77(2):107-16. Epub 2011/01/27.

[21]   Wade D, Boman A, Wahlin B, Drain CM, Andreu D, Boman HG, *et al*. All-D amino acid-containing channel-forming antibiotic peptides. Proc Natl Acad Sci USA. 1990;87(12):4761-5. Epub 1990/06/01.

[22]   Costa F, Carvalho IF, Montelaro RC, Gomes P, Martins MC. Covalent immobilization of antimicrobial peptides (AMPs) onto biomaterial surfaces. Acta Biomater. 2011;7(4):1431-40. Epub 2010/11/09.

[23]   Tan ML, Choong PF, Dass CR. Recent developments in liposomes, microparticles and nanoparticles for protein and peptide drug delivery. Peptides. 2010;31(1):184-93. Epub 2009/10/13.

[24]   Mygind PH, Fischer RL, Schnorr KM, Hansen MT, Sonksen CP, Ludvigsen S, *et al*. Plectasin is a peptide antibiotic with therapeutic potential from a saprophytic fungus. Nature. 2005;437(7061):975-80. Epub 2005/10/14.

[25]   Hancock RE, Chapple DS. Peptide antibiotics. Antimicrob Agents Chemother. 1999;43(6):1317-23. Epub 1999/05/29.

[26]   Zasloff M. Antimicrobial peptides of multicellular organisms. Nature. 2002;415(6870):389-95.

[27]   Yount NY, Yeaman MR. Emerging themes and therapeutic prospects for anti-infective peptides. Ann Rev Pharmacol Toxicol. 2012;52:337-60. Epub 2012/01/13.

[28]   Wang G, Li X, Wang Z. APD2: the updated antimicrobial peptide database and its application in peptide design. Nucl Acids Res. 2009;37(Database issue):D933-7. Epub 2008/10/30.

[29]   Fjell CD, Hiss JA, Hancock RE, Schneider G. Designing antimicrobial peptides: form follows function. Nat Rev Drug Discov. 2012;11(1):37-51. Epub 2011/12/17.

[30]   Hancock RE, Nijnik A, Philpott DJ. Modulating immunity as a therapy for bacterial infections. Nature Rev Microbiol. 2012;10(4):243-54. Epub 2012/03/17.

[31]   Giuliani A, Pirri G, Bozzi A, Di Giulio A, Aschi M, Rinaldi AC. Antimicrobial peptides: natural templates for synthetic membrane-active compounds. Cell Molec Life Sci. 2008;65(16):2450-60. Epub 2008/07/29.

[32]   Arrowsmith J. Trial watch: Phase II failures: 2008-2010. Nature Rev Drug Disc. 2011;10(5):328-9. Epub 2011/05/03.

[33]   Arrowsmith J. Trial watch: phase III and submission failures: 2007-2010. Nature Rev Drug Disc. 2011;10(2):87. Epub 2011/02/02.

[34]   Steenbergen JN, Alder J, Thorne GM, Tally FP. Daptomycin: a lipopeptide antibiotic for the treatment of serious Gram-positive infections. J Antimicr Chemother 2005;55(3):283-8. Epub 2005/02/12.

[35]   Lipsky BA, Holroyd KJ, Zasloff M. Topical *vs.* systemic antimicrobial therapy for treating mildly infected diabetic foot ulcers: a randomized, controlled, double-blinded, multicenter trial of pexiganan cream. Clinical infectious diseases : an official publication of the Infectious Diseases Society of America. 2008;47(12):1537-45. Epub 2008/11/08.

[36]   Gottler LM, Ramamoorthy A. Structure, membrane orientation, mechanism, and function of pexiganan--a highly potent antimicrobial peptide designed from magainin. Biochim Biophys Acta. 2009;1788(8):1680-6. Epub 2008/11/18.

[37]   Melo MN, Castanho MA. Omiganan interaction with bacterial membranes and cell wall models. Assigning a biological role to saturation. Biochim Biophys Acta. 2007;1768(5):1277-90. Epub 2007/03/27.

[38]   Study of Omiganan 1% Gel in Preventing Catheter Infections/Colonization in Patients With Central Venous Catheters [database on the Internet]. 2009.

[39]   Levin M, Quint PA, Goldstein B, Barton P, Bradley JS, Shemie SD, *et al.* Recombinant bactericidal/permeability-increasing protein (rBPI21) as adjunctive treatment for children with severe meningococcal sepsis: a randomised trial. rBPI21 Meningococcal Sepsis Study Group. Lancet. 2000;356:961-7.

[40]   Shaji J, Patole V. Protein and Peptide drug delivery: oral approaches. Ind J Pharmac Sci. 2008;70(3):269-77. Epub 2008/05/01.

[41]   Schuerholz T, Brandenburg K, Marx G. Antimicrobial peptides and their potential application in inflammation and sepsis. Crit Care. 2012;16(2):207. Epub 2012/03/21.

[42]   Andrä J, Gutsmann T, Garidel P, Brandenburg K. Mechanisms of endotoxin neutralization by synthetic cationic compounds. J Endotoxin Res. 2006;12(5):261-77. Epub 2006/10/25.

[43]   Dankesreiter S, Hoess A, Schneider-Mergener J, Wagner H, Mietke T. Synthetic endotoxin-binding peptides block endotoxin-triggered TNF-à production by macrophages *in vitro* and *in vivo* and prevent endotoxin-mediated toxic shock. J Immunol. 2000;164:4804-11.

[44]   Nell MJ, Tjabringa GS, Wafelman AR, Verrijk R, Hiemstra PS, Drijfhout JW, *et al.* Development of novel LL-37 derived antimicrobial peptides with LPS and LTA neutralizing and antimicrobial activities for therapeutic application. Peptides. 2005.

[45]   Li J, Nation RL, Turnidge JD, Milne RW, Coulthard K, Rayner CR, *et al.* Colistin: the re-emerging antibiotic for multidrug-resistant Gram-negative bacterial infections. Lancet Infect Dis. 2006;6(9):589-601. Epub 2006/08/26.;8:367-85.

[46]   Cruz DN, Antonelli M, Fumagalli R, Foltran F, Brienza N, Donati A, *et al*. Early use of polymyxin B hemoperfusion in abdominal septic shock: the EUPHAS randomized controlled trial. JAMA. 2009;301(23):2445-52. Epub 2009/06/18.

[47]   Martin EL, Cruz DN, Monti G, Casella G, Vesconi S, Ranieri VM, *et al*. Endotoxin removal: how far from the evidence? The EUPHAS 2 Project. Contr Nephrol. 2010;167:119-25. Epub 2010/06/04.

[48]   Kaconis Y, Kowalski I, Howe J, Brauser A, Richter W, Razquin-Olazaran I, *et al*. Biophysical mechanisms of endotoxin neutralization by cationic amphiphilic peptides. Biophys J. 2011;100(11):2652-61. Epub 2011/06/07.

[49]   Andrä J, Garidel P, Majerle A, Jerala R, Ridge R, Paus E, *et al*. Biophysical characterization of the interaction of *Limulus polyphemus* endotoxin neutralizing protein with lipopolysaccharide. Eur J Biochem. 2004;271(10):2037-46. Epub 2004/05/07.

[50]   Andrä J, Howe J, Garidel P, Rössle M, Richter W, Leiva-Leon J, *et al*. Mechanism of interaction of optimized Limulus-derived cyclic peptides with endotoxins: thermodynamic, biophysical and microbiological analysis. Biochem J. 2007;406(2):297-307. Epub 2007/05/16.

[51]   Brandenburg K, Andra J, Garidel P, Gutsmann T. Peptide-based treatment of sepsis. Appl Microbiol Biotechnol. 2011;90(3):799-808. Epub 2011/03/04.

[52]   Gutsmann T, Razquin-Olazaran I, Kowalski I, Kaconis Y, Howe J, Bartels R, *et al*. New antiseptic peptides to protect against endotoxin-mediated shock. Antimicrob Agents Chemother. 2010;54(9):3817-24. Epub 2010/07/08.

[53]   Krepstakies M, Lucifora J, Nagel CH, Zeisel MB, Holstermann B, Hohenberg H, *et al*. A new class of synthetic peptide inhibitors blocks attachment and entry of human pathogenic viruses. J Infect Dis. 2012;205(11):1654-64. Epub 2012/03/30.

[54]   Brown BK, Karasavvas N, Beck Z, Matyas GR, Birx DL, Polonis VR, *et al*. Monoclonal antibodies to phosphatidylinositol phosphate neutralize human immunodeficiency virus type 1: role of phosphate-binding subsites. J Virol. 2007;81(4):2087-91. Epub 2006/12/08.

[55]   Compton T, Nowlin DM, Cooper NR. Initiation of human cytomegalovirus infection requires initial interaction with cell surface heparan sulfate. Virology. 1993;193(2):834-41. Epub 1993/04/01.

[56]   Ruemenapf T, Thiel, H.-J. Molecular biology of pestivirus. In: Mettenleiter T.C S, F., editor. Animal Virus: Molecular Biology. Norfolk: Academic Press; 2008.

[57]   Maurer K, Krey T, Moennig V, Thiel HJ, Rumenapf T. CD46 is a cellular receptor for bovine viral diarrhea virus. J Virol. 2004;78(4):1792-9. Epub 2004/01/30.

[58]   Roessle M,, Klaering R, Ristau U, Robrahn B, Jahn D, Gehrmann T, Konarev P, Round A, Fiedler S, Hermes C, Svergun D. (2007). Upgrade of the small-angle X-ray scattering beamline at the European Molecular Biology Laboratory, Hamburg. J Appl Cryst. 40:190-4.

[59]   Mahy B. KH. Virological Methods Manual: Academic Press; 1996.

[60]   Kärber, G. Beitrag zur kollektiven Behandlung pharmakologischer Reihenversuche. Arch Exp Pathol Pharmacol. 1931;162:1.

[61]   Catania A, Colombo G, Rossi C, Carlin A, Sordi A, Lonati C, *et al*. Antimicrobial properties of alpha-MSH and related synthetic melanocortins. Scient World J. 2006;6:1241-6. Epub 2006/10/10.

[62]   Grieco P, Rossi C, Colombo G, Gatti S, Novellino E, Lipton JM, *et al*. Novel alpha-melanocyte stimulating hormone peptide analogues with high candidacidal activity. J Med Chem. 2003;46(5):850-5. Epub 2003/02/21.

[63]   Cutuli M, Cristiani S, Lipton JM, Catania A. Antimicrobial effects of alpha-MSH peptides. J Leukoc Biol. 2000;67(2):233-9. Epub 2000/02/12.

[64]   van den Berg HR, Khan NA, van der Zee M, Bonthuis F, JN IJ, Dik WA, *et al.* Synthetic oligopeptides related to the [beta]-subunit of human chorionic gonadotropin attenuate inflammation and liver damage after (trauma) hemorrhagic shock and resuscitation. Shock. 2009;31(3):285-91. Epub 2008/07/26.

[65]   Vertesy L, Ehlers E, Kogler H, Kurz M, Meiwes J, Seibert G, *et al.* Friulimicins: novel lipopeptide antibiotics with peptidoglycan synthesis inhibiting activity from Actinoplanes friuliensis sp. nov. II. Isolation and structural characterization. J Antibiot (Tokyo). 2000;53(8):816-27. Epub 2000/11/18.

[66]   Schneider T, Gries K, Josten M, Wiedemann I, Pelzer S, Labischinski H, *et al.* The lipopeptide antibiotic Friulimicin B inhibits cell wall biosynthesis through complex formation with bactoprenol phosphate. Antimicrob Agents Chemother. 2009;53(4):1610-8. Epub 2009/01/24.

[67]   Nibbering PH, Ravensbergen E, Welling MM, van Berkel LA, van Berkel PH, Pauwels EK, *et al.* Human lactoferrin and peptides derived from its N terminus are highly effective against infections with antibiotic-resistant bacteria. Infect Immun. 2001;69(3):1469-76. Epub 2001/02/17.

[68]   Lupetti A, Paulusma-Annema A, Welling MM, Senesi S, van Dissel JT, Nibbering PH. Candidacidal activities of human lactoferrin peptides derived from the N terminus. Antimicrob Agents Chemother. 2000;44(12):3257-63. Epub 2000/11/18.

[69]   Huo L, Zhang K, Ling J, Peng Z, Huang X, Liu H, *et al.* Antimicrobial and DNA-binding activities of the peptide fragments of human lactoferrin and histatin 5 against *Streptococcus mutans*. Arch Oral Biol. 2011;56(9):869-76. Epub 2011/03/09.

[70]   van der Does AM, Bogaards SJ, Ravensbergen B, Beekhuizen H, van Dissel JT, Nibbering PH. Antimicrobial peptide hLF1-11 directs granulocyte-macrophage colony-stimulating factor-driven monocyte differentiation toward macrophages with enhanced recognition and clearance of pathogens. Antimicrob Agents Chemother. 2010;54(2):811-6. Epub 2009/11/26.

[71]   Velden WJ, van Iersel TM, Blijlevens NM, Donnelly JP. Safety and tolerability of the antimicrobial peptide human lactoferrin 1-11 (hLF1-11). BMC Medic. 2009;7:44. Epub 2009/09/09.

[72]   Chen J, Falla TJ, Liu H, Hurst MA, Fujii CA, Mosca DA, *et al.* Development of protegrins for the treatment and prevention of oral mucositis: structure-activity relationships of synthetic protegrin analogues. Biopolymers. 2000;55(1):88-98. Epub 2000/08/10.

[73]   Trotti A, Garden A, Warde P, Symonds P, Langer C, Redman R, *et al.* A multinational, randomized phase III trial of iseganan HCl oral solution for reducing the severity of oral mucositis in patients receiving radiotherapy for head-and-neck malignancy. Intern J Radiat Onc Biol Phys. 2004;58(3):674-81. Epub 2004/02/18.

[74]   Giacometti A, Cirioni O, Ghiselli R, Orlando F, Kamysz W, Rocchi M, *et al.* Effects of pexiganan alone and combined with betalactams in experimental endotoxic shock. Peptides. 2005;26(2):207-16.

[75]   Haug BE, Stensen W, Kalaaji M, Rekdal O, Svendsen JS. Synthetic antimicrobial peptidomimetics with therapeutic potential. J Med Chem. 2008;51(14):4306-14. Epub 2008/06/24.

[76]   Saravolatz LD, Pawlak J, Johnson L, Bonilla H, Saravolatz LD, 2nd, Fakih MG, *et al. In Vitro* Activities of LTX-109, a Synthetic Antimicrobial Peptide, against Methicillin-Resistant, Vancomycin-Intermediate, Vancomycin-Resistant, Daptomycin-Nonsusceptible, and Linezolid-Nonsusceptible *Staphylococcus aureus*. Antimicrob Agents Chemother. 2012;56(8):4478-82. Epub 2012/05/16.

[77]   Isaksson J, Brandsdal BO, Engqvist M, Flaten GE, Svendsen JS, Stensen W. A synthetic antimicrobial peptidomimetic (LTX 109): stereochemical impact on membrane disruption. J Med Chem. 2011;54(16):5786-95. Epub 2011/07/08.

[78]   Sader HS, Fedler KA, Rennie RP, Stevens S, Jones RN. Omiganan pentahydrochloride (MBI 226), a topical 12-amino-acid cationic peptide: spectrum of antimicrobial activity and measurements of bactericidal activity. Antimicrob Agents Chemother. 2004;48(8):3112-8. Epub 2004/07/27.

[79]   Fritsche TR, Rhomberg PR, Sader HS, Jones RN. Antimicrobial activity of omiganan pentahydrochloride against contemporary fungal pathogens responsible for catheter-associated infections. Antimicrob Agents Chemother. 2008;52(3):1187-9. Epub 2008/01/09.

[80]   Fritsche TR, Rhomberg PR, Sader HS, Jones RN. *In vitro* activity of omiganan pentahydrochloride tested against vancomycin-tolerant, -intermediate, and -resistant *Staphylococcus aureus*. Diagn Microbiol Infect Dis. 2008;60(4):399-403. Epub 2008/01/08.

[81]   Elsbach P, Weiss J. The bactericidal/permeability-increasing protein (BPI), a potent element in host-defense against gram-negative bacteria and lipopolysaccharide. Immunobiology. 1993;187(3-5):417-29. Epub 1993/04/01.

[82]   De Vry CG, Valdez M, Lazarov M, Muhr E, Buelow R, Fong T, *et al.* Topical application of a novel immunomodulatory peptide, RDP58, reduces skin inflammation in the phorbol ester-induced dermatitis model. J Invest Dermatol. 2005;125(3):473-81. Epub 2005/08/25.

[83]   Gonzalez RR, Fong T, Belmar N, Saban M, Felsen D, Te A. Modulating bladder neuro-inflammation: RDP58, a novel anti-inflammatory peptide, decreases inflammation and nerve growth factor production in experimental cystitis. J Urol. 2005;173(2):630-4. Epub 2005/01/12.

[84]   Schuerholz T, Domming S, Hornef M, Dupont A, Kowalski I, Kaconis Y, *et al.* Bacterial cell wall compounds as promising targets of antimicrobial agents II. Immunological and clinical aspects. Curr Drug Targets. 2012;13(9):1131-7. Epub 2012/06/06.

[85]   Martinez de Tejada G, Sanchez-Gomez S, Razquin-Olazaran I, Kowalski I, Kaconis Y, Heinbockel L, *et al.* Bacterial cell wall compounds as promising targets of antimicrobial agents I. Antimicrobial peptides and lipopolyamines. Curr Drug Targets. 2012;13(9):1121-30. Epub 2012/06/06.

CHAPTER 2

New Antimicrobial Agents Against Common Bacteria that Cause Serious Infections

Ursula Kelly[*] and Nancy Khardori

Division of Infectious Disease, Department of Internal Medicine, Eastern Virginia Medical School, Norfolk, Virginia, USA

Abstract: Antimicrobial agents are used to treat infections involving biological agents (pathogens) other than humans and the goal is to rid the host of the pathogen. Because of the involvement of a second living agent in the Host-Pathogen-Antimicrobial agent triangle, drug therapy is affected by the biological characteristics of the pathogen including its tissue specificity and most importantly the changes it undergoes to survive while being exposed to the antimicrobial agent. It has become very clear that the drugs that are used to treat infections allow the pathogens to enhance their capability to become resistant and furthermore transfer this resistance to other pathogens. Only appropriate and judicious use of anti-infectives can prolong their effectiveness.

This chapter will provide an overview of the historical perspective and current antimicrobial agents. This will be followed by a discussion on the recently developed antimicrobial agents, those that are in development, including those with novel mechanisms of action.

Keywords: Aminoglycoside, antibiotics, antimicrobial agents, cephalosporin, fluoroquinolone, glycopeptides, lipoglycopeptide, macrolide, streptogramin, β-lactam.

INTRODUCTION

Historically, more people have died in infectious disease epidemics than all the wars. The "killing" potential of communicable diseases was known even before the germ theory of disease was established. A large number of "remedies" were tried without a known target. In the late 19th century, the Koch's postulates and microbiologic methods for *in vitro* cultivation of bacteria paved the way for

*Address correspondence to Ursula Kelly:** Division of Infectious Disease, Department of Internal Medicine, Eastern Virginia Medical School, Norfolk, Virginia, USA; Tel: 757-446-8910; Fax: 757-446-7922; E-mail: kellyum@evms.edu

scientific study of antimicrobial agents against common bacterial diseases. Sulfonamides and polymyxins were studied in the 1920's but their narrow spectrum and narrow therapeutic index respectively kept them from making history. The major characteristics that define a good antimicrobial agent are: 1) selective toxicity against pathogens while largely sparing the host cells, 2) capability to give levels in the blood, body fluids, and tissues that are sufficiently above the minimum concentrations required to inhibit/kill different types of bacteria but at the same time do not give major adverse events, and 3) chemical structure is such that it resists the large number of mechanisms available to the bacteria to become resistant to antimicrobial agents. The continued search and need for newer antimicrobials is a testament to the fact that the discovery of a "magic bullet" against bacteria has eluded generations of scientists. In the western world, the impact of infectious disease has significantly been reduced by public health infrastructure and effective and affordable vaccines. However, globally, infectious diseases remain the most common cause of morbidity and mortality.

The commercial availability of penicillin in 1944 was a landmark event and drastically reduced the fatalities from infection during the second world war. However, within a short period the capability of *Staphylococcus aureus* to produce penicillinase with ability to destroy penicillin was discovered. Since then, the war between making new antimicrobial agents and finding them rendered less and less effective by a multitude of genotypic and phenotypic resistance mechanisms used by the bacteria has been on-going. Although, this war has not been completely lost by antimicrobial agents, we have run out of novel mechanisms that can be used to make effective agents, particularly against the bacteria that have developed multiple mechanisms of resistance. This is a protective mechanism that bacteria use to survive under the "selective pressure" of wide use and misuse of antimicrobial agents. The biological capability of bacteria to transfer resistance genes between species and genera adds fuel to the fire.

This chapter will provide the reader with a short but comprehensive historical perspective on the discovery of various antibiotic classes followed by a discussion on newer agents and agents in the pipeline. Although the terms "antimicrobial agents" and "antibiotics" are often used synonymously, antimicrobial agents include antibiotics that are made by living agents, such as penicillin, while chemotherapeutic agents such as sulfonamides are synthesized. For the purpose of clarity we will use the term "antimicrobial agents" in this chapter.

1940-1980: THE GOLDEN ERA OF ANTIBIOTIC DISCOVERY

β-Lactams

Penicillins

A strong foundation is the key to future growth and development in any field. In the antimicrobial world we have a foundation in Penicillin, a 6-aminopenicillanic acid that consists of a thiazolidine ring and a β-lactam ring and a side chain [1]. A compilation of works in the early 1900's from Britain's Fleming, Florey, and Chain led to the initial clinical trials of Penicillin in streptococcal and gonococcal infections, and just a few short years later it's commercial production and widespread use is credited with saving countless lives in World War II [1, 2]. It is the safety profile of penicillin that allowed it to become the first most widely accepted antibiotic, as other antimicrobial agents had been previously used to treat infection but not mass produced. It was this pivotal event that furthered the field of antimicrobial agents in the 1940's and 1950's, bringing in antibiotics that we still use today including the tetracyclines, aminoglycosides, and glycopeptides [3]. Further modification of the parent pencillin led to the development of penicillinase resistant penicillins such as oxacillin, aminopenicillins such as ampicillin, carboxypenicillins such as carbenacillin, and ureidopenicillins such as piperacillin. Although all penicillins bind to the targeted penicillin binding proteins the further mechanism of activity depends of the type of cell wall. Today, despite having a broader selection of antibiotics to choose from, penicillins remain an integral part of our anti-infectious arsenal. Penicillins are bactericidal against susceptible gram-positive and gram-negative organisms, and the natural penicillins remain relevant and continue to be the treatment of choice for syphilis, a re-emerging disease in the developing world [4, 2]. Penicillin G is the prototype β-lactam antibiotic and shares its mechanism of action is the foundation for all the β-lactam antibiotics and requires that the bacterium has a cell wall with a peptidoglycan layer. During active bacterial division the penicillin binds to the enzymes transpeptidase, carboxypeptidase, and endopeptidase (penicillin-binding-proteins or PBP's) and prevents the crosslinking of peptidoglycan yielding an osmotically unstable cell wall structure. This then activates cell lysis and subsequent death [1].

The great power of antibiotics carries with it the weight of responsibility in their proper application. The process of selective forces inherent in the use of an antimicrobial agent on a pathogen has led to antibiotic resistance. The first and most clear example of this is in the discovery of β-lactamases in the 1940's [5]. The diverse and large family of β-lactamases present in gram-positive, gram-negative bacteria hydrolyze the β-lactam ring leading to loss of activity [5, 6, 1]. Since then it has become clear that other mechanisms of resistance are at work and they include PBP alterations such as decreased affinity to the β-lactam antibiotic, increased number of PBP's and the attainment of resistant PBP's [7]. In gram-negative organisms the lipopolysaccharide layer can alter in a way that decreases the ability of the antibiotic to penetrate the organism and particularly in organisms such as pseudomonas an efflux system can push the agent out of the cell [7, 8]. The need to overcome these complex interactions between resistance and the need for effective agents furthered the development of additional classes of antibiotics. By changing the penicillin side chain the first penicillinase-resitant penicillins were created in the 1940's, methicillin, oxacillin, and nafcillin. Following this, the need for gram negative coverage prompted the development of additional penicillin antibiotics, including aminopenicillins, ampicillin and amoxicillin. Further additions were the carboxypenicillins and ticaricillin and the ureidopenicllins including piperacillin, which brought with them anti-pseudomonal activity. The antimicrobial spectrum of these extended spectrum penicillins was further modified/enhanced by the addition of β-lactamase inhibitors to β-lactam agents. These combinations include amoxicillin-clavulanate, ampicillin-sulbactam, ticaricillin-clavulanate, and piperacillin-tazobactam. These are effective against many gram-negative organisms, except notably those with the production of extended-spectrum β-lactamases [2].

Cephalosporins

Introduced in the 1960's as Cephalosporin C, this large family of β-lactam antibiotics is structurally different from penicillins in that they have a dihydrothiaizine ring, rather than a thiazolidine ring is fused to the β-lactam ring [2, 7]. At the time of introduction, the first generation cephalosporins (cefazolin, cephalexin) were stable to the predominating β-lactamases and as they penetrated the cell wall of gram-negative bacilli more rapidly than penicillins they were used with increasing frequency. As a consequence of the production and clinical application of cephalosporins, gram-negative organisms producing a Class C cephalosporinase emerged, resulting in previously uncommon pathogens becoming the leading nosocomial pathogens [9]. Despite this, the first generation cephalosporin, cefazolin, remains a relevant antibiotic, frequently used for methicillin-susceptible *S. aureus* infections and surgical prophylaxis [2, 10].

The second generation cephalosporins, (cefoxitin, cefotetan) are more stable to cephalosporinases and plasmid-mediated β-lactamases, with increased activity against gram-negative organisms [7]. From this group cefotetan and cefoxitin have activity against anaerobes as well [2]. Resistance to this class of antibiotics was identified to be due to various species-specific inducible β-lactamases [7].

Third generation cephalosporins include ceftazidime and ceftriaxone, and in general are more active against the gram-negative organisms and less active against gram-positive cocci [7]. Ceftazidime is active against *Pseudomonas aeruginosa*, and so is cefepime, which is a member of the fourth-generation class of cephalosporins [2]. The side-chain modifications give the agents their antimicrobial spectrum and variations in pharmacokinetics and stability to the different resistance mechanisms [9].

Monobactams

Aztreonam is the only available monobactam. It is distinguished from penicillins and cephalosporins by the addition of a sulfonic acid group on the N1 position nitrogen of the β-lactam ring, which activates the β-lactam ring, with a similar mechanism of action as the penicillins and cephalosporins [11]. It is active only against gram-negative organisms [2]. The weak immunogenicity of monobactams lend them for use against susceptible bacteria in patients with serious hypersensitivity to penicillin and cephalosporins. The predominantly gram-negative activity makes them comparable to aminoglycosides without the burden of nephrotoxicity [11].

Carbapenems

With a fundamentally similar mechanism of action as other β-lactam antibiotics, the carbapenems, imipenem and meropenem have the broadest spectrum of antimicrobial activity. Interestingly, they are small in size, and capable of gaining access to the cell through porin channels of gram-negative bacterial cell walls [11]. The broad activity is also due to the high affinity of the carbapenems to penicillin-binding proteins [7]. Although weakened by metallo-β-lactamases , especially those produced by *Klebsiella pneumoniae*, they are bactericidal against susceptible gram-positive, gram-negative aerobes and anaerobes including *P. aeruginosa* [2, 7]. A newer carbapenem, ertapenem, is an exception and does not have activity against *P. aeruginosa* and Enterococci [2].

Aminoglycosides

Initially this class of antibiotics (streptomycin) was used to treat tuberculosis in the 1940's. Gentamicin was introduced in the early 1960's for aerobic gram-negative organisms, particularly *P. aeruginosa*. The aminoglycosides, tobramycin, amikacin, gentimicin, irreversibly bind to the 30s subunit of bacterial ribosomes inhibiting protein synthesis by interfering with reading of the genetic code. They have a concentration-dependent killing effect and are inactivated in anaerobic conditions (such as an abscess). Interestingly, the uptake of aminoglycosides is enhanced in the presence of cell wall synthesis inhibitors, such as β-lactams and glycopeptides. The toxicities of the aminoglycosides are significant and well known. Nephrotoxicity comes in the form of non-oliguric decrease in glomerular filtration rate at least a week in to therapy, and is usually reversible. This is in contrast to the irreversible vestibular and auditory ototoxicity, which affects the elderly disproportionately and is difficult to detect in the ill and hospitalized patient. Despite the considerable side effects profile it is still a relevant class of antimicrobial agents due to its continued activity against bacteria resistant to other classes [2, 12].

Tetracyclines

Available since the 1950's, the tetracycline class of antimicrobial agents include minocycline and doxycycline. They inhibit bacterial protein synthesis by binding to the 30s ribosomal subunit and block the attachment of transfer RNA to messenger RNA [13]. They are bacteriostatic against susceptible aerobic and anaerobic organisms including methicillin-resistant *S. aureus* (MRSA), and are particularly useful in the treatment of rickettsial diseases and atypical pulmonary infections [2, 13]. These agents have been underutilized in the past 2 decades because of the availability of newer drugs. Their retained activity against gram-positive organisms, including methicillin-resistant *S. aureus* and overall good tolerance has allowed us to bring them back as oral drugs for *S. aureus* infections. They have been especially useful for treating mild-moderate community-acquired skin infections caused by methicillin-resistant *S. aureus.* Bacterial resistance due to efflux of the tetracycline class has been overcome by the development of tigecycline, which will be discussed later in this chapter.

Macrolides and Ketolides

The macrolides are bacteristatic agents whose use began in the 1950's when the parent macrolide, erythromycin, became available. They bind to the 50S ribosomal subunit in a reversible fashion, thereby inducing dissociation of peptidyl transfer RNA from the ribosome during the elongation phase. They are frequently used to treat pulmonary infections as concentrations found in alveolar macrophages are significantly higher than that found in the serum [14]. Erythromycin has activity against streptococci and methicillin-susceptible *S. aureus*, but its major role is in treating atypical pulmonary infections, including *Mycoplasma pneumoniae* and *Legionella pneumophilia, Bordetella pertussis* and other Bordetella species. Erythromycin has a high rate of gastrointestinal intolerance and needs frequent dosing. Advances in macrolide development made available clarithromycin and azithromycin both of which have less gastrointestinal intolerance and can be given once or twice a day. In addition they have activity against *Haemophilus influenzae*, Chlamydia species and non-tuberculous Mycobacteria. Further delineating the role of the group, clarithromycin appears more active against *Legionella pneumophilia* and *Chlamydophila pneumoniae* and azithromycin is more against *Moraxella catarrhalis* and *Mycoplasma pneumoniae* [14]. A recent study from a children's hospital evaluated 2251 upper respiratory samples from 2007-2010, which showed an overall rate of macrolide resistance in *Mycoplasma pneumoniae* of 8.2%. However, when the data was evaluated in

1-year intervals, 2007-2008 and 2009-2010, it showed a rise in the rate of resistance from 3.3% to 15.8% [15]. The responsible resistance mutation is one most commonly found in Europe and Asia, and has been associated with exposure to subtherapeutic levels of azithromycin [15].

Ketolides, such as telithromycin have the same mechanism of action as the macrolides and became available in 2005 for the treatment of community-acquired pneumonia and sinusitis [2]. However, in 2007, secondary to safety concerns the FDA withdrew it from the market [16].

Lincosamides

In 1962 the prototype lincosamide, lincomycin, was isolated from *Streptomyces lincolnesis*. The modern day clindamycin was developed 4 years later. Lincosamides bind to the 50s subunit of bacterial ribosomes and inhibit protein synthesis. As the site of action is the same between the macrolides and lincosamides, these agents should never be used together as they can act in an antagonistic way. This also is the basis of cross-resistance between macrolides and lincosamides. The spectrum of activity of clindamycin includes susceptible gram-positive organisms, except enterococci, and anaerobic gram-positive and gram-negative organisms, except *Clostridium difficile* and *Fusobacterium varium*. Aerobic gram-negative organisms should be considered resistant to clindamycin [2].

Glycopeptide and Lipoglycopeptides

The first glycopeptide, vancomycin, was released to the market in 1958. Followed by the lipoglycopeptide, teicoplanin, 30 years later. The latter is approved for use in several countries in Europe and Asia but has not been approved by the FDA in the US [2, 17]. Vancomycin and teicoplanin are chemically similar antibiotics derived from naturally occurring products that are active against gram-positive organisms including staphylococcus, streptococcus, and enterococcus [2, 18, 19]. The glycopeptide/lipoglycopeptide agents act by inhibiting peptidoglycan synthesis *via* binding of peptide precursors, preventing transpeptidation in susceptible bacteria [2, 17]. Although the antimicrobial spectrum and mechanism of action of vancomycin and teicoplanin are identical the pharmacokinetics differs

[19]. Teicoplanin exhibits a longer half-life making once or twice a day dosing possible and it can be used *via* administration by intramuscular injection which notably is not adequate for CSF penetration [17, 19]. Vancomycin acheives higher levels of free serum concentrations, and is preferred for CNS infections [19]. Both act on bacteria at an earlier stage than the β-lactam class and therefore there is no cross-resistance between these two classes. Up until recently this class was uniformly active against all major gram-positive organisms. As in resistance to β-lactam class, there are several mechanisms by which bacteria have become resistant to the glycopeptide class. Essentially, plasmid carried and genetically encoded resistance occurs *via* production of low-affinity precursor binding sites or removal of the binding site, which can be transferred between species [20]. This was the mechanism described among Enterococci (VRE) in 1988 and in staphylococcus species (vancomycin-resistant *S. aureus*) in 1996. This mechanism renders the organisms fully resistant glycopeptides. Another mechanism of resistance has been described among staphylococci, generally referred to as vancomycin-intermediate *S. aureus* (VISA) which involves a cell wall barrier to vancomycin entry [18, 20].

Fluoroquinolones

Nalidixic acid was introduced in 1962 is the parent fluoroquinolone [21]. As a class, fluoroquinolones have a bactericidal mechanism of action that involves targeting of topoisomerase II (DNA gyrase) and IV, bind to these enzymes, and disrupt DNA replication, repair, and transcription which results in cell death [22]. Similar to cephalosporins, the fluoroquinolones have structural modifications which alter their spectrum of activity and help to group them in to 'generations' [22]. Nalidixic acid being the first generation, ciprofloxacin is the second generation, and levofloxacin and moxifloxacin are the third generation. The earlier, ie. first and second generation fluoroquinolones have primarily gram negative activity whereas the third generation fluoroquinolones have much improved activity against gram-positive organisms [2]. The need for the third generation fluoroquinolones was related to the appearance of penicillin resistance in *Streptococcus pneumoniae* and because of extensive penetration in to lungs, these agents are very useful for treating bacterial pneumonia for which *S. pneumoniae* is a predominant cause. All fluoroquinolones have activity against *Moraxella catarrhalis, Haemophilus influenzae*, and *Legionella pneumophila*.

Sulfamethoxazole-Trimethoprim

Sulfonamides have been used since 1930's, but in 1974 the combination antimicrobial agent trimethoprim (TMP)/sulfamethoxazole (SMX) became widely available and remains one of the most commonly prescribed agents in the world [2]. The combination blocks bacterial folate synthesis by inhibiting two sequential steps in folate production. In general, it exhibits bactericidal killing effects against susceptible aerobic gram-positive and gram-negative bacteria, including community-acquired *S. aureus, Stenotrophomonas maltophilia*, and *Burkholderia cepacia* [2].

1980's-2000: The Slowdown

In the decades leading up to the new millennium growth and development in the antimicrobial agents slowed significantly. Very few new classes of antimicrobials were introduced for systemic use between 1970 and 2000 [16]. Meanwhile the resistance of gram-positive and gram-negative organisms continued to grow. As previously noted, resistance of gram-positive organisms has extended to vancomycin. In response to concern for this growing problem, antibacterial agents with activity against gram-positive organisms including those that had developed resistance to vancomycin, were brought to market. However, discovery of agents to combat highly resistant gram-negative organisms has lagged behind. As a consequence we have turned to the use of some older agents such as polymyxins that were at one point thought to have undesireable side effects.

Streptogramins

A complicated antimicrobial agent, the streptogramin prototype quinupristin/dalfopristin is the manufactured combination of two naturally occurring products produced by *Streptomyces pristinaespiralis*. This agent was developed as a follow-up drug on pristinomycin used in oral formulation in many European countries. The function of the combination is as follows: Dalfopristin inhibits protein synthesis at the elongation phase of replication, and causes a conformational change of ribosomes which allows for increased binding affinity to quinupristin. Quinupristin then blocks the bond formation between the proteins, thereby causing release of incomplete protein chains. The effect of this synergistic relationship can be bactericidal against certain organisms, though not in the enterococcal species. It has been used since 1999 in patients with infections due to resistant gram-positive organisms not susceptible to alternate agents, or when the patient is intolerant to those agents. The spectrum of activity against gram-positive organisms includes methicillin-susceptible *S. aureus*, methicillin-resistant *S. aureus*, *S. pneumoniae, Corynebacteria, Legionella pneumophilia, Moraxella catarrhalis, Neisseria meningitides,* and *Enterococcus faecium* (including vancomycin-resistant enterococcus), but is notably not active against *E. faecalis* [2, 23].

Recalling The Past

Polymyxins

Polymyxins B and polymyxin E (colistin) were studied as effective antimicrobial agents back in the 1920's. Prior to the availability of daptomycin they were the only antibacterial agents with activity at the level of the cell membrane. This target decreased their selectivity and caused adverse events at a time when safer classes of antimicrobial agents were on the horizon. Because of this, they were not used much up until the emergence of gram-negative organisms resistant to all β-lactams including carbepenems such as *Acinetobacter baumanii*. They are now used frequently to treat infection due to multi-drug resistant gram-negative organisms and those that have not responded to other classes, with close monitoring of renal function. Other nephrotoxic agents including aminoglycoside antimicrobial agents must be avoided in patients receiving polymyxins.

Fosfomycin

Discovered in 1969, fosfomycin is a bactericidal, phosphonic class antibiotic [24]. It is active against both gram-positive and gram-negative organisms but it has found its niche as an oral agent for resistant gram negative complicated and uncomplicated lower urinary tract infections [25]. It inhibits cell wall synthesis by blocking the first step in production, stopping the formation of N-acetylmuramic acid, a critical component of peptidoglycan [25]. Fosfomycin is approved in European countries and the US, however its intravenous formulation used for severe infections in some of the European countries is not available in the US. The oral formulation is primarily used for treatment of uncomplicated and complicated cystitis.

Rifampin

Rifampin has an extensive spectrum of activity against a number of bacteria and fungi and excellent pharmacokinetics, however the rather quick development of resistance to this drug has kept it from being used as a single agent against bacterial infections. In the past decade because of the ongoing resistance among gram-positive organisms and failures in deep-seated infections, particularly those associated with devices, rifampin has been used as a second agent based on some older studies that showed synergy between cell wall active agents and rifampin against *S. aureus*. However, caution must be exercised in using this drug judicially and optimally to retain it as one of the first line and most efficacious drugs against *Mycobacterium tuberculosis*.

2000-2012: The Latest Arrivals (Table 1)

Table 1: Antimicrobial Agents Approved by the FDA in the Last 12 Years (2000-2012).

Year	Drug	Class	Novel Mechanism of Action
2000	Linezolid	Oxazolidinone	Yes
2001	Ertapenem	Carbepenem	No
2003	Daptomycin	Lipopeptide	Yes
2003	Gemifloxacin	Fluoroquinolone	No
2005	Tigecycline	Tetracycline	No
2007	Doripenem	Carbapenem	No
2009	Telavancin	Lipoglycopeptide	No
2010	Ceftaroline	Cephalosporin	No
2011	Fidaxomicin	Macrolide	No

Oxazolidinones

After 20 years of work in development, the prototype oxazolidinone, linezolid, was approved for use in 2000 by the US FDA [26]. Oxazolidinones act in a bacteriostatic or bactericidal fashion (organism and concentration dependent) against many common gram-positive organisms including methicillin-susceptible *S. aureus*, methicillin-resistant *S. aureus*, and *S. pneumonia*, and also have activity against *Legionella spp*, and *Chlamydia pneumoniae* [2]. The mechanism of action includes inhibition of protein synthesis at the initiation phase, by binding to the 23s portion of the ribosomal 50s subunit [26]. The official approval includes use for both complicated and uncomplicated skin and soft tissue infections, community and nosocomial-acquired pneumonia, and infections due to vancomycin-resistant Enterococci, though resistance to linezolid is a concern in this bacterial population [27]. It has the attractive feature of being available in an oral formulation as well as IV formulation, but long tern use is associated with myelosuppression, particularly thrombocytopenia [28].

Lipopeptides

Initially discovered in the 1980's, daptomycin, the naturally occurring cyclic fermentation byproduct of *Streptomyces roseosporus*, was pulled from clinical trials secondary to safety concerns particularly skeletal muscle toxicity [2, 29]. It did not come to life until 1997, when the need for new agents against resistant gram-positive organisms prompted its re-evaluation for use. Its activity against gram-positive organisms is by binding to the cell membrane in a calcium-dependent fashion, inserting its lipid tail which results in depolarization of the membrane. The subsequent chain of events that leads to cell death includes release of intracellular potassium and cessation of DNA, RNA, and protein synthesis [2, 29]. It is unable to penetrate the cell membrane of gram-negative organisms rendering it ineffective against those microbes [29]. Currently it is approved for use against methicillin-susceptible *S. aureus*, methicillin-resistant *S. aureus*, vancomycin-susceptible strains of Enterococcus, and Streptococci including *S. agalactiae* and *S. dysgalactiae*, in complicated skin and skin structure infections. Although it has good *in vitro* activity against vancomycin-resistant enterococcus the clinical data provided to the US FDA was not enough to warrant this indication. However, post-marketing daptomycin has been used in patients with vancomycin-resistant enterococcus infections. Clinical trials have shown that it is more effective at higher doses, despite the approved dosing of 4-6mg/kg, because the risk of skeletal muscle toxicity increases with increased concentration [29]. Furthermore it is not approved for use in primary pulmonary infections as daptomycin interacts irreversibly with surfactant, causing sequestration and inactivity of the antibiotic [30].

Third-Generation Fluoroquinolones

The third-generation fluoroquinolones, moxifloxacin, gatifloxacin, and gemifloxacin are mentioned previously in the 'Fluoroquinolone' section of this chapter. It is notable that although the drug class has its beginning in the 1960's there has been continued evaluation and modification leading to agents with improved activity against gram-positive organisms especially *Streptococcus pneumoniae*. In 1999 the FDA approved of moxifloxacin and gatifloxacin for acute sinusitis, chronic bronchitis, and community-acquired pneumonia. Just a few short years later in 2003 gemifloxacin was approved for the same indications [16, 22]. It is a common clinical practice to use these agents as monotherapy for various suspected polymicrobial infections, including diabetic foot and wound infections. This off-label use must be done with caution as there is increasing resistance to fluoroquinolones amongst common pathogens including *S. aureus* and *E. coli* [22]. The newer agents of the third generation fluoroquinolones show improved activity against *Chlamydia pneumoniae* and *Mycoplasma pneumoniae*, as well as *Bacteroides fragilis* [22].

Glycylcycline/Tigecycline

In 2003 and 2005 the emergence of *Klebsiella pneumoniae* carbapenemases (KPC) and metallo-β-lactamases (MBL) were documented on the rise in the US. These are capable of hydrolyzing β-lactams easily and are not impacted by the currently available β-lactamase inhibitors, yielding the carrier pathogen multi-drug resistant [31]. Tigecycline was added to our antimicrobial armamentarium in 2005, when it was approved by the FDA about a decade after its initial discovery. A bacteriostatic derivative of the tetracycline antibiotics, tigecycline is the first glycylcycline that acts by binding to the 30s ribosomal subunit of bacteria, preventing elongation of peptide chains. This leads to inhibition of bacterial protein synthesis [2]. The benefit of the glycylcycline class against the world of multi-drug resistant pathogens is its ability to maneuver past many of the bacterial resistance mechanisms. It has a bulky moiety which yields steric hindrance, prevents the drug efflux *via* bacterial pump mechanisms and has high affinity binding to the 30s subunit despite target site modifications [32]. These characteristics are what give it a broad spectrum of activity against clinically important multi-drug resistant organisms including vancomycin-resistant enterococci, methicillin-resistant *Staphylococcus aureus*, Strep sp, and anaerobes [2, 32]. It is notably not effective against *Pseudomonas, Provedencia* and *Proteus* species, and it does not achieve antimicrobial level the urine.

Ertapenem/Doripenem

These carbepenems were approved in 2001 and 2007, respectively. They have the same mechanism of action as other carbepenems. Doripenem has the same spectrum of activity as imipenem and meropenem. However, ertapenem lacks activity against *Pseudomonas aeruginosa* and Enterococci. All of them have activity against extended β-lactamase producing organisms. The pharmacokinetic advantage of ertapenem lies in its once a day dosing and choice of intravenous or intramuscular administration.

Semisynthetic Lipoglycopeptides

Following an over 50 year history of clinical use of vancomycin we saw a rise in the prevalence of drug resistant gram positive organisms, including methicillin-resistant *S. aureus*, vancomycin-intermediate *S. aureus*, vancomycin-resistant *S. aureus*, and vancomycin-resistant enterococci. This has placed the introduction of the new lipoglycopeptides, telavancin, oritavancin, and dalbavancin in the spotlight, and they constitute the bulk of new antimicrobial development in the past 12 years (Table **2**). They are synthetic derivatives of vancomycin that structurally have the lipophilic side chains that place them in the lipoglycopeptide group [33]. These agents are bactericidal as opposed to vancomycin which is bacteriostatic or slowly bactericidal [33]. They not only interfere with peptidoglycan crosslinking (as does vancomycin) they display an increased binding affinity to their target sites. In addition, by the effect of their lipophilic side chains and tight binding to the cell membrane, they disrupt the cell membrane causing cell lysis [33].

In 2009, the US FDA gave approval to telavancin for use in complicated skin and soft tissue infections. In clinical trials telavancin has been found to be non-inferior to vancomycin in patients with normal renal function, yet inferior to vancomycin in patients with renal failure [34]. However, telavancin has been shown to be effective against vancomycin-intermediate *S. aureus*, though not vancomycin-resistant *S. aureus* [35].

Currently in phase III development for the treatment of bacterial skin and skin structure infections, oritavancin has activity against methicillin-resistant

S. aureus, vancomycin-intermediate *S. aureus*, vancomycin-resistant *S. aureus*, and vancomycin-resistant enterococcus [36]. There are two particular features about oritavancin that standout among the developing drugs for resistant gram-positive organisms. The first, that it can sterilize biofilms of methicillin-sensitive *S. aureus*, methicillin-resistant *S. aureus*, and vancomycin-resistant *S. aureus* [36]. The second is that the extended half-life of the drug allows for a single infusion for a 2-week treatment course, potentially eliminating both concerns about compliance issues and the need for indwelling catheters in many patients, reducing both cost and complication rates [37].

Dalbavancin with its once a week dosing would offer an attractive alternative to the current IV therapies available for methicillin-resistant *S. aureus*, though it is not yet US FDA approved, and does not appear to be effective against vancomycin-resistant enterococcus.

Table 2: Telavancin, Dalbavancin, Oritavancin activity against drug resistant *S. aureus/Vancomycin-resistant enterococci*

Organism	Dalbavancin	Telavancin	Oritavancin
MSSA	+	+	+
MRSA	+	+	+
VISA	+	+	+
VRSA	-	-	+
VRE	+/-	+/-	+

Source: Partially adapted from Zhanel GG, Calic D, Schweizer F, *et al.* New lipoglycopeptides: a comparative review of dalbavancin, oritavancin and telavancin. Drugs. 2010; 70(7):859-86.

Ceftaroline

A so called 'fifth generation' cephalosporin, ceftaroline recently received FDA approval for treatment of complicated skin and soft tissue infections as well as community acquired bacterial pneumonia. It has a broad spectrum of activity against aerobic and anaerobic gram positive and aerobic gram negative bacteria [38]. Importantly, it includes activity against methicillin-resistant *S. aureus*, vancomycin-intermediate *S. aureus*, and vancomycin-resistant *S. aureus*, and there is *in vitro* data suggesting that it will be effective against *E. faecalis* [39]. The spectrum for gram-negative activity is variable for Enterobacteriaceae and it does not have independent activity against *P. aeruginosa* [39]. Like other members of the cephalosporin class it has bactericidal action by inhibiting cell wall synthesis *via* binding to penicillin-binding proteins. As referenced above, β-lactam antibiotics bind to PBPs of susceptible organisms. However, the characteristic *mecA* gene of methicillin-resistant *S. aureus,* alters the PBP2, known as PBP2a, creating a low binding affinityfor β-lactam antibiotics. This is what gives drug-resistant *S. aureus* cell wall stability in the presence of β-lactam antibiotics [38]. However, unique to ceftaroline is the characteristic of binding affinity to PBP2a, making it effective against methicillin-resistant *S. aureus* [40]. Human clinical trials have shown it to be non-inferior to vancomycin and ceftriaxone in skin and skin structure infections and community-acquired bacterial pneumonia, respectively, and data in rabbits suggest efficacy in the treatment of endocarditis [39]. As to where it will fit in the clinical armamentarium is still to be seen, but the low side effect profile and comparable cost will make it an alternative agent for resistant gram-positive organisms.

Fidaxomicin

Recent data clearly indicate that the incidence and mortality of *Clostridium difficile* infections (CDI) continues to rise, with a near doubling in numbers over a recent 5 year period [41]. Standard treatment algorithms have included metronidazole or oral vancomycin. Despite adequate therapy, CDI is associated with a 20%-30% recurrence rate, which increases with every subsequent relapse [42]. It has been proposed that the depletion of normal gut flora by metronidazole and oral vancomycin may play a role in opening the door to overgrowth of residual spores or reinfection from the environment [42]. Fidaxomicin is a macrolide antibiotic with bactericidal activity against gram-positive anaerobes, studied with particular interest in *C. difficile* [43]. Fidaxomicin stops DNA transcription *via* RNA polymerase inhibition, halting protein synthesis [41]. It has high enteric concentrations, poor systemic absorption and unlike metronidazole and oral vancomycin spares *Bacteroides* and other fecal flora [42, 43]. Randomized, double-blinded studies have shown that fidaxomicin is non-inferior to vancomcyin in the treatment of CDI, but superior in preventing early recurrence [42].

CONCLUSIONS

Our extensive review of English literature leads us to the conclusion that common infectious agents will continue to pose difficulties in treatment and eradication. As of now, no antimicrobials with novel mechanisms of action have reached clinical trials and there remain clear gaps in the anti-infection armamentarium. At the same time, novel mechanisms of resistance continue to be reported. Outside of novel mechanisms of action, the ideal approach to generation of more effective antimicrobials would be to overcome the mechanisms of resistance as has been done in the case of β-lactamase inhibitor combinations and glycylcycline. The use of biological response modifiers in the management of infectious diseases is starting to become an attractive option. The modulation of the host response by these agents would be expected to interfere with the cytokine induced pathogenesis. The cost and safety issues associated with these agents will probably limit their use to severely ill patients. As of now, polyclonal intravenous immunoglobulin is used in toxin-mediated infections such as toxic shock syndrome caused by *S. aureus* and *Streptococcus pyogenes*. Monoclonal antibodies against endotoxin and tumor necrosis factor were studied for severe sepsis but failed to get approval for sepsis syndrome. Activated protein C, as a biological response modifier, did get approved for severe sepsis even though the benefit was marginal. It was withdrawn from the market just over 10 years later because of poor risk/benefit ratio. Monoclonal antibodies against specific pathogens are being studied. A recent study showed that the addition of monoclonal antibodies against *C. difficile* toxins to antimicrobial agents reduced the recurrence rate to 7-8% compared to 25-32% in the placebo group [44]. The most difficult hurdle in the path of monoclonal antibodies is the need for expensive technology and the subsequent high cost.

While the future of antimicrobial therapy to treat infectious diseases at this point in time does not seem to be very bright the continued advances in vaccine technology for prevention and hopefully even treatment of infectious diseases will offer a more physiological way of managing infectious diseases. The short life span and short term goals of antimicrobial therapy with the collateral damage of inducing resistance will continue to contrast the long life span (potentially indefinite) and long term impact on human health with the collateral advantage of providing herd immunity.

ACKNOWLEDGEMENT

Declared None.

CONFLICT OF INTEREST

The author(s) has confirmed that there is no conflict of interest.

REFERENCES

[1]   Wright, A. The Penicillins. Mayo Clin Proc 1999; 74:290-307.
[2]   Khardori, N. Antibiotics: Past, Present, Future. Med Clin North Am. 2006; 90(6):1049-76.
[3]   Rolinson, G. Forty years of β-lactam research. J Antimicrobial Chemother. 1998; 41:589–603.
[4]   Thomas, D, *et al*. The public health response to the re-emergence of syphilis in Wales, UK Int J STD AIDS. 2011; 22:488—492.
[5]   Moellering, R. Meeting the challenges of β-lactamases. J Antimicrobial Chemother. 1993; 31, *Suppl. A*. 1-8
[6]   Livermore, D. β-Lactamase-mediated resistance and opportunities for its control. J Antimicrobial Chemother. 1998; 41, Suppl. D, 25–41.
[7]   Essack, S. The Development of β-Lactam Antibiotics in Response to the Evolution of b-Lactamases. Pharm Res.2001; 18(10):1391-1399.
[8]   Moya, B, *et al*. Pan-β-lactam Resistance Development in *Pseudomonas aeruginosa* Clinical Strains: Molecular Mechanisms, PBPs Profiles and Binding Affinities. Antimicrob. Agents Chemother. 2012;56(9):4771-8.
[9]   Medeiros, A. Evolution and Dissemination of B-Lactamases Accelerated by Generations of B-Lactam Antibiotics. CID. 1997; 24(Suppl 1):S19-45.
[10]   Dellinger, E, *et al*. Quality Standard for Antimicrobial Prophylaxis in Surgical Procedures. CID. 1994; 18:422-427.
[11]   Hellinger WC, Brewer NS. Symposium on Antimicrobial Agents. Part VII. Carbapenems and Monobactams: Imipenem, Meropenem, and Aztreonam. Mayo Clin Proc 1999;74:420-434.
[12]   Edson RS and Terrell CL. Symposium on Antimicrobial Agents. Part VIII. The Aminoglycosides. Mayo Clin Proc 1999;74:519-528.
[13]   Smilack, JD. Symposium on Antimicrobial Agents. Part X. The Tetracyclines. Mayo Clin Proc. 1999; 74(7): 727-729.
[14]   Zuckerman JM, Qamar F, Bono AR. Macrolides, Ketolides, and Glycylcyclines: Azithromycin, Clarithromycin, Telithromycin, Tigecycline. Infect Dis Clin N Am. 2009; 23:997–1026.
[15]   Yamada M, Buller R, Bledsoe S, *et al*. Rising Rates of Macrolide-Resistant Mycoplasma pneumoniae in the Central United States. Ped Infect Dis J. 2012; 31(4):409-411.
[16]   Colson, A. Policy Responses to the Growing Threat of Antibiotic Resistance. Policy Brief 6. The Antibiotic Pipeline. Extending the Cure. 2008.
[17]   Nicolau DP, Nightingale CH, Quintiliani R. Focus on teicoplanin: A new glycopeptides antibiotic. Hosp Formul. 1992; 27:675-683.

[18]   Mohr JF, Murray BE. Point: Vancomycin Is Not Obsolete for the Treatment of Infection Caused by Methicillin-Resistant *Staphylococcus aureus*. Clin Infect Dis. 2007; 44 (12): 1536-1542.

[19]   Cohen R, Barre J, Varon E. A Comparative Microbiological and Pharmacokinetic Activity of Vancomycin and Teicoplanin. Pathol Biol (Paris). 1992; 40(8):831-844.

[20]   Courvalin P. Vancomycin Resistance in Gram-Positive Cocci. Clin Infect Dis. 2006; 42(Supplement 1): S25-S34.

[21]   Oliphant CM, Green GM. Quinolones: A Comprehensive Review. Am Fam Physician. 2002; 65(3): 455-465.

[22]   Saravolatz LD and Leggett J. Gatifloxacin, Gemifloxacin, and Moxifloxacin: The Role of 3 Newer Fluoroquinolones. CID. 2003; 37(9):1210-1215.

[23]   Gurk-Turner, C. Quinupristin/dalfopristin: the first available macrolide-lincosamide-streptogramin antibiotic. Proc (Bayl Univ med Cent). 2000; 13(1): 83-86.

[24]   Raz R. Fosfomycin: an old—new antibiotic. Clin Microbiol Infect. 2012; 18: 4–7.

[25]   Michalopoulos AS, Livaditis IG, Gougoutas V. The revival of fosfomycin. Int J Infect Dis. 2011; 15:e732-e739.

[26]   Shaw KJ, Barbachyn MR. The oxazolidinones: past, present, and future. Ann NY Acad Sci. 2011; 1241:48-70.

[27]   Aksoy DY and Unal S. New antimicrobial agents for the treatment of gram-positive bacterial infections. Clin Microbiol. Infect 2008; 14:411-420.

[28]   Eliopoulos GM. Current and new antimicrobial agents. Am heart J. 2004; 147(4): 587-92.

[29]   Carpenter CF and Chambers HF. Daptomycin: Another Novel Agent for Treating Infections Due to Drug-Resistant Gram-Positive Pathogens. CID. 2004; 38(7):994-1000.

[30]   Silverman JA, Mortin LI, VanPraagh AD, *et al*. Inhibition of Daptomycin by Pulmonary Surfactant: *In Vitro* Modeling and Clinical Impact. J Infect Dis. 2005; 191(12): 2149-2152.

[31]   Doan T, Fung HB, Mehta D, *et al*. Tigecycline: A Glycylcycline Antimicrobial Agent. Clin Therap. 2006; 28(8): 1079-1106.

[32]   Peterson LR. A review of tigecycline-the first glycylcycline. Int J Antimicro Agents. 2008; 32: S215-S222.

[33]   Guskey MT, Tsuji BT. A comparative review of the lipoglycopeptides: oritavancin, dalbavancin, and telavancin. Pharmacotherapy. 2010; 30(1): 80-94.

[34]   Vibativ [package insert]. Theravance, Inc. 2012.

[35]   Zhanel GG, Calic D, Schweizer F, *et al*. New lipoglycopeptides: a comparative review of dalbavancin, oritavancin and telavancin. Drugs. 2010; 70(7):859-86.

[36]   Zhanel GG, Schweizer F, Karlowsky JA. Oritavancin: Mechanism of Action. CID. 2012; 54(S3): S214-219.

[37]   Tice, A. Oritavancin: A New Opportunity for Outpatient Therapy of Serious Infections. 2012; 54(S3):S239-S243.

[38]   Jorgenson MR, DePestel DD, Carver PL. Ceftaroline Fosamil: A Novel Broad-Spectrum Cephalosporin with Activity Against Methicillin-Resistant *Staphylococcus aureus*. Ann Pharmaco. 2011;45:1384-1398.

[39]   Saravolatz, LD. Ceftaroline: A Novel Cephalosporin with Activity against Methicillin-resistant *Staphylococcus aureus*. CID. 2011; 52(9): 1156-1163.

[40]   Ceftaroline Fosamil for the Treatment of Community-acquired Bacterial Pneumonia and Complicated Skin and Skin Structure Infections. FDA Briefing Document for Anti-Infective Drugs Advisory Committee Meeting September 7, 2010.

[41] Lancaster JW, Matthews SJ. Fidaxomicin: The Newest Addition to the Armamentarium Against *Clostridium difficile* Infections. Clin Ther. 2012;34(1): 1-13.

[42] Cornely OA, Miller MA, Louie TJ, *et al.* Treatment of First Recurrence of Clostridium difficile Infection: Fidaxomicin *Versus* Vancomycin. CID. 2012;55 (Suppl 2):S154-S161.

[43] Goldtein EJ, Babakhani F, Citron DM. Antimicrobial Activities of Fidaxomicin. CID. 2012;55(S2):S143–8.

[44] Lowy, I, Molrine, DC, Leav, BA, *et al.* Treatment with Monoclonal Antibodies against *Clostridium Difficile* Toxins. N Engl J Med. 2010; 362(3): 197-205.

Send Orders for Reprints to reprints@benthamscience.net
Frontiers in Clinical Drug Research: Anti-Infectives, Vol. 1, 2014, 53-157

CHAPTER 3

Antisense Oligonucleotides-Based Therapeutics for Pathogenic Bacteria and Viruses

Hui Bai[1,2], Yu You[3,4], Xiaochen Bo[1,*], Shengqi Wang[1,*] and Xiaoxing Luo[5,*]

[1]Department of Biotechnology, Institute of Radiation Medicine, Academy of Military Medical Sciences, Beijing, 100850, China; [2]No. 451 Hospital of Chinese People's Liberation Army, Xi'an, 710054, China; [3]The Military General Hospital of Beijing, Beijing, 100700, China; [4] The 309th Hospital of Chinese People's Liberation Army, Beijing, 100091, China and [5]Department of Pharmacology, School of Pharmacy, Fourth Military Medical University, Xi'an, 710032, China

Abstract: Human pathogenic bacteria and viruses are significant etiology of different types of infectious diseases that cause extremely high morbidity and mortality worldwide. Continuous failure of anti- pathogen/infective agents and therapies, as well as the paucity of postexposure therapeutics greatly facilitates the emergence and dissemination of pathogenic bacterial isolates and virus phenotypes with multi-drug resistance (MDR) or pan-drug resistance (PDR). Additionally, the potential use of these bacteria and viruses in acts of bioterrorism poses tremendous threat to global security. Novel counterstrike strategy of using single-stranded antisense oligonucleotides (ASOs) as prospective gene silencers has been a major area of anti-infective study, leading a potential revolution in the development of antibacterial and antiviral therapeutics by addressing the targets that are "undruggable" for traditional pharmaceutical approaches. Given 30 years of technology advances in elucidation of antisense mechanism, characterization of ASOs chemical modification, and refinement of delivery systems, ASOs based anti-infective strategy displays advantageous features of conceptual simplicity, straightforward designing and quick drug identification methods. The steric-blocking ASOs offer improving sequence-specific anti-infective effects *in vitro* and in animal models of fatal infections, which enables themselves candidates for pre-clinical and clinical tests. This chapter puts together and discusses the important advances in the field based on the above mentioned technologies and the latest development of potential targets and therapeutic AS-ODNs that have reached clinical trials with antibacterial or antiviral protocols.

Keywords: Acyl carrier protein, antisense antibacterial, antisense antiviral, antisense oligonucleotides therapy, cell penetrating peptide, host factors,

***Address correspondence to Xiaoxing Luo:** Department of Pharmacology, School of Pharmacy, Fourth Military Medical University, Xi'an, China; E-mail: xxluo3@fmmu.edu.cn
Shengqi Wang: Beijing Institute of Radiation Medicine, the Academy of Military Medical Sciences, Beijing, China; E-mail: sqwang@bmi.ac.cn
Xiaochen Bo: Beijing Institute of Radiation Medicine, the Academy of Military Medical Sciences, Beijing, China; E-mail: boxc@bmi.ac.cn

liposome, locked nucleic acid, peptide nucleic acid, phosphorodiamidate morpholino oligomers, phosphorothioate, RNA polymerase σ^{70}, RNase H activation, steric-blokcing mechanism.

INTRODUCTION

In spite of the availability of some effective therapeutics (including small-molecule chemicals of antibiotics and antivirals, as well as vaccines), bacteria and viruses are major pathogens that continue to cause globally tremendous human disease burdens and frequent outbreaks of new strains-related medical care disasters. However, traditional pharmaceutical industries in anti-infective development seem unable to keep pace with the emergence of highly pathogenic bacteria and viruses in way of providing adequate prevention measures and therapeutic countermeasures.

On the one hand, recent years have witnessed the incidence of multidrug resistance (MDR) or pan-drug resistance (PDR) bacteria escalating in a manner of global dimension, frequent prevalence and alarming magnitude. Despite the gram-positive bacteria (especially the *Staphylococcus aureus*) still in need of efficient therapies, the circulating isolates of gram-negative bacteria (GNB) species, including *Enterobacteriaceae* (especially the *E. coli*), *Klebsiella pneumonia*, *Pseudomonas aeruginosa* and *Acinetobacter baumannii* have created big problems for treatment of nosocomial infection, because they carry highly transmissible elements encoding multiple resistance genes (*e.g.*, broad-spectrum beta-lactamases and metallo-beta lactamase) that often fail many on-market antibiotics as well as new drugs in early development. On the other hand, effective vaccines or postexposure therapeutics for many pathogenic viruses are at an infant stage and are hardly unsatisfied. The newly emerging viruses, *e.g.*, the severe acute respiratory syndrome coronavirus (SARS-CoV) and the avian influenza viruses that can infect humans, are highly infectious, acutely pathogenic, and endanger public health as evidenced by lethal outbreaks with no instant and effective therapeutics. Meanwhile, a number of known viruses, *e.g.*, the human immunodeficiency virus (HIV), dengue virus, hepatitis B and C virus, and West Nile virus, claiming millions of lives every year still lack efficient or accredited drug treatments. In addition, certain highly pathogenic RNA viruses, like the filoviruses and Marburg virus, may also impact public safety and national security through the intentional use as bioweapons. Thus, new approaches that can effectively prevent bacterial and virus propagation and favor in offering timely therapeutic countermeasures are constantly and urgently needed.

Among several antisense technologies, antisense oligonucleotides (ASOs) are considered a very attractive means of fulfilling the above aims. The ASOs-based anti-pathogen strategy offers the prospect of drugs that are typically designed to bind to complementary RNA by Watson–Crick hybridization to arrest translational processes either by inducing cleavage mechanisms involving endogenous cellular nucleases such as ribonuclease H (RNAse H) or by sterically blocking enzymes involved in target gene translation (Fig. **1**). For antibacterial purposes, as bacteria lack RNA interference (RNAi) mechanism, ASOs have been studied as sequence-specific RNA silencers to realize bacterial growth inhibition or restore susceptibility to antibiotics through selective down regulation of growth essential genes or resistance mechanism related genes. For antiviral purposes, ASOs can alternatively be designed to specifically target viral genomic regions or RNA secondary structures involved in transcription or genomic-replication processes, which represent a promising type of ASO based gene-silencers that interfere with the virological processes to inhibit production of infectious virus particles in infected cells, potentially reducing cytopathic effects.

The concept of ASO-based therapeutics was first introduced by the study of Paterson *et al.* in 1977 [1], where they used single-stranded complementary exogenous nucleic acids to inhibit translation of an mRNA in a cell-free system, which led to the credence of alien nucleic acids-altered gene expression. Theoretically, ASOs-mediated inhibition of bacterial or virus replication is appealing in part because of the low likelihood of target sequence homology between pathogen and host, thus minimizing the potential for toxicity from cross-reactivity. In practice, the availability of a wider range of oligonucleotide analogs has stimulated renewed interest in their application as potential ASO-based antibacterial and antiviral agents. Nowadays, structurally modified ASOs are widely known for their potential inhibitory effects on bacterial or virus gene expression *in vitro* as well as *in vivo*. However, progress has been relatively slow as numerous new challenges rose while trying to implement this entirely new therapeutic strategy into a viable treatment (*i.e.,* synthetic feasibility and scale-up, cost of synthesis, uptake problems, selectivity, immunological effects, *etc.*). Major challenges of target identification and efficient delivery existing in early preclinical development of ASO-based therapies can be comparable for both pathogenic bacteria and viruses, although in their own uniquely troublesome ways. Furthermore, problems in clinical trials including off-target effects, resistance and pharmaco-issues (*e.g.,* systematic toxicity, bio-distribution, and pharmacokinetics) together comprise an active area of ASO-based antibacterial and antiviral drug development research.

In this chapter, we provide an overview of the antibacterial and antiviral antisense field, highlighting specific studies of interest (*e.g.,* antisense mechanism in bacteria and viruses, development of ASOs constructs, effective targets identification, difficulties with ASOs' efficiency, off-target effects, toxicity, delivery, and stability in small-animal models for preclinical testing, refinement of efficient delivery systems, and problems in clinical trials, *etc.*) over the past three decades, using our experience with targeting RNA polymerase σ^{70} as a reference point for development of promising broad-spectrum antibacterial ASOs, and antisense targeting of host genes as a reference point for development of potential adjuvant therapeutics for antiviral drugs.

ANTISENSE TECHNOLOGIES

Currently on-market drugs are mostly small chemical molecules that target proteins such as enzymes and receptors, representing a considerably small subset of total cellular proteins. Antisense technologies, which encompass ASOs, small interfering RNA (siRNA), ribozymes, and decoy RNAs or aptamers, involve the binding of complementary oligonucleotides (ONs) to target RNA through base pairing, however differ substantially in their downstream mechanisms of action and the functional outcomes they produce (Fig. **1**). Since the first introduction of concept in 1977, antisense technologies have received increasing attention as the nucleic acid-based therapeutics with highly-specific gene-silencing effects, which could be therapeutically useful in way of manipulating a wide range of cellular targets and human disorders [2,3].

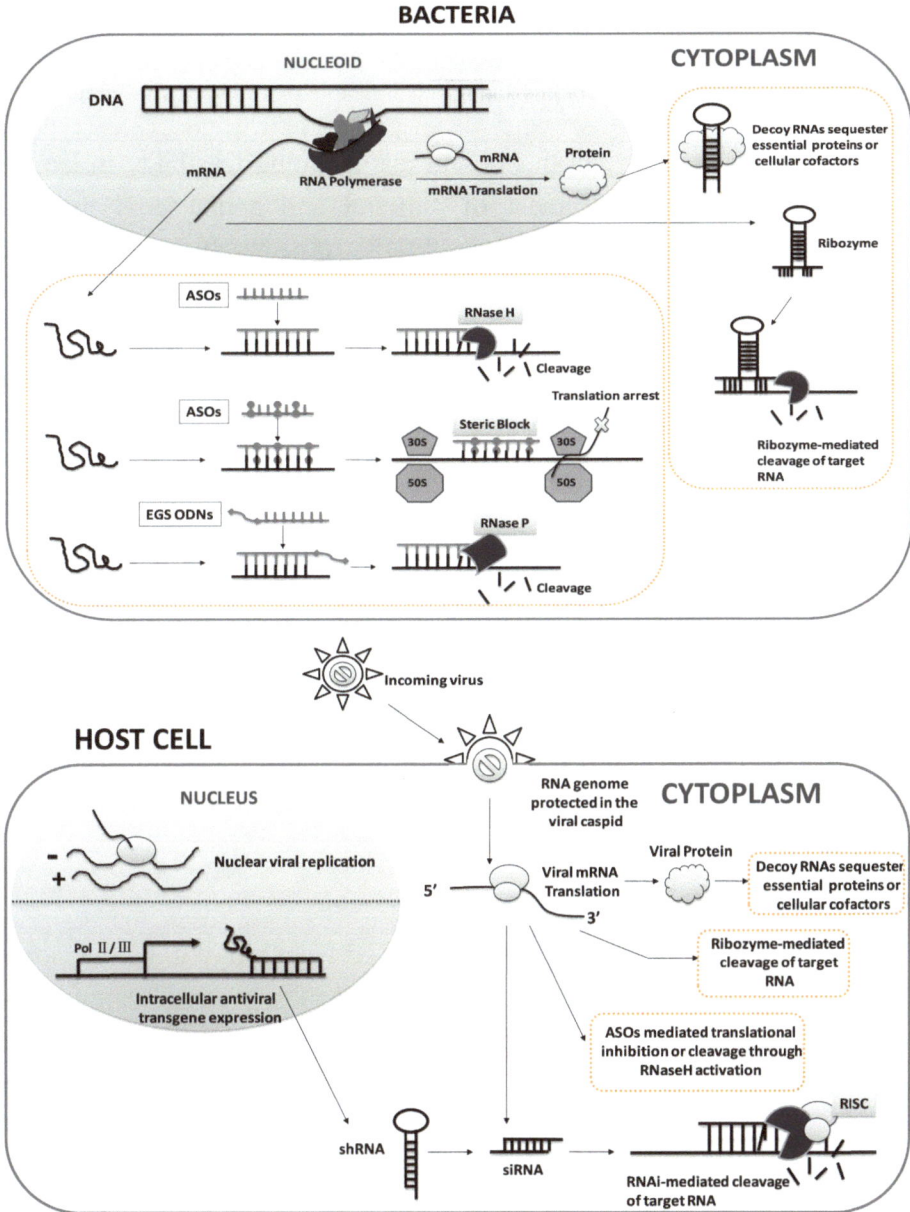

Figure 1: Nucleic acid-based anti-pathogen (bacteria and viruses) strategies. Anti-pathogen nucleic acids can either be transfected/delivered into cells (*e.g.,* siRNA or ASOs) or expressed intracellularly (shRNA, ribozymes or RNA decoys). Bacteria lack the RNAi mechanism. Viral transcripts complementary to the siRNA/ shRNA are cleaved upon assembly of the RISC machinery, which is unable to target RNA genomes protected within viral capsids or shielded from RNAi attack in subcellular compartments (*e.g.,* the nucleus or the virus-induced vesicles). Modified ASOs have a high affinity for their target sequence in both bacteria and viruses, and

inhibit gene expression by steric hindrance of the ribosome or through induction of mRNA cleavage by recruitment of RNaseH. Binding of the ribozymes to the target sequence in both bacteria and viruses should also trigger cleavage of the RNA. Decoy RNAs bind and sequester essential bacterial proteins or viral proteins and host cell factors that support viral replication.

With advantageous simplicity in theory and manipulating feasibility in lab, ASOs-based strategy for the development of antiviral and antibacterial therapeutics/ therapies has started early in 1978 [4] and 1991[5], respectively. Significant findings in either regard have been reported to show its intriguing future promises in curing infectious diseases, both scientifically and commercially. As there has been no report of RNAi mechanism in bacteria, the majority of antisense antibacterial researches have focused on utilizing ASOs-based strategy [6]. However, with adaptability in mammsalian cells, plants, insects and viruses, the RNAi technology is the latest in a long line of nucleic acid-based drug candidates for antiviral purposes, which will be discussed only briefly as it has been extensively reviewed in the literature [7,8]. Decoy RNAs or small RNA/DNA aptamers [9], which are more structurally complex and interact directly with proteins rather than complementary RNA, and ribozymes [9], which catalyse RNA cleavage and ligation reactions as well as peptide bond formation, their application studies in antiviral and antibacterial antisense field are much less and not covered in this chapter.

Antisense Oligonucleotides (ASOs)

Mechanism of Action

Antisense approach is the first nucleic acid-based technology to be developed for inhibiting specific gene expression. ASOs are single-stranded (deoxy)ribonucleotide oligomers (~20 nucleotides) with a nucleotide sequence designed to be complementary to a target mRNA transcript [10]. ASOs bind to given targets (gene or mRNA) with an apparently high degree of fidelity and exquisite specificity. The effect of gene silencing or "knock down" that happen after the binding can be broadly categorized as cleavage-dependent or occupancy-only mechanism.

As shown in Fig. **1,** cleavage-dependent mechanism includes degradation of RNA:mRNA duplexes by double-strand RNA (dsRNA)-specific RNAses (a natural means of transcriptional regulation), degradation of stable DNA:RNA heterodimers through the activity of RNase H, and degradation *via* the action of RNase P (only if external guide sequences are coupled to the oligonucleotide). Occupancy-only mechanism, also known as translation arrest, features as that antisense oligonucleotide:RNA heteroduplexes inhibit translation by steric blocking the ribosomal maturation and polypeptide elongation process. The specific gene-silencing mechanism ASOs-based therapeutics depends on the structural chemistry and special designs [11], which will be elaborated in the following part.

Three Generations of Synthesized Constructs

Oligonucleotide degradation by cellular nucleases, especially while using unmodified ONs, is evidently a major concern for the antisense research. Thus, many modifications concerning the backbone, phosphodiester bond, and sugar ring, have been introduced and investigated, giving births to three generations of nucleic acid analogs that are conjugated to form ASO oligomers (Fig. **2**). Chemical modification of the oligonucleotides was found to improve the stability and intrinsic affinity of ONs to targeted mRNA. However, the overall efficacy and safety issues (*e.g.,* poor cellular uptake, genomic toxicity, off-target effect, poor pharmacokinetics properties, *etc.*) involved in specific synthetic ASO type as therapeutics for systemic applications still waits for better solutions, which will be briefly summarized in this part and further discussed in the context.

First Generation: The first example of improvement in antisense oligonucleotide efficacy *via* chemical modifications was established in 1978 by Zamecnik and Stephenson [4], whose pioneering work introduced chemical modifications at the 3' and 5' ends of a 13 nucleotide-long oligonucleotide complementary to a target sequence in Rous sarcoma virus RNA to reduce its degradation by cellular nucleases, therefore improved its antiviral activity *in vitro*. It took another 10 years for the introduction of phosphorothioate (PS) modification to take place, which was definitely a major advancement in the field by replacing the oxygen atom in the DNA backbone with a sulphur atom (Fig. **2**). As the most widely used first generation ASO construct, PS-modified oligonucleotides are poor substrates for nucleases and thus show increased stability, increased serum half-life, and

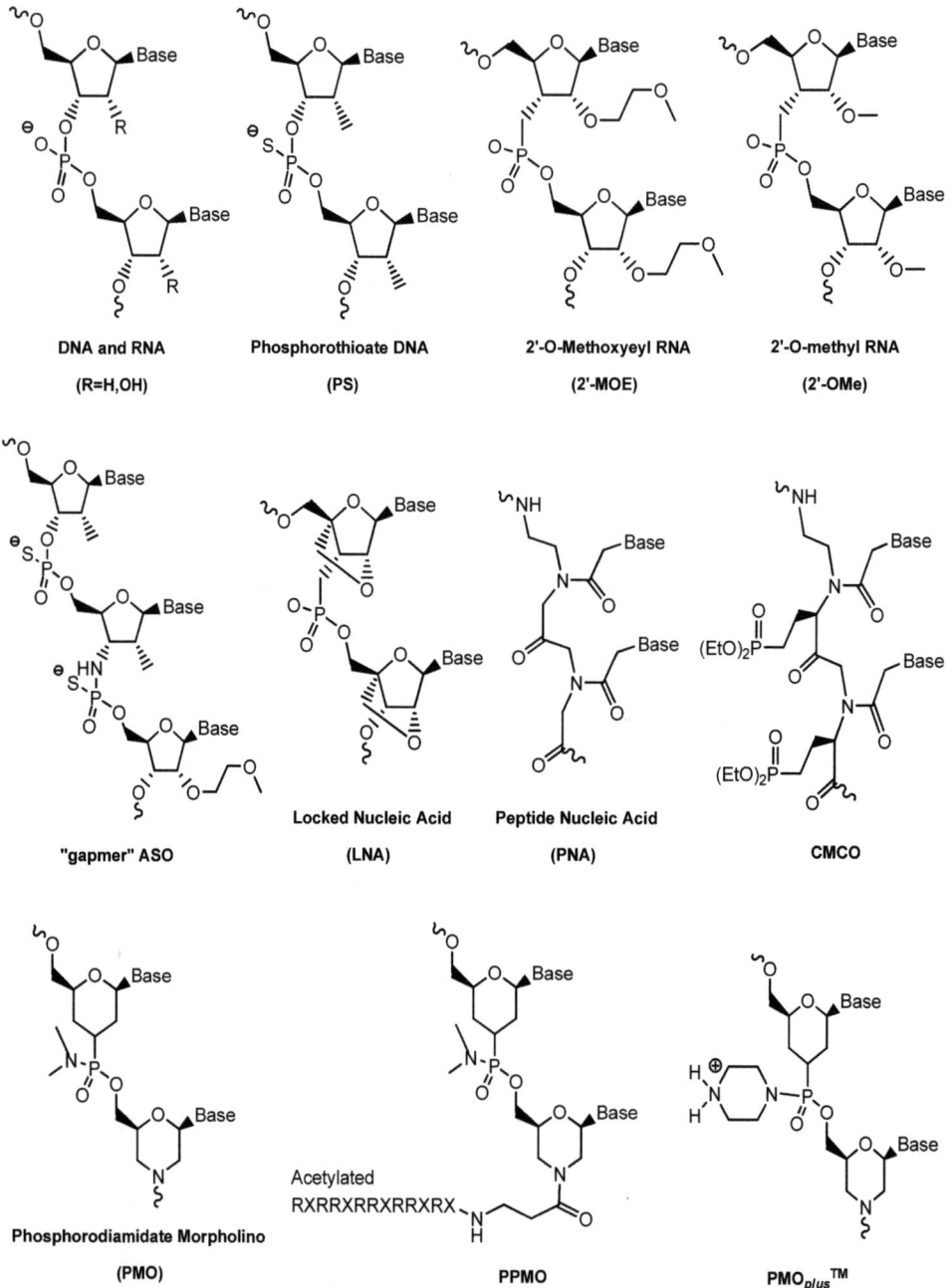

Figure 2: Oligonucleotide chemistries. Phosphorothioate (PS) backbones, as well as 2′-O-methoxyethyl (2′-MOE) and 2′-O-methyl (2′-OMe) substituents are negatively charged oligonucleotides that increase resistance to degradation and promote protein binding to target RNA (see top panel). Locked nucleic acid (LNA) modification markedly increases the binding of the oligonucleotide to the targeted mRNA(see middle panel). Peptide nucleic acid (PNA)

modification uses non-charged achiral peptide (polyamide) backbone, resulting in a good structural mimics of DNA resistant to degradation (see bottom panel). Introduction of phosphonic ester residues into a PNA backbone (CMCO) improves the solubility of the PNAs and the penetration of the compounds throug the cell membrane (see bottom panel). In phosphorodiamidate morpholino oligomers (PMOs), ribose (RNA) or deoxyribose (DNA) is replaced with morpholine rings, and the phosphorothioate or phosphodiester (RNA) groups are replaced with uncharged phosphorodiamidate groups, resulting in a compound that is neutral and very resistant to degradation (see bottom panel). Positively charged piperazine residues in positively charged PMOs (PMOplus), or positively charged arginine-rich peptides in peptide-conjugated PMOs (PPMOs), dramatically improve the intracellular uptake of the oligomers (see bottom panel).

greater tissue retention. Advantageously, it allows RNase H cleavage of the target sequence. However, the negative-charged character of PS-modified oligonucleotides shows decreased affinity for their target mRNA, are not taken up efficiently by cells, and causes nonspecific interaction with serum and intracellular proteins, which leads to strong immunostimulatory response or unexpected spurious off-target effects. Further, this backbone imparts a significant, hybridization-independent toxicity profile that varies with different sequences, which includes increased coagulation time, pro-inflammatory effects and activation of the complement pathway. PS-ASOs lead to renal tubule changes and thrombocytopaenia at higher concentrations. Several PS-ASOs drug candidates progressed through various stages of drug development during 1990 to 2010, but only one drug fomivirsen made itself to market, which was approved by FDA in 1998 as a treatment for cytomegalovirus-induced retinitis in immunocompromised patients with AIDS. With known disadvantages, the development of PS-ASOs has been continued in some companies until very recently, mainly due to its low cost at large-scale synthesis.

Second Generation: It has taken longer to develop the second generation of ASOs with larger deviation from the natural structure, the representative constructs of which include 2'-*O*-methyl (OMe) derivatives or the strongly hybridizing 2'-*O*-methoxyethyl (MOE) analogs [12] (Fig. **2**). These sugar ring modifications increase the stability of the oligonucleotide and the affinity for the target mRNA when compared to PS-ONs. However, these two oligomers are no longer able to recruit the help of RNase H, and block translation *via* steric hindrance of the ribosome during translation, which means that he oligonucleotide has to be targeted to sequences that are at or near the translation initiation site for optimal activity. A few exceptions have been found when using mixed hybrids of arabinonucleic acids (ANA) or better 2'-deoxy-2'-fluoro-arabinonucleic acids (2'F-ANA) with RNA [13], and cyclohexenyl oligonucleotides (CeNA) with fully modified ONs [14], which demonstrated induced RNase H activity to different extent (Fig. **2**). Notably, the "gapmer" ASOs [15] which feature a phosphorothioate core (gap) by 2'-modified nucleotides (2'MOE wings) (Fig. **2**), have been developed to maintain the RNase H induction activity by combining the benefits of both the first and second generation modifications. The OMe- and MOE-modified ASOs also show some unwanted effect due to their negatively-charged backbone that in some cases induce toxicity by binding to serum proteins. Nonetheless, the development of OMe- and MOE-modified ASOs as potential therapeutics has also been a long-time effort in some companies, mainly due to its flexibility to hybrid with other modifications for presenting better properties.

Third Generation: The third-generation ASO reagents, mainly comprising three constructs of namely the peptide nucleic acids (PNAs), the locked nucleic acids (LNAs) and the phosphorodiamidate morpholino oligomers (PMOs) (Fig. **2**), have been developed and thoroughly studied over the last 15 years. Together, they each have a unique backbone deviating more considerably from the natural (deoxy)ribose phosphodiester bridges, which allows more specific recognizing of targeted sequence and higher binding affinity through steric blocking mechanism. These properties endue them exerting specific and potent RNA silencing effect at shorter lengths and lower concentrations [16]. It is well known that the presence of an uncharged nucleotide backbone may greatly reduce interactions with serum and cellular proteins, which limits off-target toxicities. Thus, the third-generation chemical modifications, most being electric neutral, significantly reduce such unwanted effects once found in the first and second generation constructs, which will be briefly introduced according to respective modified structure.

PNAs are amazingly good structural mimics of DNA, which consists of a non-charged achiral peptide (polyamide) backbone comprised of N-(2-aminoethyl)glycine units to which the nucleobases are attached *via* carbonyl methylene linkers [17] (Fig. **2**). PNA oligomers bind to oligo(deoxy)nucleotides obeying Watson-Crick base paring rules (in addition to self-pairing with complementary PNA oligomers). Moreover, they even can bind to targets in duplex DNA by helix invasion. In a chemical senses, PNA bridges the gap between proteins and nucleic acids with the potential of carrying genetic information. Therefore, numerous PNA constructs of further modifications have been prepared in search for their useful antisense effects [18,19]. And some of the antibacterial findings will be discussed elsewhere in the context. However, PNAs have a solubility disadvantage.

PMOs are DNA-like oligomers with six-membered morpholine rings substituting for the deoxyribose, and with neutral phosphorodiamidate intersubunit linkages (Fig. **2**). These structureal modifications also confer high stability and solubility, both desirable qualities for use in the clinic. PMOs bind to target mRNA sequences with higher affinity than other ASOs, and like PNAs, are typically designed to target ribosome entry sites, the AUG start codon, splice junctions, or sites of critical RNA secondary structure, and thus reduce target protein levels by interfering with translation or mRNA processing. Several ASO compounds of PMO constructs have been developed in lab and are currently being evaluated in clinical trials for the treatment of a variety of ailments, including cancer, diabetes and genetic-deficient diseases like the Duchenne muscular dystrophy [20], *etc.* As this chapter restricts itself however to infectious diseases, the antisense antibacterial and antiviral potential and application of PMO was highlighted elsewhere in the context.

LNAs are ribonucleotides containing a bridging extra methylene carbon between the 2'- and 4'-positions of the ribose ring (Fig. **2**), which were first described independently by the groups of Wengel and Imanishi [21]. The supplementary ring affording strongly increased affinity for RNA oligomers. Like other two modifications, the supplementary linkage eliminates the possibility for recognition of the hybrid LNA: RNA duplex by RNase H. LNAs are one of the most studied constructs over the past 10 years, mainly because the incorporation into LNA-modified ASOs *via* standard DNA synthesis protocols allows the design of gapmers with central DNA portions, restoring the possibility for recruitment of the cleaving enzyme [22]. LNA modifications were reported to cause less toxic side effects when compared to the first and second generation modifications [23]. However, LNA-modified oligonucleotides were recently shown to induce profound toxicity in hepatocytes in mice [24].

Other important issues to be addressed include the poor intracellular uptake and side effects of the above modified ASO constructs, which are major impediments to their broad utility as therapeutics in clinic. Since pure structural chemical modification can only improve the situation to a limited extent, there has been keen interest in optimizing ASOs either by further conjugation of functional groups (*e.g.,* cell-penetrating peptide conjugation) or by forming advanced delivery systems (*e.g.,* liposome encapsulated or antibody directed delivery systems), which will be respectively analyzed in the specific application of antibacterial and antiviral ASO therapeutics in the following context.

Steric-Blocking ASOs

As different antisense technologies would all provide nucleic-acid based therapeutics, the above-mentioned synthetic analogs are also adaptable to either siRNA or ribozymes to form modified oligomers, with the aim of improving their pharmaceutical properties and therapeutic potentials [25]. As cellular enzymes, *e.g.,* the RNA-induced silencing complex (RISC), need to recognize antisense compounds like siRNA, they can only be chemically modified to a limited degree, which limits our ability to enhance their pharmacological qualities. ASOs that modulate RNA function by blocking the access of cellular machinery to RNA (*e.g.,* OMe, MOE, PNA, and PMO), do not need to exploit cellular enzymes for their activity, thus, they can be subjected to more extensive chemical modifications that improve their drug-like qualities. Besides, this different mode of action leads to outcomes such as the repair of a defective RNA or the generation of a novel protein (not discussed in this chapter), which cannot be achieved using RNAi or ASOs that activate RNaseH.

ASOs-BASED ANTIBACTERIAL THERAPEUTICS

Antisense Antibacterial Strategy

With a much less complicated and homogenous genome, bacterial cells have their double-strand DNA (dsDNA) located in nucleoid, which is a low electron density zone with no nucleic membrane to strictly separate the biochemical reactions into different time and space level. Besides, DNA replication, RNA transcription and protein synthesis in bacteria are processed in cytoplasm, which allows exogenous antisense oligonucleotides to interfere with genes and/or RNAs more readily. Meanwhile, Bacteria themselves use antisense as a natural mechanism to inhibit specific gene expression. Therefore, antisense technology suits better as an effective gene modulating tool in bacteria, especially when RNA interfering (RNAi) mechanism found in eukaryotic cells has not been reported so far in bacteria. It is generally described as RNA silencing in bacteria using synthetic nucleic acid oligomer mimetics to specifically inhibit essential gene expression and achieve gene-specific antibacterial effects [26]. Advantage of RNA silencing is unique in having the potential to selectively kill target bacteria with species and even strain specificity. Of particular interest are possibilities to tailor the antibacterial spectrum, aid the use of conventional antibiotics by potentiating their activity, and reverse resistance.

Antisense oligomers have been studied as bacterial growth inhibitors for developing new types of antibiotics as early as 1991, when Rahman MA *et al.,* firstly observed the inhibited protein synthesis and colony formation in normal *E. coli* by using PEG 1000 attached methylcarbamate DNAs targeting the start codon sequence of prokaryotic 16S rRNA [5]. The hypothesis that any gene can be antisense inhibited is quite tantalizing that the potential of gene-specific modified AS-ODNs as clinically useful antibiotics has been well accepted and further explored ever since. The present-day antisense antibacterial strategy has overcome major obstacles of target identification and poor cellular uptake that once hampered this innovative approach developing into clinically applicable therapeutics [6]. Antibacterial ASOs have been developed by constructing sequence designed synthetic RNA silencers using new chemical classes, *e.g.,* PNA and PMO, and that conjugated with cell-penetrating peptide (CPP) in multiple functional ways. And their potent bactericidal effects have been displayed in a variety of pathogenic bacteria by targeting several growth essential genes *in vitro* and *in vivo*. Further, they have also accepted thorough preclinical and clinical evaluation on the aspect pharmaceutical properties as promising antisense antibiotics in the past decade.

Target Identification and Validation

The growth inhibitory activity of ASOs relies on sequence-specific inhibition of bacterial gene expression, which offers potentially immediate bactericidal or bacteriostatic therapeutic consequences high specificity [27]. Although theoretically any gene in bacteria can be antisense inhibited, the development of potent antimicrobial ASOs principally requires to identify/determine the most essential genes for bacterial survival, which generally includes measuring gene vitality and accessibility in order to validate the a candidate gene for further qualification. Specifically, fundamental criterion to be satisfied includes sufficient concentration of ASOs at the most sensitive targeting site of mRNA, the ability of ASOs to hybridize to target mRNA sequence, the capacity of the ASOs: mRNA duplex to interfere with gene expression, and sufficient biological stability of ASOs.

Collectively, with regard to target selection, researcher are dedicated to identify an ideal gene that is with small nucleotide content but effective region of coding sequences for potent antisense inhibition. Meanwhile, the antisense property of AS-ODN itself should also be taken into consideration. Compared to gene knockout technique, antisense approach itself has been proved in practice to be an effective tool for essential gene target validation in bacteria, with advantages of controllable sensitivity, larger breadth of applicability and more realistically mimic effect of a therapeutic inhibitor [28].

Target Site Selection and Design of ASOs

Normally, a bacterial gene is comprised of hundreds or thousands of nucleotides, in which subset regions of codons are responsible for encoding certain transcripts to be translated into functional products that may alter biological state or behavior. To elicit efficient gene-silencing effect, ASOs need accessibility to the most sensitive region in a target mRNA so that stable ASO: RNA(DNA) heteroduplexesor triplex (as for PNA) can be formed to effectively hamper the corresponding translation process [29].

After years of technology refinement, researchers have summarized a comparatively fixed procedure in ASOs' design to determine the most sensitive target site with in mRNA and simultaneously obtain the ASO sequence with best potency and efficacy [30]. In general, possible antisense targeting sites are those nucleotide sequences free of any double strand (*e.g.,* hairpin) in secondary structure, which are usually predicted by RNA secondary structure softwares [31]. Then, bioinformatic algorithms are used to calculate the DNA:ASOs binding parameters (*e.g.,* minimal free energy and melting temperature, *etc.*) with set lengths for ASOs, which largely varies according to oligomer types. In principle, nucleic acid analogs with stronger affinity to target RNA require shorter lengths. Customarily, policies that control 14-30 monomers for PS-ODNs and 8-16 monomers for PNA, LNA, and PMO, have been adopted for potent inhibition and achieving better hits. Then, according to the combined data, 10-30 different targeting sites/sequences with highest scores are randomly selected based on screening scale, and complementary ASOs are synthesized for *in vitro* efficacy test. Notably, many studies have demonstrated that the start codon region of the target mRNA (Tables **1** and **2**) is the most effective region especially for RNase H-independent ASOs, because this region initiates the translation and includes the Shine-Dalgarno (SD) sequence [32,33]. Thus, following studies of target identification usually focus their ASOs design in the start codon region, and less ASOs (3-10 sequences) can be synthesized to fulfill the selection process, which saves a large amount of synthesis money.

Researchers have noticed in earlier experiments of target identification that naked ASOs were poorly uptaken by live bacteria, therefore large quantity of ASOs or bacterial strains with permeable membrane were often used to elicit gene-silencing and thereafter validate antibacterial effect of naked ASOs *in vitro*. Good *et al.,* [34] have creatively proposed and developed PNAs conjugated with attached CPPs, which dramatically improve the cellular uptake of PNA without CPP's interference, as evidenced by the minimal inhibitory concentration (MIC) of conjugates lowered to micro molar ranges. Presently, *in vitro* modified MIC and MBC tests of peptide-ASO conjugates have been well-acknowledged methods for preliminary target identification and validation. Essentiality and accessibility of targeted genes are reflected by antibacterial effect specifically designed peptide-ASO conjugates with ASO sequences complementary to different sites of mRNA, which is quantitatively determined by comparing their MIC values. Conjugate with the lowest MIC value indicates the most sensitive targeting site within mRNA for antisense inhibition, whereas the minimal bactericidal concentration (MBC) values suggest if the antisense antibacterial effect is bactericidal.

Target specificity is another important factor to finally determine the gene-silencing effect of ASOs. It is generally evaluated by testing the antibacterial activity of control ASOs (*e.g.,* ASOs with mismatched or scrambled nucleotide sequences) or naked peptides. Further, RT-PCR and western blotting can be used to observe the reduction of mRNA and protein product of the particularly targeted gene/mRNA, which further demonstrates the specificity of ASOs.

Validated Targets in Bacteria

Targeting Growth-Essential Genes

Essential genes that regulate or control bacterial growth, proliferation, virulence and synthesis of important living-dependent substances are candidate targets for antisense inhibition that have attracted much researching enthusiasms. Validated essential genes among different bacterial species include (Table **1**), 23S rRNA [35], 16S rRNA, *lacZ/bla* [36], *rnpA* in *E. coli, Enterococcus faecalis,* and *Bacillus subtilis* [37], *fbpA/fbpB/fbpC* [38,39] and *glnA1* [40] in *Mycobacterium tuberculosis, inhA* in *Mycobacterium smegmatis* [41], *oxyR/ahpC* in *Mycobacterium avium* [42], *gyrA/ompA* in *K. pneumonia* [43], NPT/EhErd2 in *Entamoeba histolytica* [44,45], *gtfB* in *Streptococcus mutans* [46], *fmhB/gyrA/hmrB* [47] and *fabI* [48] in *S. aureus, acpP* in *E. coli* [49-54] *Burkholderia cepacia* [55], as well as *Salmonella enteric* serovar Typhimurium (ST) [56,57], *gyrA* in *E. coli* [58], ST, *K. pneumonia, Pseudomonas syringae, Enterobacter cloacae, Enterococcus faecalis, Mycobacterium marinum, Acinetobacter, B. subtilis,* and *S. aureus* [37,59], and *rpoD* in *E. coli, Salmonella enterica, K. pneumonia, Shigella flexneri,* and *S. aureus* [60].

Table 1: Summary of ASOs targeting growth-essential genes in antibacterial therapy.

Target (site)[a]		ASOs[b]	Test Organism[c]	Efficacy Identified	Delivery Method[d]	Ref.
rpoD	start codon region (1-10 nt)	PNA	*E. coli, S. enterica, K. pneumonia, S. flexneri*	*in vitro/in vivo* (mice)	CPP=(KFF)$_3$K or (RXR)$_4$XB	[61]
	234-243 nt		*S. aureus*	*in vitro*		[60]

Table 1: contd…

23S rRNA	P.T. center	PNA	*E. coli* AS19	*in vitro*	—	[36]
	domain II		*E. coli* Dh5α	*in vitro*	CPP=(KFF)$_3$K	[35]
	α-sarcin loop		*E. coli* AS19	*in vitro*	—	[62]
			E. coli K12 (wild-type)		—	[63]
					CPP=(KFF)$_3$K	
16S rRNA	preceding start codon region	MDNA	*E. coli* lacking outer cell wall	*in vitro*	—	[5]
			normal *E. coli*		PEG attached	
	mRNA binding site	PNA	*E. coli* AS19		—	[36]
			E. coli K12		CPP=(KFF)$_3$K	[64]
RNase P	P15 loop	LNA	*E. coli*	*in vitro*	—	[65]
		PNA			CPP=(KFF)$_3$K	
	C5 protein	EGS-ON			—	[58]
acpP (start codon region)	SD site -24 to -13 nt	PNA	*E. coli* K12	*in vitro*	CPP=(KFF)$_3$K	[32]
	-9 to 3 nt					
	-5 to 5 nt		*E. coli*	*in vitro*		[34]
			E. coli SM101	*in vivo* (mice)		[54]
			E. coli K12			
			E. coli K12			[66]
	6 to16 nt	PMO	*E. coli* AS19	*in vitro* luciferase system	—	[33]
			E. coli SM105	*in vitro*	—	[67]
			E. coli W3110 (ATCC27325)	*in vitro*	CPP$_1$=(KFF)$_3$KXB	[56]
			EPEC (E. coli E2348.69)	*Ex vivo* cocultured Caco-2 culture	CPP$_2$=RTRTRFLRRTXB CPP$_3$=(RFF)$_3$XB CPP$_4$=(RXX)$_3$B	
			S. enterica (ATCC29629)			
			E. coli W3110 (ATCC27325)	*in vivo* (mice)	CPP$_3$=(RFF)$_3$XB	[68]
			E. coli W3110 (ATCC27325)	*in vitro*	19 synthetic CPPs	[69]
				in vivo (mice)	CPP$_1$= (RX)$_6$B CPP$_2$= (RXR)$_4$XB CPP$_3$= (RFR)$_4$XB	
		PMO	*S. enterica* LT1	*in vivo* (mice)	CPP= (RXR)$_4$XB	

Table 1: contd...

		3+Pip-PMO				
		Pip-PMO	*E. coli* W3110 (ATCC27325)	*in vivo* (mice)		[70]
		Gux-PMO				
	4 to 14 nt	PMO	14 *B. cepacia* strains (5 clinical isolates+9 from ATCC)	*in vitro*	CPP= (RFF)$_4$XB	[71]
	-5 to 6 nt					
floA, floP		PNA	*E. coli* AS19	*in vitro*	CPP=(KFF)$_3$K	[72]
glnA1		PS-ODN	*M. tuberculosis*	*in vitro*	ethambutol or polymyxin B nonapeptide	[40]
fbpA,fbpB, fbpC		5'-, 3'-HP PS-ODN			—	[38]
						[39]
inhA		PNA	*M. smegmatis*	*in vitro*	CPP=(KFF)$_3$K	[73]
adk		PNA	*S. aureus* RN4220	*in vitro*	CPP=(KFF)$_3$K	[72]
fmhB, gyrA, hmrB		PNA	*S. aureus* RN4220	*in vitro*	CPP=(KFF)$_3$K	[47]
gyrA, ompA		PNA	*K. pneumoniae*	*in vitro*	CPP=(KFF)$_3$K	[43]
gyrA		EGS-ON	*E. coli* *S. typhimurium*	*in vitro*	—	[58]
		EGS-PMO	*E. coli,* *S. typhimurium,* *P. aeruginosa,* *S. aureus,* *E. faecalis,* *B. subtilis*		CPP=(RXR)$_4$XB	[37]
fabI		UM	*S. aureus*	*in vitro*	—	[48]
fabD		PNA	*E. coli* K12		CPP=(KFF)$_3$K	[72]
gtfB		PS-ODN	*S. mutans*	*in vitro*	—	[46]
oxyR/ahpC		UM	*M. avium* complex	*in vitro /* ineffective	—	[42]
NPT/EhErd2		UM	*E. histolytica*	*in vitro*	—	[44,45]
rnpA		EGS-PMO	*E. coli,* *E. faecalis,* *B. subtilis*	*in vitro*	CPP=(RXR)$_4$XB	[37]

[a]The essential genes that were targeted encode the following proteins: *rpoD*, RNA polymerase D; *acpP*, acyl carrier protein; *fabI*, enoyl-acyl carrier protein reductase; *fabD*, malonyl coenzyme A acyl carrier protein transacylase; *folP*, dihydropteroate synthase; *fmhB*, protein involved in the attachment of the first glycine to the pentaglycine interpeptide; *gyrA*, DNA gyrase subunit A; *hmrB*, ortholog of the *E. coli* acpP gene; *adk*, adenylate kinase; inhA, enoyl-acyl carrier protein reductase; *ompA*, outer membrane protein A; *gtfB*, synthesis of water-insoluble glucans; *inhA*, enoyl-(acyl carrier

protein) reductase; RNase P, P15 loop of RNase P; *gyrA*, DNA gyrase subunit A; *oxyR*, oxidative stress regulatory protein; *ahpC*, alkyl hydroperoxide reductase subunit C; *glnA1*, glutamine synthetase; *fbpA, fbpB, fbpC*, 30/32-kDa mycolyl transferase protein complex; *NPT*, Neomycin phosphorotransferase; *EhErd2*, marker of the Golgi system; *LacZ/bla*, beta-galactosidase/beta-lactamase; *rnpA*, ribonuclease P protein; P.T. indicates peptidyl transferase; SD, Shine-Dalgarno; nt, nucleotide.
[b]UM, unmodified; MDNA, ethylcarbamate DNA; PNA, peptide nucleic acid; PMO, phosphorodiamidate morpholino oligomers; Pip-PMO and n+Gux-PMO, cations (piperazine or N-(6-guanidinohexanoyl)piperazine) attached to the phosphorodiamidate linkages; LNA, locked nucleic acid; EGS, external guided sequence.
[c]*E. coli* AS19, permeable membrane; *E. coli* SM101, defective membrane; EPEC, enteropathogenic *E. coli*.
[d]CPP, cell penetrating peptide; PEG, Polyethylene glycol; "—", no delivery method used. For synthetic peptides, X is 6-aminohexanoic acid and B is beta-alanine.

Targeting Resistance Mechanisms

Developing resistance inhibitors in traditional antibiotic industry is a sound, well-validated strategy for tackling resistance problems, because they postpone the "expire date" of on-market antibiotics and expand their application. The economic and clinical value of this rationale is well recognized and demonstrated by offering new combinations to clinical practice. Thus, a few studies focused on interrupting the expression of genes involved in resistant mechanism by antisense approach, aiming to restore bacterial susceptibility to key antibiotics in clinical practice. Validated resistance genes among different bacterial species include (Table **2**), *bla* in *E. coli* [74,59,36], *marOR AB* in *E. coli* [75], *vanA* in *E. faecalis* [76], *act* in in *E. coli* [77,78], *aac(6')-Ib* in *E. coli* [79,80], *metS/murB* in *Bacillus anthracis* [81], *mecA* in *S. aureus* [82,83], *cmeA* in *Campylobacter jejuni* [84], *oprM* in *P. aeruginosa* [85], *cat* in *E. coli* [37,59,74].

Table 2: Summary of ASOs targeting resistance mechanism in antibacterial therapy.

Target	Encoding Proteins	ASOs[a]	Test Organism(s)	Efficacy Identified	Delivery Method[b]	Ref.
cat	chloramphenicol resistance	PMO	*E. coli*	*in vitro*	CPP	[37,59,74]
oprM	outer membrane efflux protein	PS-ODN	*P. aeruginosa*	*in vitro*	liposome	[85]
mecA	penicillin-binding protein 2 prime	PS-ODN	*S. aureus*	*in vitro & in vivo*	liposome	[82,83]
cmeA	CmeABC multidrug efflux transporter	PNA	*C. jejun*	*in vitro*	CPP	[84]
aac(6')-Ib	aminoglycoside 6'-N-acetyltransferase type Ib, mediate amikacin resistance	UM	*E. coli*	EGS mediated RNaseP leavage / *in vitro*	—	[80]
				in vitro	EP	[79]

Table 2: contd...

metS/ murB	methionyl-tRNA synthetase /UDP-N-acetylenolpyruvoylglucosamine reductase	UM	B. anthracis	in vitro	—	[81]
act	chloromycetin acetyl transferase	UM	E. coli	EGS mediated RNaseP leavage / in vitro	—	[78]
				in vitro	—	[77]
vanA	class A (VanA) glycopeptide-resistant related protein	UM	E. faecalis	in vitro	—	[76]
marORAB	multiple antibiotic resistance operon	PS-ODN	E. coli	in vitro	HS/EP	[75]
LacZ/bla	β-galactosidaze/ β- l actamase	PNA	E. coli AS19 (permeable membrane)	in vitro	—	[36]
		PMO	E. coli		CPP	[59,74]

[a]UM, unmodified; ASOS, antisense oligonucleotide; PS-ODN, phosphotioates oligodeoxynucleotide; PNA, peptide nucleic acid; PMO, phosphorodiamidate morpholino oligomers.
[b]CPP, cell penetrating peptide; EP, electroporation; HS, heat shock; EGS, external guide sequences; "—", no delivery method used.

Efficient Delivery Systems

Many barriers exist for the efficient transfer of genes/oligonucleotide anologs to cells, including the extracellular matrix, the endosomal/lysosomal environment, the endosomal membrane, and the nuclear envelope. Plenty of delivery systems have been proved to serve suitably for antisense approach in eukaryotic cells regardless of their types (non-viral or viral) *vs.* cell types [86]. The optimistic concept that any microbial gene could be targeted and highly organism-specific ASOs therapeutics drugs could be envisioned had been well accepted at the early stage of antisense antibacterial studies. It has not been long before researchers realize that the obvious obstacle hampering efficient antisense gene-silencing came from the stringent and more troublesome bacterial cell membrane that causes poor cellular uptake of synthetic ASOs [87].

An unmodified 10-mer oligonucleotides is 2-3 kDa, and various chemical modifications outlined above add further molecular weights to this size. In short, ASOs are macromolecules likely to be considerably larger than vancomycin, therefore require efficient delivery systems. Further, antibacteiral ASOs may require development of delivery conditions for each bacterial species, which is attributable to the distinguished membrane stringency varying among different gram-positive and/or gram-negative bacteria or highly resistance strains within certain species. A variety of delivery strategies, including electroporation, permeablilizing solvents, cationic lipid formulations (*e.g.,* liposome), and pore-forming peptides, have been applied to facilitate the entry of ASOs into bacterial cells in the laboratory. Although what exactly will work best for ASOs remains to be determined, the cell-penetrating peptide (CPP) mediated delivery of ASOs (especially peptide-PNA and peptide-PMO conjugates) outperformed other delivery systems in way of structural simplicity and intrinsic efficiency, reaching evaluations as future applicable therapeutics.

Cell-Penetrating Peptide (CPP) Mediated Delivery

CPPs are short peptides of less than 30 amino acids that are able to penetrate cell membranes and translocate different cargoes into cells. The only common feature of these peptides appears to be that they are amphipathic and net positively charged. Two CPP strategies have been described to date; the first one requires chemical linkage between the drug and the carrier for cellular drug internalization, and the second is based on the formation of stable complexes with drugs, depending on their chemical nature. Recently, the second strategy has received increasing studies for its advantages on convenient delivery of DNAs or ASOs into eukaryotic cells, especially considering the synthesis and cost issues. There is no universal method for their delivery into different gram-positive and gram-negative specie, as they all present several limitations. However, CPP-conjugated method is now extensively applied for antimicrobial ASOs both in cultured cell and *in vivo*, representing a new and innovative concept to bypass the problem of low bioavailability caused by poor cellular uptake [88,89].

Good L and Nielsen PE, who first established synthesis chemistry that conjugated CPP to the end of PNA in 2001, have realized efficient delivery of PNA through bacterial out membrane through observing its potent bacteriocidal antisense effects at micromolar ratio [63]. Later chemistry inventions make possible the conjugation CPP to other oligonucleotide analogs (*e.g.,* PS-ODNs, LNAs, PMOs) for imparting them into bacterial cells and specific intracellular targets. Further evidence has demonstrated that spacers or linkers between ASO and CPP in the direct covalent conjugate can be introduced by different chemistry (Fig. **3**), which may increase the antisense efficacy and antibacterial potency of PNA by decreasing the spatial hindrance caused by CPP. It has also been demonstrated that the release property of the chemical bond between ASO and CPP (*e.g.,* the more stable amide bond or the less stable disulfide bond) has no significant influence on the antisense efficacy of ASOs [34,90].

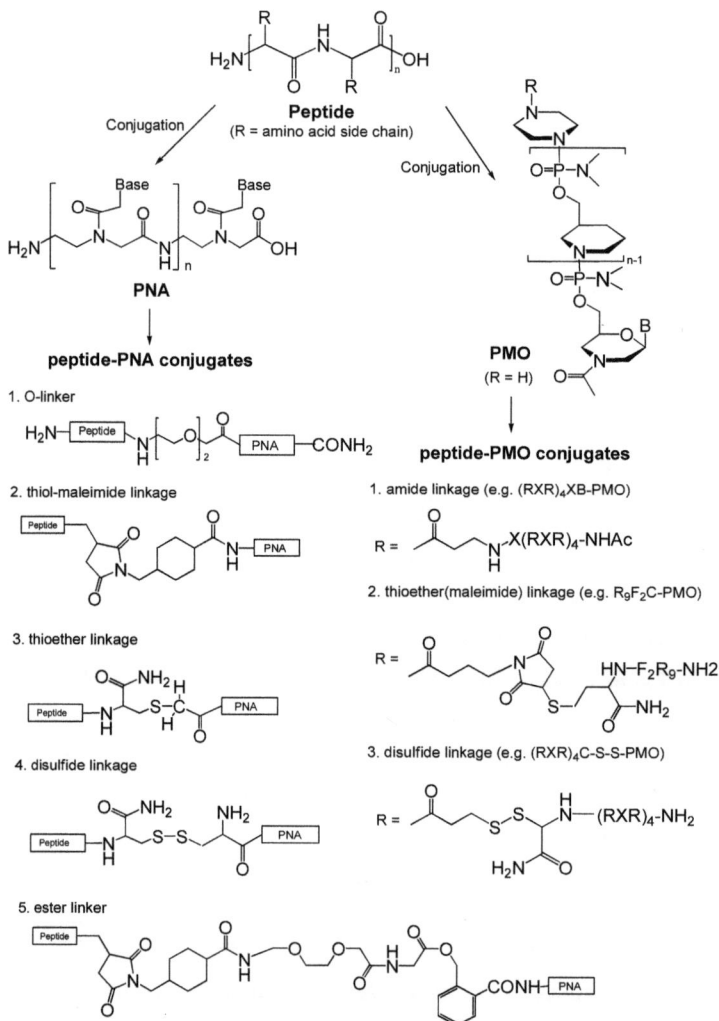

Figure 3: Structures of peptides and PNA and various types of peptide–PNA conjugate linkage. Structure of PMO and three types of peptide–PMO conjugate linkages. R=arginine, Ahx=6-aminohexanoic acid, B=beta-alanine and F=phenylalanine.

The mechanism of cell wall permeabilization of CPP itself is controversial and still under exploration. Nonetheless, the significant improvement in bacterial uptake of ASOs facilitated by conjugated CPP has been well-recognized, making it an indispensable delivery system before no advanced system is developed for antibacterial ASOs. Previous studies suggested that the repeated amphipathic motif with cationic residues followed by hydrophobic regions is an important structure for carrier efficiency of CPP. The "carrier peptide" KFFKFFKFFK, originally reported for efficient penetrating ability through brain blood barrier, is the first also the most extensively applied peptide sequence [89]. It has been proved to facilitates delivery of PNAs and PMOs into *E. coli, ST, K. pneumoniae* (however much less potent), and *S. aureus*. But it is not working for *P. aeruginosa* membrane even at higher concentrations of conjugated PNAs. Then, more efficient peptide sequences RFFRFFRFFRXB and RXRRXRRXRRXRXB (X is 6-aminohexanoic acid and B is β-alanine) have been developed, and more efficient transporting of PNAs and PMOs have been reported for *in vitro* and *in vivo* antibacterial activity tests on different bacterial species, including *E. coli, ST, K. pneumoniae, S. aureus*, and *P. aeruginosa* [69]. Notably, the membrane of two gram-negative species Acinetobacter baumanni and *S. flexneri* show highest sensitivity to $(RXR)_4XB$ (also called P7) mediated PNA-CPP conjugates. Following studies have demonstrated that gram positive bacteria *B. subtilis* and *Corynebacterium efficiens* exhibit increased susceptibility to $(RXR)_4XB$ conjugated PNAs, but in contrast, that of the gram-negative bacterium *Ralstonia eutropha* was not affected [87,64]. Recently, a unique peptide derived from human T cells, YARVRRRGPRGYARVRRRGPRRC, has been employed to conjugate with EGS-guided anti-*gyrA* PMOs, which showed at least 10- to 100-fold efficacy increase compared with previous peptides used in altering the phenotypes of as many host bacteria species as *E. coli, Salmonella enteric* serovar Typhimurium, *K. pneumonia, Pseudomonas syringae, E. cloacae, E. faecalis, M. marinum, Acinetobacter, B. subtilis*, and *S. aureus* [59].

Other Pharmaceutical Properties

A systematic overview of cargo (*i.e.,* oligonucleotides) and vector (especially the CPP, [91]) should be applied in therapeutic application of antibacterial ASOs, when properties other than chemistry are concerned. With regard to ASOs, the electric neutral PNA and PMO classes emerge and show desirable properties (especially their non-ionic backbones) as better therapeutic alternatives to other synthetic oligomers especially for antibacterial purposes. With regard to delivery strategies, non-viral methods, such as electroporation and liposome encapsulations, have been proved to be effective *in vitro* and for research purposes, but technically demanding showed for *in vivo* delivery due to toxicity, cell damage, immunogenicity, lack of tissue and cell specificity, *etc.* Thus, peptide-ASO conjugates (*i.e.,* peptide-PNAs and peptide-PMOs) offer a promising noninvasive version of gene silencers with potent antisense antibacterial activity, the pharmaceutical properties of which this part will mainly focus on.

PNAs, PMOs and Their Peptide Conjugates

PNAs and PMOs are novel classes of antisense agents that offer a better therapeutic alternative to other ASO-based antibacterial therapeutics. Although departing significantly from the sugar-phosphate backbone found in regular DNA, oligomers of both types independently (*i.e.,* with or without delivery strategies) have been found to be remarkable steric-blocking ONs for inhibiting translation and blocking miRNA activity, as demonstrated in embryos, cells and animals. Now, PMOs have been taken to pre-clinical studies for treatment of cardiovascular diseases, viral diseases and genetic disorders, such as Duchenne muscular dystrophy (DMD).

Conjugation of CPP to negatively charged ONs (*e.g.,* PS-ONs) did not result in a level of delivery into cells sufficient for biological activity. However, PNA- and PMO- CPP conjugates (covalently linked with or without spacers) confer more desirable properties on the original ODN forms [92]. Several CPPs have been developed for bacterial-specific transformation purposes (as mentioned in previous section), and they can be coupled to PNA or PMO by flexible linker types [93] (as illustrated in Fig. **3**). No general rules have yet emerged as to optimal linkage types for systemic application, since the factors affecting biological activity are often complex. Early popular labile linkers for PNA and CPP include AEEA (8-amino-3,5-dioxo-octanoic acid, a polyether spacer also known as an O-linker) and disulfide bond linkage, which were proposed to be cleavable within the reducing environment of the cell. Stable linkages, such as glycine linkage, thioether linkage and thiol-maleimide linkage, have also been reported for improved *in vivo* stability. The conjugation of CPPs to PMOs through a thioether (maleimide), disulfide or amide linker have previously been described to present similar nuclear antisense activities [51]. But, the amide linkage is advantageous with regard to synthetic procedures (*e.g.*, greater yield and less steps) and *in vivo* stability.

Toxicity

Specifically, the non-ionic character of the PNA and PMO portion of conjugates avoids potential non-specific drug interactions with bacterial cellular components (except for the target RNA sequence) which may causes severe non-specific side-effects observed with PS-ONs.

Large amount data concerning toxicity of PNA-CPP conjugates have been collected from *ex vivo* studies [94,43], in which peptide-PNAs cured cell cultures infected with bacteria in a dose-dependent manner without any noticeable toxicity to the human cells. *In vivo* inhibition of gene expression and growth have been observed for anti-*acpP* peptide-PNA in mouse after intraperitoneal *E. coli* infection. However, non-toxicity issues have not yet been seriously addressed and evaluated.

The bacteriocidal antisense effect of PMO-CPP conjugates targeting different bacterial essential genes have been evaluated in several animal bacteremia models. Amantana and *et al.*, have proved that (RXR)$_4$XB-PMOs targeting gene *acpP* had an excellent safety profile within doses for 100% survival of mice 48h after treatment [95]. Nontheless, survival rate was significantly reduced for mice treated with 2×300 mg and 2×1 mg of the same 11-base AcpP peptide-PMO, indicating toxicity at these higher doses. It is generally believed that the toxicity of conjugate is caused by peptide (RXR)$_4$XB while the PMO portion of the conjugates are essentially non-toxic. According to data from PMO-CPP conjugates targeting genes in eukaryotic cells, the degree of toxicity depends on the dose, dose frequency and route of administration [96]. Collectively, mice tolerated (RXR)$_4$XB-PMOs well with repeated intraperitoneal (*i.p.*) or intravenous (*i.v.*) injection doses of ≤15 mg/kg at diverse time intervals, showing no changes in behaviour, weight and serum chemistry, and no histopathological abnormalities detected in major organs. However, at higher doses and dosing frequency, animals experienced weight loss, despite maintaining their normal organ weights and appearances. Rats treated with a single 150 mg/kg dose appeared lethargic immediately after the injection and proceeded to lose weight, accompanied by affected kidney function. The LD_{50} of $(RXR)_4XB$-PMOs in rats was around 220-250 mg/kg.

Stability, Tissue Distribution, and Pharmacokinetics

The modified structures of PNA and PMO provide them excellent resistance to nuclease and protease activity, thereby enhanced stability in plasma, tissues, cerebrospinal fluid and urine can be observed. Rational design of conjugates, *e.g.*, the optimal linker type and the position of CPP on conjugate, has been shown to eliminate CPP's stereospecific blockade that significantly influence the antisense effect of conjugated PNA or PMO in target recognition, base matching, and binging affinity. However, stability of CPPs may be a matter of concern. The excellent stability of PNA-CPP conjugates has been confirmed for a 48h period at 37°C in rat's plasma [61]. Yet, degradation of CPP in solution and plasma has been observed for systematic delivery of CPP-2'MOE conjugate in mouse model [90]. Besides, Youngblood *et al.*, have determined that the stabilities of CPP-PMO conjugates in cells and in human serum vary according to CPP sequences, amino acid compositions and/or linkers [97]. They found that the stability of $(RXR)_4XB$ in the conjugate exhibited time- and tissue-dependent degradation, with biological stability ranked in the order of liver＞heart＞kidney＞plasma, while the PMO portion of the conjugates was completely stable in cells, serum, plasma and tissues.

Specifically, the non-ionic feature of the PNA or PMO portion of conjugates guarantees sufficient tissue concentrations required for effective oligonucleotide: RNA duplex formation, therefore enhanced affinity to target RNA sequences and increasing efficacy can be expected. Additional concern comes from the non-specificity of CPP mediated delivery of PNA or PMO in commensal environment where eukaryotic cells and prokaryotic cells both exist. Although little lethal damage to eukaryotic cells have been observed for CPP-PNA/PMO conjugates used at effective antibacterial concentrations *in vitro* and *ex vivo*, the consequences of physical disruption to normal human cell membranes caused by CPP-PNA/PMO conjugates *in vivo* have not been thoroughly evaluated, which is presumed to change the tissue distribution and pharmacokinetics properties of PMO/PNA. The application of unmodified PNAs as antisense therapeutics thus far has been limited by their low solubility under physiological conditions, insufficient cellular uptake, and poor biodistribution due to rapid plasma clearance and excretion. Although there has been no report of *in vivo* tissue distribution and pharmacokinetics properties of CPP-PNA conjugates targeting bacterial gene, limited information from PNA-CPP conjugates targeting genes in eukaryotic cells can be referred. Jia *et al.*, have determined that PNA-CPP conjugates targeting bcl-2 mRNA slowed specific tumor uptake, low uptake in blood and organs (*e.g.,* liver and spleen) except for kidney, as well as slower urinary clearance in Mec-1-bearing severe combined immunodeficiency (SCID) mice. Recently, Wancewicz *et al.*, have reported that conjugation of PNA (targeting murine phosphatase and tensin homolog) to short basic peptides (serve as solubility enhancers and delivery vehicles) allowed for rapidly distribution and accumulation of conjugate in liver, kidney and adipose tissue, while their rates of elimination *via* excretion were dramatically reduced compared to unmodified PNA. Unlike PNA-CPP conjugates, the pharmaceutical properties of PMO-CPP conjugates have been evaluated in an extensive scope as specific gene modulators [95]. In general, the conjugation of CPP to PMO enhances the PMO pharmacokinetic profile, tissue uptake, and subsequent retention [98]. Amantana *et al.*, have reported that conjugation of a PMO to the peptide $(RXR)_4XB$ increased the tissue uptake (in all organs except in brain, with greater increases being seen in liver, spleen and lungs) and retention time in these organs, while efflux of the conjugated PMO from tissues to the vascular space was slow. They have also confirmed that peptide conjugation also improved the kinetic behavior of PMO, as demonstrated by increased volume of distribution, estimated elimination half life, and area under the plasma concentration *versus* time curve.

In all, systematic studies on parmaco-issues of CPP-PNA/PMO conjugate based antisense antibacterial therapy need to be done if any possible candidate for clinical development is ever recommended [99].

Broad-Spectrum ASOs Therapeutics

A range of functional genes in bacteria have been validated as potential targets by using unmodified or CPP conjugated ASOs. Collectively, consistent efforts on antisense targeting of a small bacterial gene *acpP* (encoding the essential fatty acid biosynthesis protein) have passed the proof-of-principle phase and have gathered plenty of positive *in vivo* results against several pathogenic bacterial species (*i.e.*, *E. coli*, *ST*, and *Burkholderia cepacia*). However, the validated gene targets in different bacterial species show discouragingly low similarity in sequence and homology. Thus, the target sites within a same gene for ASO targeting in most cases varied according to species, which undermines the potential of validated gene-specific ASOs for developing promising broad-spectrum therapeutics. Antibacterial peptide-ASOs may present an unusual opportunity for developing broad-spectrum therapeutics against upgrading infections caused by multi-drug or pan-drug resistant pathogenic species (as elaborated below), where many successful compounds have failed.

Effective Shared Target Region

A challenging aspect of identifying essential genes in bacteria for broad-spectrum antisense inhibition mainly exists in the efforts to locate the exact targeting site for complementary ASOs, the antisense inhibition of which may elicit potent and specific bactericidal effect against different bacterial species (*e.g.,* among gram-negatives, gram-positives, or both). It can be inferred that prerequisite in target selection is to search for genes with high similarity and identity among as many bacterial species as possible. As time consuming as it is, an economical way is to focus on validated targets for both traditional antibiotics and antibacterial ASOs, with respect to their massively available sequence data. To name a few are 16S rRNA, *acpP*, and *gyrA*. However, the issue of target accessibility among different species still needs further investigation and validation. The ribosome has a complex structure involving rRNAs and ribosomal proteins, and therefore, inaccessibility of the target site could be one of the reasons for the ineffectiveness of anti-ribosomal PNAs. Systematic researches *in vitro* and in animal models have demonstrated that gene *acpP* opens limitless possibility for recommending the very first "antisense antibiotic" into market. Besides, *acpP* in pathogenic gram-negative species shares highly homology in sequences, making it an ideal candidate for developing antibacterial ASOs with broad spectrum. What needs to be confirmed is accessibility of an already-validated 11-nucleotide targeting site in the start codon region of *E. coli acpP* in other candidate bacterial species. Meanwhile, newly discovered gene targets for new types of protein-targeting antibiotics, *i.e.,* the bacterial cell division inhibitor (targeting bacterial cell division protein *FtsZ for* terminating bacterial proliferation [100]), and virulence inhibitor (targeting quorum sensing sensor protein *QseC* without affecting bacterial growth [94]), also show promises and potential for developing specific or broad-spectrum antibacterial ASOs based on their homology assessment. Our research focuses on validating the known target in broad-spectrum antibiotic development by antisense strategy, in which the DNA-dependent RNA polymerase (RNAP) is a candidate of great interest [101].

Universally Applicable Delivery Systems

The term "broad-spectrum" also qualifies the delivery systems for ASOs. Specifically, rational design of peptide-ASO conjugates could optimize the effect AS-DONs in way of enhancing antibacterial potency and expanding antibacterial spectrum, in which CPP choice is of equal importance. The range of sensitivities observed for different bacterial species to CPPs largely confines their application potential. Collectively, our results suggested that the peptide component of peptide-PNA conjugates may be developed for a wide range of indications to realize broad antisense antibacterial spectrum.

Proof-of Principle Studies

Our previous studies provided the first proof-of-principle evidence for identifying bacterial RNAP σ^{70} as a promising target for potent broad-spectrum antisense inhibition in both gram-negative [61] and gram-positive bacteria [60].

Peptide (RXR)$_4$XB- and (KFF)$_3$K-conjugated peptide nucleic acids (PPNAs) were developed to target *rpoD* in different pathogenic gram-negative bacteria (GNB) species (including antibiotic sensitive and resistant strains of *E. coli*, ST, *K. pneumoniae* and *S. flexneri*), which share high homology in *rpoD* sequence. The (RXR)$_4$XB- and (KFF)$_3$K- conjugated PNAs were bactericidal against different strains of MDR-GNB in concentration-dependent and sequence-selective manner. Among tested PPNAs, (RXR)$_4$XB conjugated PNA06 (targeting the 1-10 nucleotides of the start codon region) showed more potent and broad spectrum inhibition *in vitro* and *in vivo*. The results were associated with suppression of *rpoD* mRNA and σ^{70} expression, as well as σ^{70} downstream regulated genes. The treatment of PPNA06 on mono- or multiple MDR-GNB infected human gastric mucosal epithelial cells demonstrated complete inhibition on bacterial growth and no influence on morphology and growth of human cells *ex vivo*. Notably, PPNA06 did not show the induction of antibiotic resistance as compared with classical antibiotics in MDR-GNB.

By following similar procedures, we also synthesized a serial of peptide conjugated PNAs based on software predicted parameters and further design optimization. We identified a target sequence (234 to 243 nt) within the conserved region 3.0 of *S. aureus rpoD* mRNA, which showed comparably higher sensitivity to antisense inhibition. A (KFF)$_3$K conjugated 10-mer complementary PNA (PPNA2332) was developed for potent micromolar-range growth inhibitory effects against four pathogenic *S. aureu*s strains with different resistance phenotypes, including clinical vancomycin-intermediate resistance *S. aureus* (VISA) and MDR- MRSA isolates. It showed bacteriocidal antisense effect at 3.2 fold of MIC value against MRSA/VISA Mu50 with high sequence specificity. It also specifically interfered with *rpoD* mRNA, inhibiting translation of its protein product σ^{70} in a concentration-dependent manner. Its *ex vivo* bacteriocidal antisense effect was also observed for eliminating viable bacterial cells in a time- and concentration- dependent manner at ≥1 μ without showing any apparent toxicity at 10 μM. Meanwhile, the potent broad-spectrum antisense inhibition in gram-positive bacteria of this particular conjugate PPNA2332 was also observed on *Staphylococcus epidermidis* (unpublished results), which also shares high similarity in *rpoD* sequence.

ASOs-BASED ANTIVIRAL THERAPEUTICS

Antisense Antiviral Strategy

Antisense strategies have been used for a variety of purposes, such as target validation, gene function studies and experimental therapy for different diseases. A significant research on antisense technology has been contributed to counteract viral diseases. ASOs can be designed to interact with any viral RNA, provided its sequence is known, making antisense strategies a very attractive means of inhibiting viral replication. The antisense antiviral principle which applies ASOs potential therapeutics aimed to inhibit viral growth was first documented in 1978, by the observations of Zamecnik and Stephenson using a phosphodiester oligodeoxynucleotide composed of 13 nucleotides (a 13-mer) to block Rous sarcoma virus replication [4,102]. And the approval to use Vitravene (fomivirsen, ISIS 2922), which is an ASO composed of 21 PS-ONs against CMV IE2, by FDA in 1998 for local treatment of cytomegalovirus (CMV)-induced retinitis provides ample proof that antisense has the potential to emerge as a new antiviral therapeutic of this century.

As sequence-specific approaches to inhibition of virus replication, it is quite appealing of antiviral ASOs in part because of the low likelihood of target sequence homology between pathogen and host, thus minimizing the potential for toxicity from cross-reactivity. Furthermore, using ASOs to target sequence that is highly conserved between viral strains or species raises the prospect of enlarging the application potential by developing broad-spectrum antiviral therapeutics. However, the path from initial observations to the practical use of ASOs for defining novel antiviral agents has been a long one, during which the four major concerns have not changed much since first proposed in 1998 [103]: (a) choose preferred viral RNA target sequences for optimal antisense impact, (b) to alter the ASOs to enhance stability and retain hybridization capacity and favorable biological characteristics and (c) to recognize that oligonucleotides may display an array of antiviral activities unrelated to any antisense function for which they were designed.

Antiviral ASOs: A Brief Overview of Structural Chemistry

ASO based antiviral agents are usually designed to induce cleavage of, or to anneal to and block, targeted viral RNA. Damaged viral RNA, or the obstruction of viral RNA from factors necessary for the efficient progression of the viral life cycle, can reduce viral propagation. Importantly, the ensuing targeted destruction of viral RNA should interfere with viral replication without entailing negative effects on ongoing cellular processes. Practical application of antisense oligonucleotides has required modifications of DNAs or RNAs, with the aim of retaining the hybridization capacity while increasing stability.

Phosphorothioates: Once the Main Player

The earliest study using modified ASO construcs as potential antiviral therapeutics dated from 1978 by Zamecnik and Stephenson, who used a tridecamer oligonucleotide as a hybridization competitor to inhibit Rous sarcoma virus replication. From 1985, Miller and Ts'o published several papers describing inhibition of herpes simplex virus type 1 by oligodeoxyribonucleoside methylphosphonates in cell lysates and in reticulocytes. The methylphosphotriester DNA constructs, where a nonbridging oxygen is replaced by a methyl moiety, results in more hydrophobic (loss of a negative charge) ONs with a slight drop in hybridization affinity (lowered melting temperature or Tm) and a loss in RNaseH activation. However, it remains an interesting study object despite concerns from synthesis and purity.

Replacement of a non-bridging oxygen with sulfur otherwise yields phosphorothioates with strongly enhanced stability against enzymatic degradation. The loss in hybridization affinity ($0.3\text{-}1^\circ$C drop in Tm per modification) is largely compensated for by preservation of the sensitivity to RNase H cleavage of the hybrid between a PS-ON and its RNA complement. Hence in late 1990s, by this innovative modification PS one can synthesize tailored ASOs with a balance of characteristics of hybridization affinity, hydrophobicity and the capacity to recruit RNase H mediated hydrolysis of the target RNA. Most of the literature on use of ASOs as antiviral agents has used the PS-modified oligonucleotides, and the first generation of drugs that have progressed to clinical trials have been PS compounds [104,105]. Historically, PS-ONs have played a vital part in the development of antiviral ASOs and received thorough evaluations on biomedical properties before serious non-specific effects forced them to step out of stage (as mentioned in the "First generation" part).

Second and Third Generation Constructs: Morpholino ASOs as New Star

As outlined before, most of the second and third generation ASO constructs are steric-blocking ONs that are unable to recruit RNaseH and therefore are limited to inhibition of translation. The creation of gapmers (as mentioned in the "Second generation" part), once precluded with PNA or PMO constructs, are now applicable, which thanks to the strategy that endows PNA [106] or MPO [107] oligomer with conjugated 5'-phosphorylated-2',5'-linked oligoadenylate to recruit of 2-5A dependent RNase L. Numerous constructs have been prepared in search for useful antisense antiviral effects, however, the success of ASOs, especially the morpholino ASOs (*i.e.,* PMOs, peptide-PMOs, and positively-charged PMOs, as illustrated in Fig. **2**), at reducing translation of targeted cellular mRNA, along with several characteristics of RNA viruses, suggested that a steric-blocking antisense strategy may be effective against viral pathogens.

The fundamental reasons include [108]: (1) the austere genomes of RNA viruses showing little redundancy in the function of their encoded proteins, and many viral proteins executing multiple critical functions; (2) multiple proteins are sometimes coded by the same mRNA through the use of a number of molecular mechanisms; (3) many defined viral sequence elements and/or structures involved in functional regulation of translation, RNA synthesis, mRNA capping, and virion assembly are often located in RNA that does not code for protein, some of which include sequence which is highly conserved within a group of viruses. Thus in all, inhibition of viral translation could therefore be expected to have severe consequences for viral replication, and those critical motifs constitute appealing targets for sequence-specific intervention.

The morpholino ASOs, which are capable of forming a stable duplex with such sequences and thereby sterically blocking access of other biomolecules to specific regions of viral RNA, or of disrupting important RNA structures, have been the most extensively applied ASOs and proved to be useful in antiviral researches and in medicine as well. It has been suggested that PMOs readily enter primary cells [109] whereas its nonionic characteristics hinder viral uptake. The first report of antiviral activity by a PMO appeared in 2001 [110], and the *in vivo* activity of PMO antisense agents against a variety of viral pathogens has been demonstrated. However, none has yet been licensed for treatment of viral infection or other indications.

Attempts to improve the *in vivo* efficacy of PMOs have given births to peptide conjugated PMOs (PPMOs), which remarkably enhance transport across cellular membranes. The first report of antiviral activity by a PPMO appeared in 2004 [111], and by a PPMO containing the P7 CPP in 2005 [112]. Later reports have provided abundant evidence that PPMO are capable of potent and specific activity against RNA virus infections in cell cultures and in mice. Considering the number of reports and the magnitude of antiviral activity of PPMO *in vivo*, therapeutic applications appear possible, and continued drug development research warranted. However, serious problems (*e.g.,* peptide degradation, poor toleration, *etc.*) in *in vivo* applications of PPMOs are encouraging various attempts to introduce non-natural amino-acids or other chemical changes in conjugated peptide to offset these unfavorable effects. This led to the design of a new class of positively charged PMOs (PMO*plus*^{TM}), which contain positively-charged piperazine groups along the molecular backbone (Fig. **2**). It has recently been shown that PMO*plus*^{TM} are well tolerated and provide improved efficacy in numerous *in vivo* viral infection models [113,114]. Presumably, broad implications for therapeutic development of PMO*plus*^{TM} against viruses and other highly pathogenic microbes can be highly expected. The above mentioned items shall be discussed in greater detail in the following context.

Targets for ASOs-Based Antiviral Strategies

Only a limited number of proteins for each virus are available for selective inhibition, thereby a shortage of viable targets has always hampered traditional antiviral researches. Collectively, the antisense technique and/or more general genetic interference, offers the possibility of designing highly specific ligands targeted to any of the viral genes in order to inhibit viral growth. Typical targets for ASOs include genes involved in viral DNA replication, early gene expression and viral coat proteins. The goal of this part is to summarize notable findings from reports documenting abrogation or amelioration of virus infections by ASOs, with a focus on the genetic locations that provided the most sensitive targets for ASO-based intervention against various viruses.

Target Identification and Validation in General

In the early phase when researchers thought that the array of RNA targets for antisense is limited only by nucleotide sequence information, theoretically a random screening process of potential ASOs which complement the entire RNA target sequence should be carried out as a result of lacking knowledge of preferred oligonucleotide hybridization sites. Plenty of trials and errors have taught us that there are in fact multiple opportunities during RNA processing may be vulnerable to ASO binding. Favorite target sites have included RNA splice donor or receptor sites, 5' capping, 3' adenylation, translation start and termination sites, and ribozome entry. Although there is not a fully rational approach to RNA target sequence selection, there are multiple techniques which can impact on success. Computer models of RNA secondary structure, nucleotide hybridization efficiency and frequency have been compared to ASO activity. Antisense activity correlates well with hybridization strength, while RNA secondary structure does not necessarily predict optimal antisense activity. *In vitro* translation inhibition has been used to screen potential ASO inhibitors, and then the most active ASOs evaluated in antiviral assays.

Additionally, it is essential to define an antisense activity with solid supporting data when using an antisense-designed ON to probe new antiviral targets. The modifications in ASOs have resulted in a variety of non-antisense antiviral activities of ONs which were designed for the selectivity expected of antisense. Thus, assaying the antiviral activity alone is insufficient and probably misleading when evaluating an ON which has been designed as antisense to a target RNA. As a matter of fact, demonstration of both ON sequence-specificity of action and selectivity of inhibition of expression of antisense gene target should be included and fulfilled in the target identification and validation process. Conversely, credible demonstration of an antisense mechanism of action warrants the antisense to be a specificity of design and function, and its utility in defining novel antiviral targets. The criteria will continually undergo refinement as ever more sophisticated molecular techniques elucidate intracellular mechanisms of action.

ASOs Against Double-tranded DNA (dsDNA) Virus Infections

A DNA virus is a virus that has DNA as its genetic material and replicates using a DNA-dependent DNA polymerase. The nucleic acid is usually double-stranded DNA (dsDNA). Genome organization within this group varies considerably. Some have circular genomes (Baculoviridae, Papovaviridae and Polydnaviridae) while others have linear genomes (Adenoviridae, Herpesviridae and some phages). Some families have circularly permuted linear genomes (phage T4 and some Iridoviridae). Others have linear genomes with covalently closed ends (Poxviridae and Phycodnaviridae).

Hepatitis B Virus (HBV)

Persistent Hepatitis B virus (HBV) infection often leads to the development of chronic hepatitis, cirrhosis and hepatocellular carcinoma. Despite the availability of an effective HBV vaccine, the virus is still a major health problem with approximately 350 million persons infected worldwide, which is mainly due to lack of effective antiviral therapeutics. In the past 30 years, new antiviral approaches for the treatment of this disease have been developed, in which nucleic acid-based strategies are of the greatest potential to be future therapeutics. Collectively, this innovative strategy designed to inhibit HBV replication includes the use of ASO constructs, DNA-based immunization techniques to stimulate broad-based cellular immune responses, hammerhead ribozymes to cleave HBV pregenomic RNA *in vitro*, dominant negative HBV core mutant proteins as inhibitors of nucleocapsid formation within cells, and RNAi-based therapy expressing anti-HBV short hairpin RNAs (shRNAs). Besides, various novel expression vectors have been developed to deliver such DNA constructs to cells in order to optimize these antiviral effects (as listed in Table **3**). However, ASO-based antiviral strategy seems the most straightforward technique to inhibit essential viral targets as well as host-encoded products that control or assist viral replication.

HBV is a member of the Hepadnaviridae and its genome is a 3.2 kilobases (kb) double-stranded circular DNA. As a small DNA virus, HBV contains four open reading frames (ORFs) encoding two predominant transcripts and two minor transcripts. These transcripts are translated into three surface antigens, two nucleocapsid proteins, the polymerase and a transactivator-the X gene product. For antisense intervention these transcripts and the short encapsidation signal, an RNA sequence with a well-defined secondary structure, have been the targets of antisense researches (more information are summarized in Table **3**). Alternatively, the host-targeted antisense antiviral strategy used in the validation of effective host targets (*e.g.*, proteins and micro RNAs) and the development of adjuvant anti-HBV therapeutics, where examples will be listed in Table **10**.

Table 3: Summary of effective ASO target sites in double-stranded DNA viruses.

Productive ASO Target Region(s)[a]	ASOs[b]	Experimental Systems[c]	Delivery Method[d]	Ref.
1. Hepatitis B virus, HBV (Family *Hepadnaviridae*)				
cap site/SPII and initiator/gene S	AS-ODN, PS-ODN	hepatocellular carcinoma-derived cells stably transfected with HBV DNA	—	[115]
5' region of pre-S gene (795-812 nt)	PS-ODN	primary duck hepatocyte cultures and DHBV infected Pekin ducks	—	[116]
			liposome	[117]
preS1 open reading frame, S gene, and C gene	PS-ODN	hepatocellular carcinoma-derived cells stably transfected with HBV DNA	—	[118]
S gene	Antisense RNAs	HepG2.2.15 cells	lentiviral vectors	[119]
S and C genes	LNA	HBV transgenic mice	liposome	[120]
Pre-S region (951-968 nt)	PS-ODN	DHBV infected ducks	—	[121]
upstream sequence of the encapsidation site (1849-1864 nt)	AS-ODN, 2'-allyl ODN	chicken hepatoma cell line LMH	ASGP-R mediated delivery system composed of NAcGlcNH2.BSA and adenovirus mutant dl312	[122]
U_5-like region (1980-1905 nt) in pre C and C region	ligand PS-ASO	HepG2.2.15 cells	Linked with liver-targeting ligand Gal-PLL	[123]
		HBV transgenic mice		[124]
pol gene (2468-2487 nt)	poly-DNP-RNA	DHBV infected cell culture and ducks	—	[125]
pregenomic RNA	Antisense RNAs	HBV infected human hepatocellular carcinoma cells	—	[126]
core promoter (1734-1754 nt)	TFO	HepG2.2.15 cells	—	[127]
initiation sites of pre C	PS-ODN			
initiation sites of pregenomic RNA				
initiation site, DR2, and EN II in X gene	PS-ODN	HepG2.2.15 cells	—	[128]
initiation site of HBV X gene	PS-ODN	HBx transgenic mice	—	[129]
X and P regions	Antisense RNAs	HBx transgenic mice	Retrovirus vector pLXSN for constrct of 4 recombinant vector plasmids	[130]
bulge (2567-2581nt) and upper stem (2573–2587 nt) of encapsidation signal ε in DHBV	PNA	primary duck hepatocytes	peptide conjugated	[131]
	PS-ODN		—	

Table 3: contd...

2. Herpes simplex virus, HSV (Family *Herpesviridae*)					
IE 4 and 5 pre-mRNA	intron:exon junction of the splice acceptor	octamer MODN	HSV1 infection in cell culture	—	[132]
		dodecamers and psoralen derivatized ON			[133]
	splice donor site and translation initiation site	MODN	HSV1 infection in cell culture, and HSV1 ear infection in mice		[134]
		2'-OMe with methylphosphonate internucleoside connection	HSV1 infection in cell culture		[135]
		PO-ODN with minihairpins at the 3'-end	HSV1 infection in cell culture		[136]
translation start site of IE 1 mRNA		PO-ODN with 2 residues of PS-ODN at 5' and 3' end	HSV1 infection in cell culture	—	[137]
		end-capping and pyrimidine protection			[138]
IE 3 mRNA		PO-ODN	HSV-1 infection in tissue cell culture	—	[139]
IE 1 and 3 mRNA		MODN	HSV-1-infected U937 cells, and transient expression assays with infectious HIV or HIV-LTRcat activated by HSV-1	—	[140]
splicing acceptor site of IE 5 mRNA		PO-ODN, MODN, PS-ODN	HSV-1 infected Vero cells	—	[141]
		PS-ODN		lipophilic compound geraniol conjugated a 5' terminal	[142]
KSHV IE transcription activator Rta mRNA		2'-OMe modified EGS molecule	human primary effusion lymphoma cells infected with KSHV	—	[143]
internal AUG codon of UL13 mRNA of HSV-1		PS-ODN (ISIS1080), 5' or/and 3' modified PS-ODN	HSV-1 infected HeLa cells	—	[144]
5' noncoding region of HSV-1 UL29 mRNA		PS-ODN (ISIS4015)	HSV-1 (KOS strain) infection in cell culture	—	[145]
translation-start-site regions of HSV-1 ICP0 or ICP27mRNA		PMO	Rabbit skin cells or vero cells infected with acyclovir (ACV)- susceptible and resistant HSV-1 clinical isolates (strain 294.1 and 615.9)	Arg-rich peptide (P7) conjugated at 5' end	[146]
translation-start-site regions of HSV-1 unique long gene (UL)30 and UL39			Eight-week-old Swiss Webster female mice infected with HSV-1 in the eyes		[147]

Table 3: contd…

translation initiation region of Vmw65 (virion tegument protein) mRNA		ODN	Vmw65-expressing cell line and HSV-1 infection in tissue culture	—	[148]
		PS-ODN	HSV-1 (KOS) infection in cell culture		[149]
translation initiation site of the large subunit of RR1		AS-ON	HSV-2 infection and reactivation by cocultivation of latently infected ganglia in cell culture	—	[150]
vIL6 mRNA	ORF-K2 translation initiation site (−14 to 7 nt)	PMO	KHSV infected PEL cells	Arg-rich peptide (P7 or P4) conjugated at 5' end	[151]
	5' UTR (−106 to −84 and −42 to −20 nt)				
DNA polymerase		PS homo-ODN	HSV-2 (strain 333) induced DNA polymerase	—	[152]
			HSV-2 (strain 333) infected Vero cells or HeLa S3 cells		[153]
virus adsorption		GT-ODN	HSV-1 and -2 infected Vero cells	—	[154]
3. Epstein–Barr virus, EBV (Family *Herpesviridae*)					
coding region just 3' of the AUG on the EBNA-1 mRNA		PO-ODN, PS-ODN	EBV infected EBV nonproducer line Raji cells	—	[155]
EBNA-1 RNA		UM AS-ODN	EBV-immortalized lymphoblastoid cells (X50-7, 11-23, and LS)	—	[156]
			EBV infected BJAB, Louckes, and U937 cells		
translation initiation codons and their flanking sequences of BZLF1 transcript		PO-ODN, PS-ODN	EBV-producer cell line P3HR-1	—	[157]
translation initiation site of LMP1 mRNA		UM AS-ON	EBV-infected B lymphocytes (B95.8 cells)	—	[158]
				lipid-based and receptor-mediated delivery systems	[159]
4. Human cytomegalovirus, HCMV (Family *Herpesviridae*)					
mRNA of HCMV DNA polymerase gene		PS-ODN	HCMV strains AS169 and Towne infected normal human dermal fibroblast cells	—	[160]
RNA transcripts of IE1					
IE2 (170120-170140 nt)		PS-ODN (ISIS 2922), chemically modified PS-ODN (ISIS13312)	human fibroblast cell line (MRC-5) OR HRPE cell infected with HCMV		[161]
the intron-exon boundary of HCMV genes UL36 and UL37 (UL36ANTI)		PS-ODN	ganciclovir-resistant strains and clinical isolates of HCMV	—	[162,163]

Table 3: contd…

	chemically modified PS-ON (GEM132)	HCMV-infected human fibroblasts	cholesteryl moiety linked to the 3'-end	[164]
ORF UL83	AS RNA (stably expressed)	HCMV replication in astrocytoma cells stably transfected with a retroviral vector carrying an antisense UL83 cDNA	—	[165]
UL36 mRNA	AS-ODN	HCMV AD169 strain infected megakaryocytes in a semi-solid CFU-MK culture system	—	[166]
5. Marek's disease virus, MDV (Family *Herpesviridae*)				
splice donor sequence	PO-ODN	MDV infected lymphoblastoid cells	—	[167]
6. Human papillomavirus, HPV (Family *Papillomaviridae*)				
translation initiation site of the E2 mRNA for HPV 6 and 11	PS-ODN (ISIS 2105)	cells transfected with the target HPV sequences which transactivate a reporter gene	—	[168]
		bovine papillomavirus cell culture focus formation model		
PV E1 helicase transcript	PS-ODN	cell culture studies and in a mouse xenograft model of HPV replication	—	[169]
E1 translation start site	PS-ODN	cell based HPV E1-luciferase fusion assay	—	[170]
	2'-OCH$_3$ hybrid of PS-ODN	kidney xenograft nude mouse model implanted with HPV-infected human foreskin fragments	—	
HPV16 E6 RNA	AS RNA	human cervical cancer cell lines, CaSki and SiHa cells harboring HPV 16 genome	lipofectamine 2000	[171]
HPV11 E6/E7 RNA	AS-ODN	HPV11-induced papillomas in human foreskin grafts on immunodeficient mice	lipofectamine	[172]

[a]C gene, HBV core antigen (HBcAg) gene;S gene, HBV surface antigen (HBsAg) gene; cap site/SPII, the cap site of mRNA transcribed from the SPII promoter; initiator/gene S, the translation initiation site of the S gene; nt, nucleotides; pol gene, polymerase gene; IE, immediate-early; vIL6, viral Interleukin 6; ORF, open reading frame; UTR, untranslated region; RR1, ribonucleotide reductase 1; EBNA-1, Epstein-Barr virus nuclear antigen 1; BZLF1, ; LMP1, latent membrane protein 1.

[b]ASO, antisense oligonucleotide; AS-ODN, antisense oligodeoxynucleotide; PS-ODN, phosphotioates oligodeoxynucleotide; poly-DNP-RNA, poly-2'-O-(2,4-dinitrophenyl)-oligoribonucleotide; PNA, peptide nucleic acid; LNA, locked nucleic acid; MODN, methyl phosphonate oligodeoxynucleotide; PO-ODN, phosphodiester oligodeoxynucleotide; PO-ODN with minihairpins at the 3'-end, PO-ODNs constituting of a 5-dodecameric sequence, flanked at the 3'-end by octameric sequences adopting hairpin-like structures; GT-ODN, ODNs composed entirely of deoxyguanosine and thymidine, but not specifically designed to act as antisense agents,with either natural PO or PS modified internucleoside linkages; PMO, phosphorodiamidate morpholino oligomers; 5' or/and 3' modified PS-ODN, PO analogs of ISIS 1080 containing three PS linkages placed on the 3' (ISIS 1365), 5' (ISIS 1370), both the 3' and 5' (ISIS 1364) ends or with four linkages in the middle (ISIS 1400); EGS, external guided sequence that activate RNase P cleavage

of targeted mRNA; UM, unmodified; ISIS13312, a chemically modified analog of ISIS 2922 with changes including substitution of 5-methyl cytosine (5-methylC) for cytosine in the sequence and the addition of 2'-O-alkyl substitutes (2'-methylethoxy) on nucleosides on the 3' end of the oligonucleotide and nucleosides on the 5' end; GEM132, a chemically modified PS-ON Version of UL36ANTI with changes including three 2-O-methyl nucleotides at the 3-end and four 2-O-methyl nucleotides at the 5-end.

^cDHBV, duck hepatitis B virus. DHBV shares a common genomic size, organization, and mode of replication to the human HBV, and chronic infection can be readily induced by neonatal infection.DHBV infection is a useful *in vivo* model for the screening of antiviral drugs for HBV. The 2.2.15 cell is the hepatoblastoma cell line HepG2 transfected with cloned HBV DNA; Vero cells, African green monkey kidney cells; KSHV, Kaposi's sarcoma-associated herpesvirus also known as human herpesvirus 8; HIV, human immunodeficiency virus; HIV-LTRcat, an HIV-LTR-directed chloramphenicol acetyltransferase construction; Raji cells, EBV nonproducer line; BJAB and Louckes cells, EBV-negative Burkitt B-cell lines; U937 cells, EBV-negative monoblastic cell line.

^dASGP-R, asialoglycoprotein receptor; NAcGlcNH2.BSA, N-acetylglucosamine derivative of bovine serum albumin; Gal-PLL, alactosylated poly-L-lysine; Arg-rich peptides include P7, CH_3CONH-$(RXR)_4$-XB, P4, $R_5F_2R_4$, NH_2-RRRRRRFFRRRRC-$CONH_2$ (R stands for arginine, X stands for 6-aminohexanoic acid, F stands for phenylalanine, C stands for cysteine, and B stands for beta-alanine); lipid-based and receptor-mediated delivery systems, oligomers were complexed or not to the following vectors:(i) TFX reagents (TFX10®, TFX20® and TFX50®) are liposomes containing the same concentration of polycationic synthetic lipids but combined with different molar ratios of the fusogenic lipid DOPE; (ii) Superfect (SF), an activated dendrimeric transfection reagent; (iii) Transferrin-polylysine (TFPL) complex, as a carrier internalized *via* receptor-mediated endocytosis; "—", no delivery method used.

Herpes Simplex Virus (HSV)

Herpes simplex virus 1 and 2 (HSV-1 and HSV-2), also known as Human herpes virus 1 and 2 (HHV-1 and -2), are two members of the herpes virus family, Herpesviridae, that infect humans. Both HSV-1 (which produces most cold sores) and HSV-2 (which produces most genital herpes) are ubiquitous and contagious. Symptoms of herpes simplex virus infection include watery blisters in the skin or mucous membranes of the mouth, lips or genitals. Lesions heal with a scab characteristic of herpetic disease. Sometimes, the viruses cause very mild or atypical symptoms during outbreaks. However, as neurotropic and neuroinvasive viruses, HSV-1 and -2 persist in the body by becoming latent and hiding from the immune system in the cell bodies of neurons. The HSV-1 genome contains about 152,000 basepairs and most of the genes are catalogued into immediate-early, early and late genes, the expression of which is tightly and coordinately organized. ASOs have been evaluated against different sites of the HSV-1 genome, and demonstrated differing success (Table **3**).

Epstein–Barr Virus (EBV)

The Epstein–Barr virus (EBV), also called human herpesvirus 4 (HHV-4), is a virus of the herpes family and is one of the most common viruses in humans. EBV is the etiologic agent for most cases of infectious mononucleosis, particular forms of cancer (*e.g.,* Hodgkin's lymphoma, Burkitt's lymphoma, nasopharyngeal carcinoma), and certain autoimmune diseases (*e.g.,* dermatomyositis, systemic lupus erythematosus, rheumatoid arthritis, Sjögren's syndrome, and multiple sclerosis). The EBV genome is about 192 thousand base pairs in length and contains about 85 genes. EBV has two forms of infection, replicative (lytic) and latent cycle [173]. In the lytic form, nearly 100 viral genes are expressed, and replication occurs by use of virus-encoded DNA polymerase. EBV generally produces a lytic infection in the oral and pharyngeal epithelium, which is necessary for transmission of the virus to neighbouring cells. In the latent form of infection, which occurs primarily in B lymphocytes, the expression of the viral genome is restricted to six EBV-encoded nuclear antigenic proteins (EBNAs), three latent membrane proteins (LMPs), and two small non-polyadenylated RNAs (EBERs). The replication of the virus is mediated by the host DNA polymerases and the viral EBNA1 protein. Two key genes, EBNA-1 which is required for maintenance of cell transformation, and BZLF 1 which is required for reactivation of latently infected cells, have been the subjects for design and evaluation of ASOs (Table **3**).

Human Cytomegalovirus (HCMV)

Human cytomegalovirus (HCMV),alternatively known as human herpesvirus-5 (HHV-5), is a species of virus that belongs to the viral family known as Herpesviridae or herpesviruses. HCMV may be found throughout the body, but its infections are frequently associated with the salivary glands. Further, HCMV infection is typically unnoticed in healthy people, but can be life-threatening for the immunocompromised, such as HIV-infected persons, organ transplant recipients, or newborn infants. After infection, HCMV has an ability to remain latent within the body over long periods. Eventually, it may cause mucoepidermoid carcinoma and possibly other malignancies. Several researches have focused on the use of ASOs to inhibit CMV. A series of ASOs complementary to the translation start sites, coding regions, intron:exon region and 5' caps in RNAs including the DNA polymerase, and immediate early genes IE 1 and IE2 have been evaluated [174] (Table **3**).

Marek's Disease Virus (MDV)

Marek's disease virus (MDV) is an avian herpesvirus which induces lymphoproliferative disease and demyelination of peripheral nerves in chickens.

The viral DNA is a linear double-stranded molecule of 180 kb, of which the BamHI, Sinai and BglI restriction maps have been established. Kawamura *et al.* (1991) evaluated a phosphodiester 18-mer complementary to a splice donor sequence (Table **3**).

Human Papilloma Virus (HPV)

Human papillomavirus (HPV), a virus from the papillomavirus family, is capable of infecting humans and establishes productive infections only in keratinocytes of the skin or mucous membranes. While the majority of the known types of HPV cause no symptoms in most people, some types can cause warts (verrucae), or lead to cancers of the cervix, vulva, vagina, penis, oropharynx and anus.Specifically, HPV 16 and 18 infections are strongly associated with an increased odds ratio of developing oropharyngeal (throat) cancer. Recently, HPV has been linked with an increased risk of cardiovascular disease.

The HPV genome is composed of six early (E1, E2, E4, E5, E6, and E7) genes, two late (L1 and L2) genes, and a non-coding long control region (LCR). After the host cell is infected E1 and E2 are expressed first. High E2 levels repress expression of the E6 and E7 proteins.E6 and E7 are the HPV proteins associated with cancer. When the host and HPV genomes integrate, E2 function is disrupted, preventing repression of E6/E7.

HPV infection is limited to the basal cells of stratified epithelium, the only tissue in which they replicate. The virus cannot bind to live tissue; instead, it infects epithelial tissues through micro-abrasions or other epithelial trauma that exposes segments of the basement membrane. Thus, what is an extremely challenging issue for an antiviral researcher is identifying a selective inhibitor of a virus for which neither a cell culture replication assay nor a readily accessible animal infection model of disease is available. HPV provides such a challenge, and an opportunity for the antisense approach [174] (Table **3**).

ASOs Against RNA Virus Infections

While HIV remains the RNA virus with the largest impact on global public health, nonretroviral RNA viruses including dengue, measles, respiratory syncytial, hepatitis C, and influenza A viruses are currently major pathogens causing of infectious disease with high mortality and thereby create heavy burdens to global health care. Despite the availability of vaccines and success of small molecule drugs for limited viruses (*e.g.,* DNA virus and HIV), the development of safe and effective therapeutic treatment against non-retroviral RNA viruses has been comparatively scanty.

RNA virus infection requires delivery of the viral genome into cells, transcription of viral mRNA and subsequent translation of viral proteins that aid in genome replication, viral assembly and budding. With the rapid expanding of knowledge concerning RNA structure and function, as well as an increasing awareness of the pivotal role RNA played in viral diseases, many researchers apply nucleic acid-based technologies, especially the antisense oligomers, to identify new viral targets and design ASO-based therapeutics that specifically target viral single-stranded nucleic acid sequence. As an active area of antiviral drug development research, antiviral ASOs may offer the prospect of actively pursuing the greatest unmet medical needs in RNA viruses, the investigations of which shall be respectively discussed below.

Retrovirus: Human Immunodeficiency Virus Type 1 (HIV-1)

Human immunodeficiency virus (HIV) is a lentivirus (a member of the retrovirus family) that causes acquired immunodeficiency syndrome (AIDS), a condition in humans in which progressive failure of the immune system allows life-threatening opportunistic infections and cancers to thrive. The genome of HIV-1 is a small (about 9000 nucleotides) and hence molecular function is packed densely into the RNA sequence, with functionality sometimes overlapping or coincident (Fig. **4**). The replicative cycle of HIV comprises a number of essential steps, including adsorption, fusion, uncoating, reverse transcription, integration, DNA replication, transcription, translation, maturation, and budding (assembly/release), which could be considered adequate targets for chemotherapeutic intervention. Yet, most clinically available drugs are small molecules aimed at only a very limited number of such targets, predominantly the viral proteins reverse transcriptase (RT) and protease.

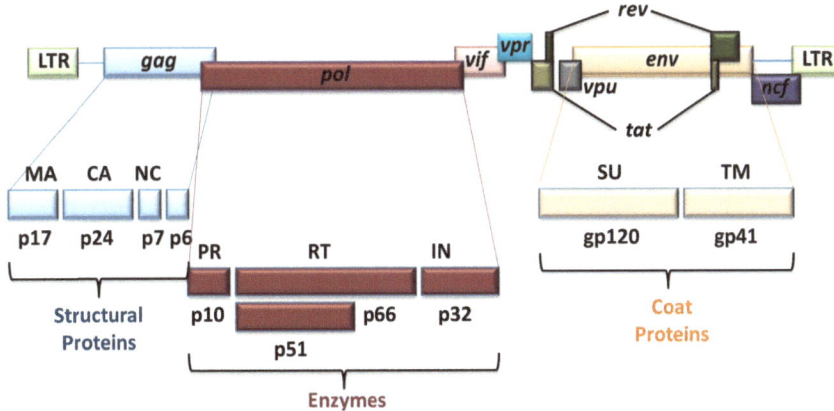

Figure 4: Genomic organization of HIV-1. HIV-1 has several major genes coding for structural proteins that are found in all retroviruses, and several nonstructural ("accessory") genes that are unique to HIV-1. The *gag* gene provides the basic physical infrastructure of the virus, and *pol* provides the basic mechanism by which retroviruses reproduce, while the others help HIV to enter the host cell and enhance its reproduction. *gag* (group-specific antigen): codes for the Gag polyprotein, which is processed during maturation to MA (matrix protein, p17); CA (capsid protein, p24); SP1 (spacer peptide 1, p2); NC (nucleocapsid protein, p7); SP2 (spacer peptide 2, p1) and p6. *pol*: codes for viral enzymes reverse transcriptase, integrase, and HIV protease. *env* (for "envelope"): codes for gp160, the precursor to gp120 and gp41, proteins embedded in the viral envelope which enable the virus to attach to and fuse with target cells. Transactivators: *tat*, *rev*, *vpr*. Other regulators: *vif*, *nef*, *vpu*. *tev*: This gene is only present in a few HIV-1 isolates. It is a fusion of parts of the *tat*, *env*, and *rev* genes, and codes for a protein with some of the properties of tat, but little or none of the properties of *rev*.

For HIV, modified ASOs may be potent inhibitors, because their sequence-specific targeting of functional RNA elements as probes facilitates the understanding the molecular biology of HIV-1 and development of possible antiviral or virucidal agents. The concept of using ONs targeting HIV-1 RNA as a therapeutic agent *via* either RNase H activation or steric-blokcing mechanism is almost 30 years old. The first experimental evidence was given by Zamecnick PC *et al.*, who utilized 12–26 long unmodified oligodeoxyribonucleotides targeting upstream of the primer binding site (PBS) to successfully inhibit HIV-1 growth in either peripheral human blood lymphocytes or transformed T cells [175]. Many different sites on the HIV-1 genome were investigated over the following years and a number of preclinical oligonucleotide candidates have been developed. Notably, inhibition of HIV-1 by different PS-ONs had been reported numerous times until further evaluations have demonstrated that most of these compounds most probably act *via* a non sequence-specific mechanism as inhibitors of reverse transcriptase, RNase H, primer extension, integrase, envelope glycoprotein gp120, and virus adsorption to cells. That is why the only ASO (a 25-mer PS-ODN, named GEM 91, targeted to the AUG initiator site of the Gag protein) that entered clinical trials was sadly forced to withdraw from later clinical trials [176]. Additionally, the infidelity of the HIV-1 RT during replication results in rapid mutation of much of the RNA sequence, and thereby resistance to drug action is obtained. Thus, later researches tend to identify new target and/or sites in a few highly functional regions in HIV-1 that have short sequences well conserved between viral isolates and rarely or never mutated, *e.g.,* the HIV-1 RNA leader sequence [177]. The historical context of targeting the HIV-1 RNA sequence with ASOs is summarized in Table **4**.

Table 4: Summary of effective ASO target sites in human immunodeficiency virus (HIV).

Productive ASO Target Region(s)[a]		ASOs[b]	Experimental Systems[c]	Delivery Method[d]	Ref.
splice acceptor sites (upstream of tRNA^{Lys} for PBS)		UM-ODN	peripheral human blood cells and transformed T-lymphocyte (H9) cells separately infected with HTLV-III	—	[175]
Human tRNA^{Lys3} including the 3' terminus, TΨC stem-loop and variable loop (41-76 nt)			synthetic tRNA^{Lys3}		[178]
Pre-PBS-PBS region for tRNA^{Lys3} primer (164-183 nt)		2'-OMe	P4 cell infected by HIV-1$_{LAI}$		[179]
exon I of *art/trs* genes (6018-6032 nt)		M-ODN, PS-ODN	immortalized T4+ T cells (ATH8 cells) infected with HTLV-III	—	[180]
		Phosphoro-selenoate ODN			[181]
ends of retrovirus RNA (R region)	Poly(A) signal (9183-9202nt)	AS-ODN	Molt-3 cells infected with HTLV-III$_B$ or HTLV-IIIcc	—	[182]
	Cap (1-20nt)				
	5' untranslated region (54-73nt)				
	upstream of the *env* initiator (7947-7966nt)				
splice acceptor site	*tat* initiator (5349-5368nt)	PS-ODN	Molt-3 cells infected with HIV-1	cholesteryl group tethered at 3'-terminal internucleoside link	[183]

Table 4: contd…

first splice acceptor site of the *tat-3* gene		M-ODN	H9 cells infected with HIV-1 (strain HTLV-III$_B$)	—	[184]
capped 5' end and splice acceptor site of *tat*		M-ODN	H9 or Molt-3 cells infected with HTLV-III$_B$ or HTLV-IIIcc	—	[185]
splicing acceptor site	5350-5380 nt	2'-OMe/PS	Molt-4 cells infected with HTLV-III$_B$	—	[186]
	5346-5380 nt	patch derivative			
env (5585-9153nt) covering the complete exon II of the tat gene near the 3' terminus of HIV genome		AS-ON	H9 cells infected with HTLV-III$_B$	CD3 monoclonal antibody-targeted liposome	[187]
U3 Enhancer Element (-93 to -83 nt)		M-ODN	U1.1A cells and 8E5cells chronically infected with HIV-1	—	[188]
tat initiator (5402-5410nt)			U1.1A cells chronically infected with HIV-1		
splice acceptor site of exon II of gene *tat*		AS-ODN	CEM cell or HeLa-T4+ cell infected with HIV-1	—	[189]
translation initiation site of *tat*		ON	cell culture in acute infection of HIV-1	poly(L-lysine) conjugated	[190]
p18/p24 junction in gag		PS-ODN	cell culture in acute infection of HIV-1	—	[191]
active site of HIV protease in *pol*					
first exon of *rev* gene			cell culture in chronic infection of HIV-1		
conserved sequence of gene *env*		ON	MT-4 cells infected with HIV-1	lipophilic groups conjugated	[192]
negative sense viral RNA					
5'-terminus of gene *rev*					
poly(A) sequences					
RRE	stem I and the initial portion of stem loop II	PS-ODN	HeLa-tat cells were transfected with pIIIAR alone and with plasmid pREV	—	[193]
	stem loop II		Molt-3 cells infected with HTLV-III$_B$		
translational initiation region of *gag* region (324-362nt)		Self-stabilized ODN	H9 lymphocytes infected with HIV-1	—	[194]
translational initiation region of *gag* region (776-802nt), and rev1 (5970-5997nt)		PS-ODN	Molt-3 cells infected with HTLV-III$_B$ (long-term: >80 days)	—	[195]
			human monocyte/macrophage and PBMC cultures infected with HIV-1 (long-term)	—	[196]

Table 4: contd…

rev and *tat* gene	AS-ODN, PS-ODN	HIV-1 acutely or chronically infected cells	immunoliposome (HLA class I molecule targeted protein-A bearing liposome)		[197]
initiation codon region of *rev* gene	α and β phosphodi-ester ON				[198]
tat gene	modifiedAS-ODN	*de novo* HIV infected CEM-SS lymphocytes	Influenza derived fusogenic peptide		[199]
gag (p24 leader, 1185-1214nt)	PS-ODN	Molt-4 cells infected with HTLV-III$_B$	—		[200]
pol (p24 leader, 1185-1214nt)					
rev (splice acceptor, 4486-4505nt)					
tat (splice acceptor, 5348-5367nt; translation start, 5550-5569nt)					
tar (1-20nt, 15-34nt, 40-59nt)					
Initiation site of *rev*	PS-ODN	ATH8 cells infected with HTLV-I and H9 cells infected with HTLV-III$_B$	—		[201]
1-20 region of HIV RNA	DNA	—	OKT4 cross-linked oligo$_{1-20}$/ Moli-1 complex		[202]
5'-LTR (262-281nt)	PS-ODN	CMT3, and HIV-1 infected H9 cells	—		[203]
U5 region (149-174 nt) and PrePBS region (224-238 nt)	ODN	HIV RT system, and HIV-1 *de novo* infected Molt-4 cells	poly(L-lysine) conjugated		[204]
					[205]
Non-regulatory region (1189-1208nt)	PS-ODN	CMT3	—		[206]
		CMT3, and HIV-1 acutely or chronically infected H9 cells	encapsulated in cationic lipid		[207,208]
DIS loop	272-280 nt	RNA, DNA/RNA	synthetic RNA 1–615 from HIV-I Mal	—	[209]
	272-290	LNA, LNA/DNA	Jurkat-Tat T-cell infected with HIV-1 subtype A		[210]
5'-end of HIV Rev mRNA	PS-ODN	HIV-1-infected H9 cells and peripheral blood T-lymphocytes	CD4 MAb-targeted liposome		[211]
		HIV-1 acutely infected Molt-3 cells and primary cells, and HIV-1 chronically infected H9 cells	DLS-liposome		[212]
stem-loop structure of TAR element at the 5' end of HIV-1 and HIV-2 viral RNAs (11-35 nt and 20-35 nt)	ODN	recombinant forms of HIV-1 or 2 RT carrying or not carrying the RNAse H activity	—		[213]
PBS and U5 region	PNA and PNA-DNA chimera	cell culture	—		[214]

Table 4: contd…

RRE		AS-ODN	HIV-1 infected macrophages	CD4 coupled pH-sensitive liposome	[215]
near the 3' end of *gag-pol* transframe region (2262-2276 nt, highly conserved)		PNA	chronically HIV-1-infected H9 cells and in PBMC infected with clinical HIV-1 isolates with the multidrug-resistant phenotype	Peptide conjugated	[216]
			human HIV-infected macrophages	erythrocytes as carriers	[217]
Srev + SDIS (245-270 nt) + SPac (295-324 nt)		AS-ODN, PS-ODN	(long-term) acute assay using HIV-1 (IIIB strain)-infected Molt-3 cells	DLS carrier system for a combination of ODNs	[218]
apical stem-loop of 39-mer TAR RNA	20-45 nt	UM	chronically and autely HIV-1$_{IIIB}$-infected CEM cells		[219]
	21-36 nt	PS-ODN, OMe, NP, PNA	HIV RT p66/51		[220]
		OMe/LNA (α-L-LNA and S-LNA)	HeLa Tet-Off/Tat/luc-f/luc-R cells	—	[221]
	25-36 nt	OMe, PNA OMe /pC OMe /LNA	a DNA template carrying the HIV promoter and TAR sequence		[222]
	22-36 nt	mr-AOMP	HIV LTR in HeLa HL3T1 cells in culture		[223]
minimal functional TAR sequence comprising the apical stem-loop and bulge regions (18-34 nt)		PNA	a reporter gene construct pHIV LTR−CAT and pCMV−Tat in cell culture	—	[224]
			chronically HIV-1-infected H9 cells	transportan-PNA conjugate	[225,226,227]
			CEM cells infected with pseudotyped HIV-1 virions expressing the firefly luciferase reporter gene	5 different MTD (penetratin-PNA conjugate with most efficacy)	[228]
stem-loops in 5'-UTR	TAR (Tat protein binding site)	2'-OMe/LNA	HeLa T4 LTR beta-Gal cells infected with HIV-1	—	[229]
	SL3 loop (primary packaging element binding the Gag polyprotein)				
PPT		ODN A	HTLV-1-transformed T cell line C81-66/45	—	[230]

Table 4: contd…

		ex vivo treatment of plasma of HIV-1-infected patients and HIV-1 isolates from Africa and drug-resistant strains		[231]
a region immediately downstream from the primer binding site (PBSD, 203-222 nt), a region encompassing the major splice donor site (SD, 278-297 nt), a region covering the *gag* initiation site (AUG, 326–345 nt)	LNA antisense gap-mer	HIV RNA	—	
		HEK 293-T cells were co-transfected with HIV-1 LAI genomic DNA plasmid, renilla luciferase plasmid and the LNAs	Lipofectamine-2000	[232]

[a]PBS, primer binding site; nt, nucleotides; *pol* gene, polymerase gene; RRE, Rev-responsive element; LTR, long terminal repeat; TAR, the *trans*-activating responsive; Srev, a 28-mer ODN complementary to 5'-end of HIV Rev mRNA; SDIS, a 26-mer ODN complementary to a highly conserved sequence localized between the primer binding site and the major splice donor site; SPac, a 30-mer ODN complementary to a sequence localized between the major splice donor site and the first ATG gag initiation codon and corresponds to the packaging signal psi (ψ); DIS, dimerization initiation site (PPT, polyp-urine tract.

[b]UM, unmodified; 2'-OMe, 2'-O-methylnucleosides; M-ODN, methylphosphonate ODN; patch derivative, 2'-OMe oligomers in which two phosphates on the average of within five from 3'-end were substituted by thiophosphates and five phosphates from 5'-end were completely substituted; PS-ODN, phosphotioates oligodeoxynucleotide; ASO, antisense oligonucleotide; MPdC, AS-ODN, antisense oligodeoxynucleotide; modified AS-ODN, AS-ODN modified with a 2-propanol-3-amino group or a ribonucleotide at the 3' end and a hexamethylene-bridged pyridyldisulfide or primary amine at the 5' end; mr-AOMP, oligo-2'-O-methylribonucleosides that have alternating methylphosphonate/phosphodiester linkages; PNA, peptide nucleic acid; NP, N3'-P5' phosphoramidate; LNA, locked nucleic acid; 2'-OMe/LNA, mixmer oligonucleotides consisting of residues of both 2'-Me and LNA; Self-stabilized ODN, oligodeoxynucleotid analog that has hairpin loop structures at 3' ends; ODN A, oligodeoxynucleotide A consists of a 25-mer antisense and a 25-mer passenger strand, connected by four thymidines (T4), phosphorothioated at each end (three bases) and in the T4 linker; pC, propynylC substituted; LNA antisense gap-mer, the LNA ASOs were designed as gap-mers with 5 LNA residues flanking a 10-mer phosphorothioate modified DNA body to enable RNase H cleavage.

[c]H9 cells, established from T cell leukemia patients chronically infected by the LAI HIV-1 isolate; HTLV-III, human T-cell lymphotropic virus type III, the etiological agent of acquired immunodeficiency syndrome (AIDS), now called human immunodeficiency virus (HIV)BH10; U1.1A cells, subcloned from Ul, a clone of promonocytic U937 cells infected with the lymphadenopathy-associated virus strain of HIV-1; 8E5cells, immortalized cells of CD4[+] T-lymphocyte lineage; PBMC, peripheral blood mononuclear cells; CEM cell, T lymphoblastic cell; Molt-4 cells, the human T lymphotropic virus type I (HTLV-I)-positive human T cell line; HL3T1, a HeLa cell line that contains a chloramphenicol acetyl transferase (CAT) reporter gene under the control of an integrated HIV-1 LTR promoter; HeLa Tet-Off/Tat/luc-f/luc-R cells, HeLa cells containing stably integrated HIV-1 Tat gene under Tet-Off promoter, firefly luciferase gene under HIV-1 LTR, and Renilla luciferase gene under cytomegalovirus (CMV) promoter; CMT3, COS-like Monkey kidney cell line stably transfected with plasmids pCMVgagpol-rre-r(containing *gag* and *pol* genes) and pCMVrev (containing the *rev* gene of HIV-1), derived from cDNA clone BH10; HIV RT system, HIV reverse transcript system.

[d]lipophilic groups conjugated, aliphatic linear structures and cholesterol were coupled to the 5'-terminal phosphate of oligonucleotides *via* glycine or propylene diamine spacers; pIIIAR, a plasmid in which the chloramphenicol acetyltransferase (CAT) gene is inserted into the *env* region (containing the RRE) of HIV-1; pRev, a plasmid expressing the Rev protein, also under the control of the long terminal repeat (LTR); fusogenic peptide, a membrane destabilized peptide derived from the Influenza hemagglutinin envelop protein with sequence GLFEAIAGFIENGWEGMIDGGGYC; OKT4, anti-CD4 monoclonal antibody (MAb); Moli-1, an MAb against oligo$_{1-20}$ obtained from spleen cells from BALB/c mice immunised with the oligo-avidin complex; transportan, a 27-amino-acid peptide with sequence GWTLNSAGYLLGKINLKALAALAKKIL; MTD, membrane transducing peptides used include the 16mer penetratin (RQIKIWFQNRRMKWKK), the 13mer tat-peptide (GRKKRRQRRRPPQ), Transportan-27 (GWYLNSAGYLLGK(e-Cys)INLKALAALAKKIL) and its truncated derivatives Transportan-21 (AGYLLGK(e-Cys)INLKALAALAKKIL)and Transportan-22 (GWYLNSAGYLLGK(e-Cys)INLKALAAL); "—", no delivery method used.

Non-Retroviral Positive-Sense Single Stranded RNA Virus

Non-retroviral positive-sense single-stranded RNA viruses ((+) RNA viruses), including the Flaviviridae, Picornoviridae, Caliciviridae, Togaviridae, Arteriviridae, Coronaviridae, and Hepeviridae virus families, constitute a broad and prevalent group of pathogens that threaten human health and life worldwide [233]. While effective vaccines have been developed for some, such as poliovirus and hepatitis A, others such as coxsackievirus, severe acute respiratory syndrome coronavirus (SARS-CoV), and West Nile virus, have no accredited drug treatments.

Without constituents of the RNA-dependent RNA polymerase (RdRp), the virion of (+) RNA viruses use genomic RNA as mRNA, upon entry into the cytosol of a newly in-fected cell, to produce the viral replicase proteins necessary to carry out viral RNA synthesis and continue the infectious cycle.

Family *Flaviviridae*

Flaviviruses contain a ~11 kb, plus-sense orientation, single stranded RNA genome consisting of a 5' untranslated region (UTR), a single long open reading frame (ORF), and a 3' UTR. The single ORF encodes a single long polyprotein, which is cleaved to produce three structural, and seven non-structural proteins once generated [234]. The family *Flaviviridae* includes three genera of important human pathogens: Hepacivirus [including hepatitis C virus (HCV)], Flavivirus [including yellow fever virus (YFV), dengue virus, Japanese encephalitis (JEV), West Nile virus (WNV), St. Louis encephalitis, Murray Valley encephalitis virus (MVEV), and tick-borne encephalitis viruses (TBE)], and Pestivirus [including bovine viral diarrhea virus (BVDV)]. Treatment options against flaviviral disease are extremely limited, with vaccines for humans available only for YF, JE, and TBE viruses, and no effective drugs yet commercially available. Studies utilizing antisense technology have provided insights into flaviviral molecular biology and the basis for the development of novel therapeutic approaches. Oligomers of various antisense structural types, targeted to different locations in the flaviviral RNA genome, have now been used to successfully suppress viral gene expression and thereby inhibit flavivirus replication. Findings of many of the significant reports that have appeared on the topic of ASOs-based strategies for the development of antiviral therapeutics for flaviviruses are summarized in Table **5** (HCV) and **6** (Flaviviruses). Most of the ASOs tested targeted sites in the viral untranslated regions (UTRs). The 5'-terminal region of the genome and highly conserved sequence in the 3' UTR were repeatedly shown to be especially productive targets.

Hepatitis C virus (HCV), first identified in 1989, is of a 9.6 kb long RNA sequence that encodes a single polyprotein cleaved into three structural proteins (core, E1, and E2) and seven nonstructural (NS) proteins (p7, NS2, NS3, NS4A, NS4B, NS5A, NS5B) (Fig. **5**). Progress on basic understanding of HCV virology has facilitated the development of the new anti-HCV agents [235]. HCV enters host hepatocytes through endocytosis, and undergoes endosomal fusion followed by cytoplasmic translation of single-stranded RNA into polyprotein, then polyprotein cleavage. The viral NS3/4a protease cleaves the HCV polyprotein at downstream junctions. Viral proteins then combine on altered host membranes to form a replication complex for production of viral RNA. NS4B and NS5A appear to play key roles in replication.

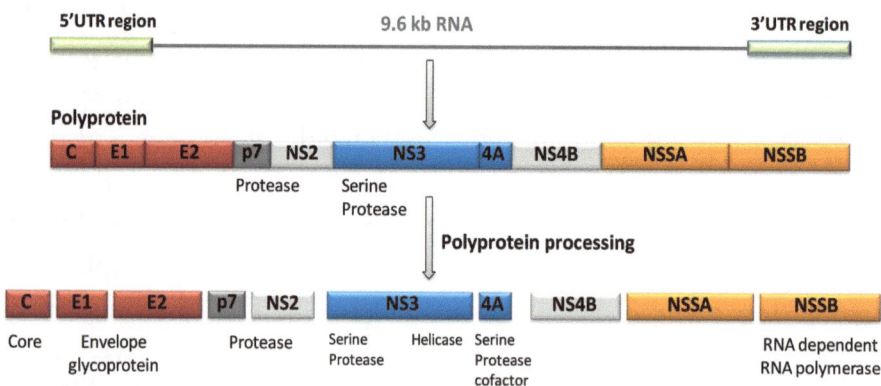

Figure 5: Hepatitis C virus (HCV) genomic organization. Schematic of the HCV RNA genome (*top*), the HCV polyprotein prior to processing (*middle*), and the cleaved HCV polyprotein (*bottom*). HCV RNA encodes a single polyprotein that is cleaved into three structural proteins (core, E1, and E2) and seven nonstructural proteins (p7, NS2, NS3, NS4A, NS4B, NS5A, NS5B). NS2 cleaves at the NS2/NS3 junction, whereas the NS3/4A complex cleaves remaining junctions, indicated by thin arrows.

Current standard therapy for HCV consists of pegylated interferon alpha (pegIFNα) and ribavirin (RBV), depending on viral genotype. HCV treatment is complicated by major side effects leading to treatment discontinuation, and is associated with a <50% response rate. Until now, there has been limited effective treatment for non-responders to standard therapy. Recent years have witnessed therapeutic options for HCV expanding [236]. Initial traditional drug development focuses on the NS3/4A serine protease and the NS5B RNA-dependent RNA polymerase (RdRp) with well-defined virus-specific enzymatic functions. However, drugs targeting other HCV components and host proteins that interact with HCV (Table **10**), though in earlier phases of development, have attracted more attention, especially for new antiviral strategies like the antisense technology [237]. ASOs, ribozymes, and more recently, siRNAs have been widely used to control gene expression, and several clinical trials are in progress. The potential to use ASOs as tools to control HCV infection, either by promoting an RNase H mediated cleavage of viral genomic RNA or by interfering with the assembly of a translation initiation complex, have shown evidence of many promising viral targets [238] (Table **5**). The antisense molecules developed for use in anti-HCV therapy have mainly been directed against the 5' untranslated region (UTR) region, and more specifically against the ribosome entry site (IRES) domain. Extensive knowledge of internal structure and conservation among HCV genotypes have rendered the HCV IRES (and, in particular, its IIId loop) particularly attractive [239] target for antisense inhibition. Of all the assayed antisense oligonucleotides, ISIS 14803 has probably shown the most promising ability to block HCV translation and replication in preclinical investigations in cell, which have encouraged clinical trials of this agent currently in phase II (Table **5**).

Table 5: Summary of effective ASO target sites in Hepatitis C virus (HCV).

Productive ASO Target Region(S)[a]	ASOs[b]	Experimental Systems[c]	Delivery Method[d]	Ref.
5' NCR (38-65, 134-175, 312-339 nt), core open reading frame (341-377 nt)	ODN	synthetic HCV RNA	—	[240]
core region containing initiator codon AUG (342-361 nt), upstream of initiator AUG (321-340 nt), and downstream of initiator AUG (346-363 nt)	PS-ODN	cell free protein synthesis system, and HCV infected MT-2C cells	—	[241]

Table 5: contd…

5' NCR (from HCV 355 to 364-374 nt)		DNA, 2'-F, P=S ODN	*in vitro* translation assay and in a human H8Ad-17c cell-based HCV core protein expression assay	—	[242]
5' NCR	core coding region (371-388 nt)	PS-ODN	RRL *in vitro* translation assay, and HepG2 *in vitro* cell culture system	— and lipofection method	[243]
	3' end of the NCR (326-348 nt)				[244]
		B-ODN, M-ODN, S-ODN			[245,246,247]
	(326-342 nt)	modified AS-ODN	—	cholesterol or bile acids conjugated at 5'-end of ODN for liver specific drug targeting	[248,249]
NS3 protease RNA (1916-1950 nt)		AS-ODN	human hepatocellular carcinoma cell line HuH-7 or the human kidney cell line HEK293	—	[250]
stem loop of 5' NCR (260-279 nt)		PS-ODN (ISIS6095)	H8Ad17 immortalized hepatocytes	—	[251]
initiator codon region of 5' NCR (330-349 nt)		PS-ODN (ISIS6547)			
(340-359 nt)		2'-Methoxye-thoxy ON (ISIS11155)			
initiator codon region of 5' NCR (330-349 nt)		PS-ODN (ISIS6547), 5-methylcytidine-modified PS-ON (ISIS14803)	VHCV-IRES-infected BABL/c mice	—	[252]
			HCV-positive patients with hepatocellular carcinoma		[253]
domain IV of the IRES region in 5' NCR (containing the translation initiation codon, 345-365 nt) (ISIS14803)		MPO	HCV IRES luciferase reporter messenger RNA (mRNA) translation in living mice		[254]
		5-methylcytidine-modified PS-ON	Phase I dose-escalation clinical study : 24 patients with HCV genotype 1 chronic HCV	—	[255]
			Phase I: clinical trials: 28 chronic HCV patients.		[256,257]
stem loop IIId of IRES region in 5' NCR (260-273 nt,)		2'-OMe	bicistronic RNA construct in RRL	—	[258]

Table 5: contd...

IRES region in 5' NCR (197-207 nt, 314-330 nt, 333-349 nt)	PNA, chimeric LNA	CV-1 (monkey kidney) cells transfected with reporter plasmid pRL-HL	lipid-mediated transfection of PNA:DNA duplexes and LNAs	[259]
stem loop IIId (264-282 nt) and initiator codon region (326-348 nt) of IRES region in 5' NCR	PS-ODN	HepG2 cells infected with native HCV RNA genomes in a replication competent system or native viral particles from HCV type 4 positive serum	—	[260]
IV loop region of IRES (340-349 nt)	PS-ODN, PNA*	Huh-7.5 cells electroporated with a full-length HCV genome construct, HCV IRES-driven translation in a RRL assay	—	[261]
		HCV replicon-harboring cells and JFH1 infected Huh-7.5 cells	lipid-mediated transfection of PNA:oligonucleotide duplex	
stem loop III region at the 5'-end of X-RNA (1-17 nt)	PNA	*in vitro* RdRp assay, and JFH1 HCV infected HuH-7 cells	CPP (HIV Tat peptide and arginine-rich peptide) conjugated at N-terminus of PNA *via* an O-linker	[262]
25-40 nt of IRES (1-350 nt) containing the distal and proximal miR-122 binding sites	LNA	subgenomic HCV replicon assay	—	[263]

[a]NCR, non coding region; IRES, internal ribosome entry site.
[b]AS-ODN, antisense oligodeoxynucleotide; PS-ODN, phosphotioates oligodeoxynucleotide; S-ODN, B-ODN, M-ODN, polar phosphorothioates (S), non-polar methyl- (M) or benzylphosphonates (B) modifications partially made in the 23-mer ODN that contained only six modified nucleotides which are located at the ODN termini or are scattered along the molecule; modified AS-ODN, at each end of the 17-mer AS-ODN sequence three phosphodiesters were modified as phosphorothioates, methyl- or benzylphosphonates; 2'-O Me, 2'-O-methyloligoribonucleotides; PNA, peptide nucleic acid; PNA*, phosphorothioates linkages for some PNA monomers ; LNA, locked nucleic acid.
[c]MT-2C cells, a human T-lymphotropic virus type I infected cell line, which is cloned by the limited dilution method from MT-2 cells and shows more efficient HCV replication than an uncloned population of MT-2 cells; RRL, rabbit reticulocyte lysate; HepG2 *in vitro* cell culture system, a hepatoblastoma cell line transfected with a plasmid expressing the HCV-luciferase fusion RNA; H8Ad17c cells, Neomycin-selected H8Ad17 immortalized hepatocytes transfected with an expression vector containing the HCV type II 59 NCR, core, and the majority of the envelope coding sequence (HCV nucleotides 1 to 1357) expressed from the immediate-early promoter of human cytomegalovirus (HCMV); RdRp, RNA-dependent RNA polymerase; IRES, internal ribosome entry site;
[d]CPP, cell penetrating peptide; HIV Tat peptide, Tat48–60 (GRKKRRQRRRPPQ), arginine -rich peptide (YGRRRRRRRRR); O-linker, AEEA, 8-amino-3,5-dioxo-octanic acid; "—", no delivery method used.

West Nile Virus (WNV) is one of the Japanese encephalitis (JE) antigenic serocomplex of viruses, in the family *Flaviviridae*, which is transmitted by urban culex mosquitoes. Ever since the first described case of in 1937 in West Nile

Province, Uganda, the virus had spread throughout other regions of Africa, Europe, and Southwestern Asia, before entering the United States in 1999 and later rapidly moved into Canada, Central America, Mexico, and the Caribbean. The WNV produces one of three different outcomes in humans, including an asymptomatic infection, a mild febrile syndrome termed West Nile fever, and a neuroinvasive disease termed West Nile meningitis or encephalitis. No treatment yet exists for the serious neuroinvasive form of WNV infections. Recent years have presented many researches of targeting important steps/functional RNA regions by ASO-based therapeutics (Table **6**).

Dengue Virus (DENV) is a mosquito-borne RNA virus of four antigenically distinct serotypes, anyone of which is the cuase of a complex of clinical syndromes called Dengue fever-dengue hemorrhagic fever (DF-DHF). When a person who has been previously infected with dengue gets infected for the second, third or fourth time, the previous antibodies generally developed from a long-lasting immunity to the old strain of DENV now interfere with the immune response to the current strain, leading paradoxically to more virus entry and uptake, which is in part through an immunologic process known as antibody-dependent enhancement. It is therefore essential that a prospective vaccine and highly desirable that a prospective therapeutic be effective against all four DENV serotypes, among which ASO-based therapeutics have showed promising results (Table **6**).

Japanese Encephalitis Virus (JEV), the major cause of acute viral encephalitis in humans, is of single-stranded, plus-sense viral genome, which is used for translation and minus-strand RNA synthesis, and it complementary minus-strand viral RNA contain various sequences and RNA secondary structures conserved in flaviviruses, providing potential targets for antisense agents (Table **6**).

Yellow Fever Virus (YFV), is a 40 to 50 nm wide enveloped RNA virus causing an acute viral hemorrhagic disease called yellow fever because of the increased bleeding tendency (bleeding diathesis).The WHO estimates that yellow fever causes 200,000 illnesses and 30,000 deaths every year in unvaccinated populations, and today nearly 90% of the infections occur in Africa. A safe and effective vaccine against yellow fever has existed since the middle of the 20th century, and vaccination programs in affected areas are of great importance due to no effective therapy. The number of cases of yellow fever has been increasing since the 1980s, making it a re-emerging disease of highly potential also to be used as biological weapon.

Tick-Borne Encephalitis Virus (TBEV) is the virus associated with Tick-borne encephalitis. TBEV has three subtypes: Western European subtype (formerly Central European encephalitis virus, CEEV; principal tick vector: Ixodes ricinus), Siberian subtype (formerly West Siberian virus; principal tick vector: Ixodes persulcatus), and Far Eastern subtype (formerly Russian Spring Summer encephalitis virus, RSSEV; principal tick vector: Ixodes persulcatus). Several events that occurred during the final decades of the twentieth century and the beginning of the twenty-first century suggest a rise of tick-borne infections worldwide [264].

St. Louis Encephalitis Virus (SLEV) is an emerging arbovirus in South America, with human SLEV encephalitis cases reported in Argentina and Brazil. Genotype III strains of SLEV were isolated from mosquitoes during the largest SLEV outbreak ever reported in South America (Córdoba, Argentina, 2005). These strains are related to a non-epidemic genotype III SLEV strain isolated in 1979 in Santa Fe Province, Argentina. There is currently no clear explanation for the reemergence of SLEV in Argentina. SLEV can cause fatal neurological infection and currently there is neither a specific treatment nor an approved vaccine for these infections.

Table 6: Summary of effective ASO target sites in *Flaviviruses*.

Productive ASO Target Region(s)[a]	ASOs[b]	Experimental Systems[c]	Delivery Method[d]	Ref.
1. West Nile virus (WNV)				
5'-terminal (first 20 nt at 5' end) or the 3'-terminal element involved in a potential genome cyclizing interaction (3' CSI)	PMO	full-length WNV containing a luciferase reporter (RlucWN)	Arg-rich peptide (R$_5$F$_2$R$_4$) conjugated at 5' end	[265]
		WNV reporting BHK cell line containing persistently replicating replicons with dual reporter genes (RlucNeoRep)		
		Vero cells infected with an epidemic strain of WNV isolate 3356		[265,266]

Table 6: contd…

		Five-week-old female C3H/HeN (C3H) mice inoculated with WNV		[266]
G-rich region of RCS2/CS2, and CS1 near 3'-terminus	PS-ON with partial deoxyguano-sines	Vero76 cells infected with 13 different WNV strains that contain common RNA sequence	G-quartets formation	[267]
2. Dengue virus (DENV)				
5'-terminal (first 20 nt at 5'End)	PMO	Vero cells infected with DENV serotypes 1-4	Arg-rich peptide $(R_5F_2R_4)$ conjugated at 5'end	[268]
the 3'-terminal element involved in a potential genome cyclizing interaction (3'CSI)		Vero cells infected with DEN-2 (New Guinea)		[265]
top of the terminal 3' stem-loop (3' SLT)	PMO	BHK 21 clone 15 cells infected with DENV serotypes 1-3	Arg-rich peptide (R_9F_2C) conjugated at 5' end	[269]
5'SL- and 3' CS-region (22-24 mer)	PMO	AG129 mice infected with DENV-2 (strain New Guinea C)	Arg-rich peptide $(R_5F_2R_4)$ conjugated at 5'end	[270]
3. Japanese encephalitis virus (JEV)				
the 3'-UTR loop structure in the cis-acting element on the plus-strand RNA	PNA	BHK 21cells infected with the Nakayama strain of JEV	HIV Tat peptide covalently linked at the N terminus of PNA with an O-linker	[271]
4. Yellow fever virus (YFV)				
3'-terminal element involved in a potential genome cyclizing interaction (3'CSI)	PMO	Vero cells infected with YFD 17D	Arg-rich peptide $(R_5F_2R_4)$ conjugated at 5'end	[272]
5. Tick-borne encephalitis viruses (TBE)				
NS3 protein gene sequence	ON	porcine embryo kidney cell culture infected with TBE	—	[273]
		white mice infected with TBE	—	[274]
6. St. Louis encephalitis virus (SLEV)				
3'-terminal element involved in a potential genome cyclizing interaction (3'CSI)	PMO	Vero cells infected with SLEV Kern217.3.1.1	Arg-rich peptide $(R_5F_2R_4)$ conjugated at 5'end	[265]

[a]RCS2/CS2, complementary sequence at the 5'- and 3'- terminus respectively for cyclization of viral RNA by base pairing; CS1, an intervening 22 nucleotide sequence following CS2 for cyclization mediation; NCR, non coding region.
[b]PMO, phosphorodiamidate morpholino oligomers; PS-ON, phophorothioate oligonucleotide ; PNA, peptide nucleic acid.
[c]BHK cells, baby hamster kidney cells; Rluc, Renilla luciferase, RlucNeoRep, neomycin phosphotransferase; Vero cells, African green monkey kidney cells (ATCC CCL-81); Vero76 cells, African green monkey kidney cells (ATCC CCL-1587); DENV serotypes 1-4, dengue (DEN) virus strains of DEN-1 (strain 16007), DEN-2 (16681), DEN-3 (16562) and DEN-4 (1036).
[d]Arg-rich peptide $(R_5F_2R_4)$, CH3CONH-(RAhxR)$_4$-Ahx-βAla, designated P7 (R stands for arginine, Ahx stands for 6-aminohexanoic acid, F stands for phenylalanine, C stands for cysteine, and βAla stands for beta-alanine) was covalently

conjugated to the 5' end of the PMOs through a non-cleavable piperazine linker; G-quartets formation, four consecutive guanosine residues (GGGG) at the terminus of AS-O(D)N that enhance the termini stability and cellular uptake; HIV Tat peptide, Tat48–60 (GRKKRRQRRRPPQ), Tat57–49 (RRRQRRKKR); O-linker, AEEA, 8-amino-3,5-dioxo-octanic acid; "—", no delivery method used.

Family *Picornaviridae*

A picornavirus is a virus belonging to the family *Picornaviridae*. Picornaviruses are non-enveloped, positive-stranded RNA viruses with an icosahedral capsid. As a small RNA virus, the genome RNA is unusual because it has a protein on the 5' end used as a primer for transcription by RNA polymerase. The family *Picornaviridae* includes several genera of important human pathogens: Enterovirus [including poliovirus (PV), coxsackievirus groups A and B (CVA and CVB), and the numbered EVs such as EV71], Rhinovirus [including human Rhinovirus (HRV)], Aphthovirus [including foot-and-mouth disease virus (FMDV)], Cardiovirus [including Encephalomyocarditis virus (EMCV)], and Hepatovirus [including Hepatitis A virus (HAV)].

The picornavirus genome is an approximately 7.5kb singles tranded RNA molecule of positive polarity possessing a covalently linked protein (VPg) at the 5' terminus and a 3'-terminal poly(A) tail. The 5' and 3' untranslated regions (UTRs) flank the single open reading frame and are known to have various important roles in viral translation, RNA synthesis, and virion assembly. Picornavirus replication takes place in the cytoplasm of the host cell, and no event of the viral life cycle is known to occur in the nucleus, the activity of which though is affected by viral proteins. Viral genomic RNA is translated into a single large polyprotein, which is subsequently cleaved into 10 mature proteins by virus-encoded proteases. The 5' UTRs of all picornaviruses are unusually long, averaging about 700 nucleotides (nt), and contains an internal ribosome entry site (IRES), a region that mediates the initiation of viral RNA translation several hundred nucleotides downstream of the 5' terminus in a cap-independent manner. Inspite of apparently identical functional roles, the four classes of picornaviral IRES configurations defined on the basis of RNA secondary structure and biological properties share little similarity in sequence or secondary structure.

The small molecule compounds in development for effective control and treatment of picornaviruses caused diseases and morbidity thus far appear to have limited promise, due variously to high toxicity or low efficacy, stability, or solubility. Anti-picornaviral antisense agents are part of a broader group of nucleic acid-based molecules developed for sequence-specific inhibition of translation and/or transcription of the target sequence through induced nuclease activity or physical hindrance [275,276]. A number of ASOs designed to inhibit picornavirus infections have generated positive preclinical results but are still early in the development process (Table **7**). Specifically, antisense phophorothioate DNA has been used against coxsackievirus type B3 (CVB3) *in vitro* and *in vivo*, antisense RNA has been used against FMDV, and PMO-peptide conjugates have been used against several picornaviruses all with some success.

Table 7: Summary of effective ASO target sites in *Picornaviruses*.

Productive ASO Target Region(s)[a]	ASOs[b]	Experimental Systems[c]	Delivery Method[d]	Reference
1. Coxsackie virus groups A and B (CVA and CVB)				
pyrimidine-rich tract (S(CB1)), structural protein VP(1) (S(CB4)) and initiation of translation (S(CB2))	ODN	CVB3 infected HeLa cells	—	[277]
proximal terminus of the 5' UTR (1-20 nt)	PS-ODN	CVB3 infected HeLa cells	lipofectin transfection	[278]
IRES region (557-576 nt)				
translation initiation codon AUG region including 9 nt of the 5' UTR (733-752 nt)				
3' proximal terminus of the 3' UTR (7380-7399 nt)		cardiomyocytes (HL-1 cell line) infected with CVB3		[279]
		CVB3 myocarditis mouse (Adolescent A/J mice) model	—	
3' portion of IRES region (570-590 mer)	PMO	CVB3 infected HeLa cells and cardiomyocytes (HL-1 cell line), and CVB3 myocarditis mouse (Adolescent A/J mice) model	Arg-rich peptide (P7) conjugated at 5'end	[280]
2. foot-and-mouth disease virus (FMDV)				
Second AUG of polyprotein gene	DNA and RNA	BHK 21 cells infected with FMDV	cytoplasmic microinjection	[281]
domain 5 of the IRES (1015-1035 nt), first and second AUG of polyprotein gene (1036-1056 nt and 1121-1142 nt)	PMO	BHK-21 cells infected with seven distinct FMDV serotypes (A24Cru, A12 strains 119ab, O1 Campos, O/Taiwan/2/99, C3 Resende, Asia1-1pak 54, SAT1/SAR/9/81, SAT-2/Zim/7/83,	Arg-rich peptide (R₉F₂ C) conjugated at 5'end	[282]

Table 7: contd...

3. poliovirus (PV)				
IRES stem-loop 5 (551-574 nt)	PMO	clinical isolates of PV1 infected HeLa cells, and 10-week-old poliovirus receptor transgenic (cPVR) mice	Arg-rich peptide (R_9F_2C) or P7 conjugated at 5'end	[283]
4. human Rhinovirus (HRV)				
3' base of IRES stem-loop 5 (541-562 nt)	PMO	HRV type 14 infected HeLa cells	Arg-rich peptide (R_9F_2C) or P7 conjugated at 5'end	[283]

[a]IRES, internal ribosome entry site; NCR, non coding region; UTR, untranslated region.
[b]ODN, oligodeoxynucleotide; PS-ODN, phophorothioate oligodeoxynucleotide; PMO, phosphorodiamidate morpholino oligomers.
[c]BHK 21 cells, baby hamster kidney cells of strain 21; A24Cru, foot-and-mouth disease virus strain A24 Cruzeiro/Brazil/1955.
[d]Arg-rich peptide (R_9F_2C), NH_2-RRRRRRRRRFFC-$CONH_2$; P7, CH3CONH-$(RAhxR)_4$-Ahx-βAla, (R stands for arginine, Ahx stands for 6-aminohexanoic acid, F stands for phenylalanine, C stands for cysteine, and βAla stands for beta-alanine); "—", no delivery method used.

Nidoviruses

Nidoviruses are positive-strand RNA viruses named for their strategy of replication, which includes the formation of a nested set of messenger RNAs in the 3' position. These viruses have the largest and most complex genomic RNA (~30 kb) known, the category of this order includes families of coronaviruses (including SARS-CoV and Mouse hepatitis virus (MHV)), toroviruses that contain members infecting humans, arteriviruses (including equine arteritis virus (EAV) and Porcine reproductive and respiratory syndrome (PRRS)), and roniviruses [including Hepatitis A virus (HAV)] that only infect some animals.

These genomes share the structure of eukaryotic mRNA, so the viruses can use some host cell proteins during replication and gene expression which occurs in the cytoplasm of the host cell. Unlike many viruses, they do not have any polymerase in the virus particle, as the genome can be read directly as mRNA when it first enters the host cell. This order of viruses can be distinguished from other RNA viruses by a constellation of seven conserved domains. Family *Coronaviridae* and *Arteriviridae* produce a nested set of mRNAs by the process of discontinuous subgenomic (sg) mRNA synthesis The sg mRNAs share common sequence with the genome at their 5' and 3' ends. This common sequence is attractive for ASOs targeting, as both the genome and all sg mRNA may be targeted with a single appropriately positioned ASO.

Human coronaviruses are the second most frequent cause (after rhinoviruses) of the common cold, and like rhinoviruses, they occasionally also cause more serious respiratory disease — particularly in small children, the immunosuppressed, and the elderly. The SARS epidemic, which started quietly as an atypical pneumonia in Guangdong Province, China, in late 2002, and spread in early 2003 to most of the world's continents. Death occurred in approximately 10% of infected individuals following flu-like symptoms, which included fever, malaise and dry cough, leading to acute respiratory distress with diffuse alveolar damage. An intensive research effort identified a novel coronavirus (SARS-CoV) as the causative agent of SARS [284, 285]. It is only the most recent and newsworthy evidence of the capacity of a coronavirus to cause an overwhelming disease with a unique pathogenesis. Thus, it has resulted in a large infusion of research money into the field utilizing different strategy to combat this troublesome virus (detailed information for ASO-based anti-SARS researches are summarized in Table **8**).

Table 8: Summary of effective ASO target sites in Nidoviruses, Alphaviruses, and Vesiviruses.

Productive ASO Target Region(s)[a]		ASOs[b]	Experimental Systems[c]	Delivery Method[d]	Ref.
1. SARS-associated coronavirus, SARS-CoV (Family *Coronaviridae*)					
TRS region in 5' UTR (53-72 nt, 56-76 nt)		PMO	SARS-CoV (Tor2 strain) infected Vero-E6 cells	Arg-rich peptide (R_9F_2C) or $(R_5F_2 R_4)$ conjugated at 5'end	[286,287]
-1 PRF (encompassing the loop3, 13458-13472 nt)		PNA	SARS-CoV replicon replication assay and interferon β reporter assayin HEK293 cells	Tat_{57-49} covalently linked at the N terminus of PNA with an O-linker	[288]
2. Equine arteritis virus, EAV (Family *Arteriviridae*)					
Genomic 5'-UTR	5' terminus (1-20 nt)	PMO	EAV infected Vero-E6 cells	Arg-rich peptide (R_9F_2C) conjugated at 5'end	[289]
	Conserved hairpin (53-72 nt)				
	Leader TRS (199-218 nt)				
	leader TRS and replicase initiator AUG (207-228 nt)				
5'-terminal region (1-24 nt)		PMO	HeLa cells persistently infected VB strain of EAV	Arg-rich peptide (P7) conjugated at 5'end	[290]

Table 8: contd…

3. Porcine reproductive and respiratory syndrome, PRRS (Family *Arteriviridae*)				
5' terminus (9-28 nt)	PMO	PRRSV (strains ATCC VR2385 and NVSL97-7895) infected porcine alveolar macrophages	Arg-rich peptide (P7) conjugated at 5'end	[291]
hairpin loop region of 5' UTR (54-74 nt)				
ORF6 (AUG translation initiation codon included, 14055-14074 nt)				
ORF7(AUG translation initiation codon included, 14560-14581 nt)				
ORF4 (AUG translation initiation codon included, 12922-12941 NT)		porcine alveolar macrophages cells infected with heterologous strains in the North American PRRSV genotype		[292]
5' terminus (9-28 nt)		PRRSV infected porcine alveolar macrophages cells collected from piglets		[293]
		3-week-old piglets infected with PRRSV (strain VR2385)		[294]
4. Mouse hepatitis virus, MHV (Family *Coronaviridae*)				
beginning of the nonstructural ORF1a polyprotein gene covering two possible initiation codons in favorable Kozak consensus environments (PP)	PMO	MHV-A59 infected murine astrocytoma (DBT) cells	Arg-rich peptide (R_9F_2C) conjugated at 5'end	[295]
initiation codon (H1)				
internal region (H2) of the hemagglutinin-esterase gene				
initiation codon of the nonstructural protein 2a gene (2A)				
genomic and subgenomic TRS region (51-72 nt), pp1ab AUG region (200-220 nt), 5' terminus (2-22 nt)	PMO	17Cl-1 cells infected with MHV-1 or DBT cells infected with all other MHV strains (MHV-2, MHV-3, MHV-4 (JHM), MHV-A59, and MHV-Alb139)	different Arg-rich peptides conjugated at 5'end of PMO with an amide or thioether linker	[296]
		Five-week-old male C57BL/6 mice infected with MHV-A59	mannosylated PMO-peptide conjugates	

Table 8: contd…

5. Sindbis virus, SINV (Family *Togaviridae*)				
5′ terminus and the AUG translation start site region of the genomic	PMO	BHK cells with *in vitro* transcribed SinLuc RNA, and BHK cells infected with wild type SINV	Arg-rich peptides (P3 or P7 AMCA) conjugated at 5'end	[297]
6. Venezuelan equine encephalitis virus, VEEV (Family *Togaviridae*)				
5′ terminus and the AUG translation start site region of the genomic	PMO	Vero cells infected with VEEV (TC-83)	Arg-rich peptides (P3 or P7) conjugated at 5'end	[297]
		VEEV (highly virulent ZPC738 strain) infected NIH Swiss mice	Arg-rich peptides (P7 or AMCA) conjugated at 5'end	
7. Vesivirus (Family *Caliciviridae*)				
AUG translation start site regions of ORF1 (7-26 nt)	PMO	porcine kidney (PK-15) and Vero cells infected with vesivirus isolates SMSV-13 and PCV Pan-1	scrape-loading delivered	[298]
5' region of feline calicivirus (isolated from the liver of a cat that died with hemorrhage and hepatitis and sequenced)	PMO	112 kittens of various sex and age in 4 trials involving 3 outbreaks of naturally developing caliciviral disease	—	[299]

[a]TRS, transcription-regulating sequence; NCR, non coding region; ORF, open reading frame; UTR, untranslated region; PRF, programmed -1 ribosomal frameshifting.
[b]ODN, oligodeoxynucleotide; PS-ODN, phophorothioate oligodeoxynucleotide; PMO, phosphorodiamidate morpholino oligomers; PNA, peptide nucleic acid.
[c]Vero cells, African green monkey kidney cells; BHK cells, baby hamster kidney cells; SinLuc RNA, the recombinant fulllength Sindbis virus which expresses the luciferase (luc) reporter protein (SinLuc) as a tool for monitoring anti-SINV compounds.
[d]Arg-rich peptides include (R_9F_2C), NH_2-RRRRRRRRRFFC-CONH$_2$, P7, CH3CONH-$(RXR)_4$-XB, and peptides having two to four repeats of the RXR peptide motif conjugated to a PMO through one of two linkers, either a thioether bond between the HS group of the cysteine (C) and the PMO or an amide bond between the carboxyl group of beta-alanine and the PMO (R stands for arginine, X stands for 6-aminohexanoic acid, F stands for phenylalanine, C stands for cysteine, and B stands for beta-alanine); mannosylated PMO-peptide conjugates, three acetylprotected and 9-fluorenylmethoxy carbonyl-modified mannosyl serine residues sequentially appended to the N terminus of the $(RX)4B$ or $(RXR)4XB$ peptide; Tat57–49, HIV-1 Tat peptide (RRRQRRKKR); O-linker, AEEA, 8-amino-3,5-dioxo-octanic acid; P3, oligoargrine peptide; AMCA, 7-amino-4-methylcoumarin-3-acetic acid; "—", no delivery method used.

The genus ***Alphavirus*** in the family Togaviridae contains members [including Venezuelan equine encephalitis virus (VEEV), Eastern equine encephalitis virus (EEEV), Western equine encephalitis virus (WEEV), O'nyong-nyong virus and Chikungunya virus] that threaten human health, both as natural pathogens and as potential biological weapons. Of these, VEEV is the most important human pathogen, with several recent outbreaks consisting of hundreds of thousands of cases occurring mostly in Latin America. SINV has been extensively used as a model alphavirus because of its low pathogenicity to humans, easy propagation in a variety of cell lines, and molecular biology that is considered representative of the genus. No therapeutic for alphavirus-induced disease exists, though veterinary vaccines of varying quality against EEEV, WEEV, and VEEV are commercially available with limited availability for ordinary people. Recently antisense PMO-peptide conjugates have been shown to be effective against the alphaviruses SINV and VEEV in cell cultures and in murine models (Table **8**).

Caliciviridae are human or non-human pathogenic viruses with a high diversity. Certain members of this family have unusually broad host ranges, and some are zoonotic (transmissible from animals to humans).The family Caliciviridae includes several genera of important pathogens: Lagovirus [including rabbit hemorrhagic disease virus (RHDV)], and human pathogenic Norovirus and *Vesivirus*. All viruses in this family possess a non segmented, polyadenylated, positive sense single stand RNA genome of ~7.5-8.5 kb in length, enclosed in an icosahedral capsid of 27-40 nanometers in diameter, and Virtually, every calicivirus replicate is a mutant, which is due to lack of effective genetic repair mechanisms in the replicative processes. Hence, traditional therapeutics dependent on s protein epitopes lack the required diversity of conformational specificity that would be required to reliably detect, prevent or treat infections from these mutant clusters (quasi-species) of Caliciviruses [300]. Antisense technology using PMO oligomers shows promise in overcoming these current diagnostic and therapeutic problems inherent with newly emerging viral diseases (Table **8**).

Non-Retroviral Negative-Sense Single Stranded RNA Virus

For the replication of non-retroviral negative-sense RNA viruses ((-) RNA viruses), infecting viral genomic RNA must first be transcribed into mRNA. Initial mRNAs of (-) RNA viruses are produced with the use of RdRp present in the infecting virion, and nascent transcripts are translated to produce proteins required to engage the infectious cycle. Replication requires the production of full-length antigenomic positive-sense RNA (cRNA), which in turn serves as a template for (-) strand genome synthesis. The genomic RNA and cRNA of (-) RNA viruses is known to be tightly complexed with viral protein, forming nucleocapsid, at all stages of the viral life-cycle. In contrast, viral mRNA is not complexed in ribonucleoprotein form, and is therefore the RNA species most likely accessible to duplexing by ASOs.

Filoviruses

The family Filoviridae is the taxonomic home of several related viruses that form filamentous virions. Two members of the family that are commonly known are ***Ebola virus (EBOV) and Marburg virus (MARV)***. Both viruses, and some of their lesser known relatives, cause severe disease in humans and nonhuman primates in the form of viral hemorrhagic fevers. All accepted members of the family (all ebolaviruses and marburgviruses) are Select Agents, World Health Organization Risk Group 4 Pathogens (requiring Biosafety Level 4-equivalent containment), National Institutes of Health/National Institute of Allergy and Infectious Diseases Category A Priority Pathogens, Centers for Disease Control and Prevention Category A Bioterrorism Agents, and listed as Biological Agents for Export Control by the Australia Group.

EBOV and MARV have a single stranded, negative-sense 19 kb RNA genome that codes for seven proteins. Viral replication and transcription depend on a complex of nucleoprotein (NP), VP30, VP35 and the RNA polymerase L. VP24 and VP40 are involved in budding of mature virions from the cell surface while glycoprotein (GP) is found embedded in the lipid bilayer surrounding the nucleocapsid. The filoviruses are listed as Category A bioterrorism agents by the CDC and must be handled under biosafety level 4 conditions. Both viruses cause severe hemorrhagic fever in humans with mortality rates of 30–90% [301]. Currently, there are no vaccines or therapeutics for treating filovirus infections in humans [302]. Two preclinical studies have been published investigating PMOs as a treatment for EBOV infection, with both showing efficacy in animal models (Table **9**).

Respiratory Syncytial Virus (RSV)

Human respiratory syncytial virus (RSV), a member of subfamily Pneumovirinae in the family *Paramyxoviridae* (Order: Mononegavirales), is a major virus that causes lower respiratory tract infections and hospital visits during infancy and childhood. After an RSV virion enters a cell, the nonsegmented genomic RNA is transcribed into 5' capped and 3' polyadenylated mRNAs that each code for one of the 11 individual RSV proteins. The RSV genome is replicated through the production of a complementary positive-sense full-length replicative intermediate (the antigenome). The viral 'L' protein is a major component of the RNA-dependent RNA polymerase complex responsible for all viral RNA synthesis, and likely also plays roles in the capping and polyadenylation of viral mRNA. Currently, there is no vaccine for the prevention of RSV infection, and available therapeutics are of limited utility. However, there have been several reports over the past few years describing studies in which RSV infections have been diminished through ASOs-based approaches (Table **9**).

Influenza A Virus (FLUAV)

Influenza A viruses (FLUAV) contain eight segments of single stranded, negative sense genomic RNA which code for 11 known proteins (designated PB1, PB1-F2, PB2, PA, HA, NP, NA, M1, M2, NS1, and NS2). Multiple subtypes of FLUAV exist based on antigenic variation of envelope associated hemagglutinin (HA) and neuramidase (NA) proteins (*i.e.*, H1N1, H3N2 and H5N1). Influenza viruses are etiological agents of deadly flu that continue to pose global health threats, and have caused global pandemics that killed millions of people worldwide. Specifically, the H5N1 is a highly virulent avian influenza virus that can also infect humans, the pandemic consequences of which can be truly devastating. The availability of neuraminidase inhibitors and attenuated vaccines improves our ability to defend against influenza, but their benefits can be significantly limited by drug-resistance and virus mutations. Nucleic acid-based drugs may represent a promising class of antiviral agents that could play a role in the prevention and treatment of influenza [303]. Since the 1990s, ASOs targeting FLUAV RNA sequences that are highly conserved across viral subtypes and considered critical to the FLUAV biological-cycle, have been investigated as potential therapies for FLUAV infection (Table **9**).

Table 9: Summary of effective ASO target sites in negative-sense single stranded RNA viruses.

Productive ASO Target Region(s)[a]	ASOs[b]	Experimental Systems[c]	Delivery Method[d]	Ref.
1. Ebola virus, EBOV (Family *Filoviridae*)				
near or overlapping the AUG translational start site of VP24, VP35, and RNA polymerase L	PMO	ZEBOV infected Vero E6 cells	—	[304]
		C57BL/6 mice or female Hartley guinea pigs infected with lethal ZEBOV (pre- and post-exposure treatment)		
		female rhesus macaques infected with lethal ZEBOV (pre-exposure treatment)		
translational start site region of VP35 positive-sense RNA (3136-3115 nt)	PMO	Vero E6 cells and C57BL/6 mice infected with lethal ZEBOV (pre-exposure treatment)	—	[305]
		Vero E6 cells and C57BL/6 mice infected with lethal ZEBOV (pre- and post-exposure treatment)	Arg-rich peptide (R₉F₂C) conjugated at 5'end	
5' to, overlapping, or immediately downstream of the translational start codon of VP24	PMO	Vero E6 cells and C57BL/6 mice infected with lethal ZEBOV (pre-exposure treatment)	Arg-rich peptide (P7) conjugated at 5' or 3' end	[113]
	PMO*plus*		—	
eVP24 (10331-10349 nt) and eVP35 (3133-3152 nt) transcripts	PMO*plus* (five piperazines linkages)	rhesus monkeys infected with lethal ZEBOV (post-exposure treatment, combination therapy)	—	[306]
2. Marburg virus, MARV (Family *Filoviridae*)				
NP24 (73-95 nt)and VP24 (10204-10224 nt) transcripts	PMO*plus*	cynomolgus monkeys infected with MARV	—	[306]
3. Respiratory Syncytial Virus, RSV (Family *Paramyxoviridae*)				
start site for the NS2 and P genes of the RSV genomic RNA	PS-ODN	HEp-2 cells infected with RSV strain A2	—	[307]
the 'gene-start'sequence (GS) present at the 5' terminus of the L gene mRNA to 13 nt into the coding sequence	PMO	HEp-2 cells and A549 cells infected with RSV strain A2	Arg-rich peptide (P7) conjugated at 5' end	[308]
		BALB/c mice with infected with RSV strain A2 (pre-exposure treatment)		
4. Influenza A virus, FLUAV (Family *Orthomyxoviridae*)				

Table 9: contd…

loop-forming site of the polymerase 3 mRNA	ON	influenza A/PR8/34 (H1N1) virus infected MDCK cells	undecyl residue conjugated at the 5' terminal phosphate group	[309]
the juxtaposed sequences in the 5'-terminus of the molecule around and upstream of the initiation codon	ON	*in vitro* translation of the influenza virus M1 protein mRNA	—	[310]
AUG initiation codons of PB2 and PA	ODN, PS-ODN	CAT-ELISA assay	—	[311]
AUG initiation codon sites of the NP protein and the RNA polymerase (PB1, PB2, and PA) genes	PS-ODN	CAT-ELISA assay	liposome encapsulation	[312]
PB1, PB2, PA and NP genes	PS-ODN	FLUAV infected MDCK cells	liposome encapsulation	[313]
surrounding the translation initiation codons of the viral PB2 or PA genes	PS-ODN	FLUAV infected mice	cationic liposome (Tfx-10) encapsulation	[314]
AUG initiation codons of PB1, PB2, PA and NP genes	PS-ODN	FLUAV infected MDCK cells and mice	liposome encapsulation	[315]
3'/5' terminal sequences	PS-ODN	influenza A/JingFang/86-1(H1N1) infected MDCK cells	—	[316]
AUG initiation codons of PB2	PS-ODN	influenza A/PR/8/34 (H1N1) and B/Norway/1/80 viruses infected female BALB/c mice	— cationic liposome (DMRIE-C) encapsulation	[317]
AUG initiation codons of PB2 and PA	PS-ODN	influenza A/PR/8/34 (H1N1) infected MDCK cells and mice	cationic liposome (Tfx-10) encapsulation	[318]
AUG translation start site region of PB1 mRNA and 3'-terminal region of NP viral genome RNA	PMO	Vero cells infected with FLUAV strains of A/PR/8/34 (H1N1), A/WSN/33 (H1N1), A/Memphis/8/88 (H3N2), A/Eq/Miami/63 (H3N8), A/Eq/Prague/56 (H7N7), A/Thailand/1(KAN-1)/04 (H5N1)	Arg-rich peptide (P7) or $(R_5F_2 R_4)$ conjugated at 5' end	[319]
viral mRNA encoding the heamagglutinin protein	AS-ON	influenza A/PR/8/34 virus infected mice (post exposure treatment)	liposome encapsulation	[320]

Table 9: contd…

the translation start site region of PB1 or NP mRNA or the 3'-terminal region of NP viral RNA	PMO	MDCK cells and female BALB/c mice infected with highly pathogenic mouse-adapted influenza A virus (SC35M; H7N7)	Arg-rich peptide (P7) conjugated at 5' end	[321]
5'-terminal conserved sequence	PS-ODN	influenza H5N1 infected MDCK cells and mice	—	[322]
the AUG translation start site region of the polymerase subunit PB1 mRNA	PMO	FLUAV A/Eq/Miami/1/63 (H3N8) infected MDCK cells and female BALB/c mice	Arg-rich peptide (P7) conjugated at 5' end	[323]
the 22 terminal nucleotides at the 3' end of the NP virion RNA				
packaging signals in the 5' end of the viral PB2 RNA (2279-2293 nt)	PS-ODN	MDCK cells infected with eight H1N1 strains, eight local H3N2 strains, four H7N3 avian strains, and seven local B strains	liposome encapsulation	[324]
5' ends of segments 2 or 3 (complementary to the 3'-coding regions of PB1 and PA, respectively)	PS-ODN	MDCK cells infected with strains of H1N1 subtype (A/Taiwan/1/86, A/PR/8/34, A/Roma/2/08, an oseltamivir resistant isolate A/Parma/24/09, and the oseltamivir sensitive new pandemic variant A/Italy/05/09), local strains of H3N2 subtype (A/Firenze/1/03, A/Firenze/5/03, A/Firenze/4/07), and two avian strains of H7N3 subtype (A/Mallard/Italy/33/01, A/Turkey/Italy/214845/02)	liposome encapsulation	[325]
5. Arenavirus (Family *Arenaviridae*)				
the 5'termini of genomic segment L (1-20 nt) and S (1-19 nt)	PMO	Vero-E6 cells infected with four distantly related arenaviruses (LCMV,PICV,TCRV, and JUNV)	Arg-rich peptides (P7) conjugated at 5'end	[326]
		LCMV-infected C57/Bl6 mice (pre-exposure treatment)		
6. Measle virus, MeV (Family *Paramyxovirus*)				
52 nucleotides at the 5' end of the NP mRNA	antisenseRNA (endogenous expression)	MeVEd infected HeLa cells (transfected with eukaryotic expression vector pRc/CMV)	—	[327]
AUG start codon of NP in genomic MV RNA (91-113 nt)	PMO	Vero, Vero/hSLAM, CHO/SLAM and BHK/sr/T7 cells infected with MeVEd, and other genotypes	Arg-rich peptides (P7) conjugated at 5'end	[328]

Table 9: contd…

7. Vesicular stomatitis virus, VSV (Family *Rhabdoviridae*)				
the major ribosome protected initiation site around the 5' proximal AUG codon of VSV NP mRNA	oligodeoxyribonucleotides	L929 cells infected with Indiana strain of VSV	poly(L-lysine)-conjugated	[329]
	PS and chimeric AS oligodeoxyrib-onucleotides	VSV infected J774E cells	Macrophage-targeting short polyguanylic acid stretches conjugated at 3' end	[330]
8. Borna disease virus, BDV (Family *Bornaviridae*)				
GP mRNA (934-951 nt)	PNA	Vero cells persistently infected with BDV	16 amino acid residue and NLS conjugated PNA	[331]
		Primary hippocampal neurons (isolated from 1-day-old Sprague-Dawley rats) acutely infected with BDV		

[a] eVP, ZEBOV viral protein; NP, nucleoprotein; ORF, open reading frame; GP, glycoprotein;
[b] PMO, phosphorodiamidate morpholino oligomers; PMO*plus*, positively charged PMOs containing piperazine linkages within the molecular backbone; PS-ODN, phophorothioate oligodeoxynucleotide; ON, oligonucleotide; PNA, peptide nucleic acid.
[c] Vero cells, African green monkey kidney cells; ZEBOV, Zaire Ebola virus; MARV, Lake Victoria Marburg virus; A549 cells, Human A549 pulmonary type II epithelial cells; CAT, chloramphenicol acetyltransferase protein expressed in the clone 76 cell line, which expresses the influenza virus RNA polymerase and nucleoprotein (NP) genes in response to dexamethasone; MDCK cells, Madin-Darby canine kidney cells; eight H1N1 strains include A/Taiwan/1/86 and the local isolates A/Firenze/1/04, A/Firenze/1/06, A/Firenze/11/06, A/Firenze/13/06, A/Firenze/15/06,A/Firenze/17/06, A/Firenze/23/06; eight local H3N2 strains include A/Firenze/1/03, A/Firenze/5/03, A/Firenze/4/07,A/Firenze/5/07, A/Firenze/7/07, A/Firenze/8/07, A/Firenze/10/07, A/Firenze/11/07; four H7N3 avian strains include A/Mallard/IT/33/01, A/Mallard/IT/43/01, A/Turkey/IT/214845/02, A/Turkey/IT/220158/02; seven local B strains include B/Firenze/1/04, B/Firenze/2/06, B/Firenze/3/06, B/Firenze/4/06, B/Firenze/5/06, B/Firenze/6/06, B/Firenze/14/06; LCMV, lymphocytic choriomeningitis virus Arm53b; PICV, Pichinde virus; TCRV, Tacaribe virus; JUNV, Junín virus strain Candid#1; MeVEd, measles virus Edmonston wild-type strain; and other MeV genotypes include MeV genotype G1 Berkeley/83, MeV genotype B3-2 Ibadan/97 or MeV genotype D3 Chicago1/89; J774E cells, a murine monocyte/macrophage cell line.
[d] Arg-rich peptide P7, CH$_3$CONH-(RXR)$_4$-XB, R$_5$F$_2$R$_4$, NH$_2$-RRRRRRFFRRRRC-CONH$_2$ (R stands for arginine, X stands for 6-aminohexanoic acid, F stands for phenylalanine, C stands for cysteine, and B stands for beta-alanine); poly(L-lysine)-conjugated, ε-amino groups of lysine residues of poly(L-lysine)(Mr, 14,000) were coupled at 3' ends of oligodeoxyribonucleotides through a N-morpholine ring; NLS, nuclear localization signal of the simian virus 40 T antigen (PKKKRKV); 16 amino acid residue and NLS conjugated PNA, PNA was coupled to 16 amino acid residues derived from the third helix of the pAntp of Drosophila, the PNA construct was linked covalently to an NLS sequence. At the N terminus of the PNA, a spacer of two lysines (K-K) was introduced to prevent spatial binding hindrance of the following NLS sequence, and one cysteine was introduced to enable linkage with the pAntp mannosylated PMO-peptide conjugates, three acetylprotected and 9-fluorenylmethoxy carbonyl-modified mannosyl serine residues sequentially appended to the N terminus of the (RX)4B or (RXR)4XB peptide; Tat57–49, HIV-1 Tat peptide (RRRQRRKKR); O-linker, AEEA, 8-amino-3,5-dioxo-octanic acid; P3, oligoargrine peptide; AMCA, 7-amino-4-methylcoumarin-3-acetic acid; "—", no delivery method used.

Arenavirus

Arenavirus is a genus of virus that infects rodents and occasionally humans. At least eight Arenaviruses are known to cause human disease. The diseases derived from Arenaviruses range in severity. Aseptic meningitis, a severe human disease that causes inflammation covering the brain and spinal cord, can arise from the

Lymphocytic choriomeningitis virus (LCMV) infection. Hemorrhagic fever syndromes are derived from infections such Guanarito virus (GTOV), Junin virus (JUNV), Lassa virus (LASV) causing Lassa fever, Machupo virus (MACV), Sabia virus (SABV), or Whitewater Arroyo virus (WWAV).

The unusual process of arenavirus gene expression may offer several opportunities for ASOs-based therapeutic intervention. All arenaviruses have a single-stranded, ambisense RNA genome containing two segments, "L" and "S," each of which encodes two proteins. The "L" segment encodes polymerase (L) and matrix (Z) protein and the "S" segment nucleoprotein (NP) and glycoprotein (GPC). All arenaviruses contain nearly identical complementary sequences at the 5' and 3' termini of each genomic segment, which indicates the potential target site for broad-spectrum antisense inhibition among different strains within the family (Table **9**).

Measles Virus (MeV)

Measles virus (MeV) is a highly contagious human pathogen causing measles, which is characterizedby high fever, conjunctivitis,coryza, cough, the presence of Koplik spots on the buccalmucosa and a maculopapular rash. Vaccination programs have dramatically reduced the incidence of measles on a global scale, however, measles is still responsible for an estimated 245,000 deaths each year, due to currently no antiviral compounds available for the treatment.

MeV is a member of the genus Morbillivirus, family *Paramyxoviridae*, which has a monopartite RNA genome of 15,894 nucleotides in length. The genome consists of a short highly conserved 3' leader RNA region followed by coding regions for the six structural proteins. Each mRNA transcription unit is separated by defined stop-start sequence motifs and the genome ends with a short highly conserved 5'-trailer RNA region. The genome encodes, in a 3' to 5' order, the nucleocapsid (N) protein, phosphoprotein (P), matrix (M) protein, fusion (F) protein, hemagglutinin (H) protein and the large (L) protein. Antisense peptide-PMO conjugates have been studied as steric-blocking RNA silencers for inhibiting the replication of MeV in cell cultures (Table **9**).

Vesicular Stomatitis Virus (VSV)

Vesicular stomatitis Indiana virus (VSIV, often still referred to as VSV), a virus in the family Rhabdoviridae, is zoonotic and leads to a flu-like illness in infected humans. The genome of the virus is a single molecule of negative-sense RNA that encodes five major proteins: G protein (G), large protein (L), phosphoprotein, matrix protein (M) and nucleoprotein (NP). However, the NP mRNA has been the only target validated by ASOs to inhibiting VSV replication and cytopathic effect (CPE) in cell cultures (Table **9**)

Borna Disease Virus (BDV)

Borna disease virus (BDV) is the etiological agent of Borna disease which causes neurological symptoms in a wide variety of animal species, and there is considerable evidence that BDV also infects humans. BDV is a nonsegmented negative-strand RNA virus which consists of an 8.9kb RNA genome including at least six ORFs. It has the property, unique among members of the order Mononegavirales that infect animals, of transcribing and replicating its genome in the nucleus. Characterization of amino acid sequences of several BDV strains and isolates from various parts of the world revealed that BDV is a remarkably conserved virus with 84 to 95% amino acid identities among all gene products. Antisense peptide-PMO conjugates have been studied as steric-blocking RNA silencers for inhibiting the replication of MeV in cell cultures (Table **9**).

From Bench to Bedside

Along with productive target site identification and validation in different pathogenic viruses, limited but rigorous studies have also explored and evaluated ASOs' mechanism of action, efficacy of various delivery systems, pharmacologic properties, and the generation and characterization of resistant virus, which are of equal importance to the transformation from antisense oligomers to clinically applicable antiviral ASO therapeutics. These studies provide guidance for the better design of antisense agents against RNA or DNA viruses for systematic application in humans.

The *in vivo* activity of antisense agents against a variety of viral pathogens has been demonstrated, and a number of PS-ODNs and neutral charge PMOs have entered human clinical trials. However, none has yet been licensed for treatment of viral infection or other indications. The purpose of this part is to summarize notable findings from the reports documenting issues currently preventing the further development of antiviral ASOs, with a focus on the specific challenges (*i.e.,* tissue-specific delivery, time of treatment, virus resistance to ASOs, and several phamarco-considerations) posed against viral pathogens.

Delivery Strategies for Antiviral ASOs

The poor intracellular uptake of ASOs is a major impediment to their use as therapeutics. This is the main reason why the antiviral drug fomivirsen3 is currently the only approved antisense drug (although the drug was discontinued in 2004 as the market for the drug diminished). Efficient, tissue-specific conjugation of ASOs to cell-penetrating peptides (CPP, please refer to **"Efficient delivery systems"** in the **"ASOs-Based Antibacterial Therapeutics"** for detailed information) is an increasingly common approach to improve cellular uptake. With efficient delivery of ASOs into cells made possible by the advent of peptide-ASO conjugate (PASO), the use of PASO-technology in the arena of virology research has intensified. Recently, a number of publications describing inhibitory activity by PASO (especially the PPMO) against RNA virus infections of cell cultures and mice have appeared (see the **"delivery method"** section in Tables **3** to **9**).

The enhanced efficacy of PASOs has been hypothesized to result from the enhanced affinity resulting from ionic interactions between the net positive charge of the peptides and the negative charge of the complementary RNA and through enhanced pharmacokinetic properties and cellular uptake. However, analyses have shown that while the ASO portion of PASO conjugates are stable in cells, serum or tissue homogenate, the peptide portion is susceptible to degradation [332,333]. This degradation impacts the effectiveness of PASOs by preventing the efficient escape of the conjugated molecule from endosomal or lysosomal vesicles [333]. Besides, PASOs are more poorly tolerated *in vivo* compared to naked ASOs, as illustrated by the dose-dependent reductions in weight, behavioral alterations, and mild liver histopathology observed following repeat administration of 200–300 lg doses of a PMO conjugated to an arginine-rich peptide in mice [266].

Various attempts have been taken to introduce non-natural amino acids or other chemical changes in the conjugated peptide to offset toxicity and improve cellular uptake and intracellular distribution, though with varying degrees of success [334,333]. Meanwhile, a few efficient and tissue-specific delivery systems have been evaluated in viral infection cultures, including antibody-coupled pH-sensitive liposome encapsulation, lipid-mediated carrier system, and tissue-targeting ligand conjugated system. However, with conceptual advantages, these complex delivery systems that lack *in vivo* evaluation data stands less superiorities in reality, considering factors of systematic application in humans, manufacturing, quality control, and costs. Delivery strategies with simpler modification to ASOs, higher efficiency, and excellent *in vivo* tolerance are more welcomed, in which the poly-(L-lysine) and PMO_{plus}^{TM} (which positively-charged piperazine groups along the molecular backbone, see Fig. **2** and [335]) are representative candidates.

Virus Resistance to ASOs

It is well established that viruses can escape the effects of antiviral therapeutics through genomic alterations. Not all nucleic acid–based antiviral strategies are equally sensitive to passive and active viral escape strategies. Some viruses can escape from RNAi inhibition through an active countermeasure (or induced viral resistance) or a passive countermeasure (that the replication cycle of a virus provide an intrinsic protection against RNAi attack). Although no data available on virus escape from antiviral ribozymes, point mutations in the target sequences will prevent efficient binding of the ribozyme and therefore result in escape. Decoy and aptamers that bind and sequester viral factors are not very likely to suffer from escape, although detailed studies are lacking.

Inhibition of gene expression by antisense approaches does not rely on recruitment of a cellular silencing mechanism. Instead, the inhibitory effect is largely based on strong binding of antisense oligonucleotides to the target mRNA, which can block splicing events or the elongating ribosome through steric hindrance. In addition, unmodified DNA or phosphorothioate-modified oligonucleotides may induce RNA cleavage by activation of cellular RNase H, the enzyme that specifically recognizes DNA RNA duplexes. This is a concern for ASO-based therapeutics, for which antiviral resistance could conceivably occur through the propagation of minority variants containing nucleotide polymorphisms that prevent the efficient binding of the antisense agents to the target nucleotides.

As early as 1997, Bell *et al.,* constructed an eukaryotic expression vector designed to express antisense RNA complementary to 52 nucleotides at the 5' end of the MVnucleoprotein mRNA and demonstratd that transfection of HeLa cells with this construct result in no resistance to Measles Virus (MV) infection [327]. In 1998, Mulamba GB and *et al.,* [336] firstly reported that a human cytomegalovirus (HCMV) mutant isolated for resistance (10-fold) to the antisense oligonucleotide fomivirsen (ISIS 2922) exhibited cross-resistance to a PS modified derivative of fomivirsen with an identical base sequence, although no changes in the mutant's DNA corresponding to the fomivirsen target sequence were found. Several reports document the rapid selection of RNA viruses containing compensatory mutations that enabled these viruses to overcome PMO antiviral activity following serial passage in cultured cells, including specifically SARS-CoV [287,286], WNV [266], poliovirus type 1 (PV1) [337], and FMDV [282]. Nucleotide variations that alter the complementarity of the target RNA with the antiviral antisense agents were responsible for conferring resistance in SARS-CoV, FMDV, and PV1; however, in WNV, resistance determinants were identified outside of a PPMO target site when virus was serially passaged in the presence of a PPMO targeting the untranslated 30 conserved sequence I (30 CSI) [266]. WNV isolated from brains of infected mice treated with 30 CSI PPMO showed no mutations compared to original virus challenge stock [266]. Likewise, sequence obtained from virus isolated from the muscle of PV1 PPMO-treated mice did not show mutations [337]. Rigorous deep-sequence-based investigations are in need to characterize the occurrence and prevalence of potentially resistant minority quasispecies in virus challenge stocks and in viruses isolated from infected humans.

Time of Treatment

A central issue in ASOs-based drug development against a virus is the question of how soon after the onset of infection a drug candidate would need to be administered in order to produce significant clinical benefit. The level of antiviral activity by ASOs administered at various times in relation to the time of virus inoculation has been addressed. Studies of PPMO in cell cultures and *in vivo* [108] indicate that the antiviral activity of PPMO diminishes in concert with the period of time that elapses after infection and before treatment. However, several *in vivo* studies showed that PPMO treatment beginning after infection did cause a significant reduction in viral replication or severity of disease [108]. Several reports contend that the above demonstrated ability of a candidate ASO compound *via* a nontoxic regimen in an *in vivo* setting, especially against rapidly pathogenic viruses, is sufficient to merit its consideration for further exploration as a potential drug, regardless of the time-of-infection/time-of-treatment issue in the mouse models employed.

Pharmaco-Considerations

ASOs along with its delivery component must have a favorable pharmacokinetic and safety profile, if to be considered for development as an antiviral pharmaceutical agent in humans. Important issues need to be addressed include *in vivo* toxicity, biodistribution, and pharmacokinetic behavior, the similar problems of which shall not be discussed repeatedly here as detailed information can be referred to that also exists for antibacterial ASOs. Notably, the pharmaceutical properties and pharmacological behavior of antiviral PMOs have been the most extensively assessed and reported (please refer to [108] for detailed information). It has been reported that various PMO produced no serious drug-related adverse events in over 10 phase I and II human clinical studies [108]. However, the number of reports of significant antiviral efficacy by PMO is low, making the evaluation of the pharmacologic behavior of PPMO imperative.

Normally, *in vivo* PPMO toxicity evaluation was first carried out in non-infected mice, and a safe PPMO dose regimen thereafter evaluated for antiviral activity. Antiviral regimens of PPMO were shown to cause few ill effects when delivered *i.v., i.p.* or *i.n.* to non-infected mice. Generally, PPMOs are generally well tolerated by animal recipients, although toxicity can be observed as the dose increases [338]. The nature of the treatment regimen may affect PPMO toxicity. For example, Burrer and *et al.,* [296] observed no treatment-associated toxicity when PPMOs were administered to healthy mice. However, when PPMO treatment followed viral challenge, significant toxicity was observed. Besides, conjugation of PMO to the arginine-rich peptide $(RXR)_4$ increases elimination half-life and tissue retention [338], whereas no consistent correlation between peptide conjugation and increased efficacy has been revealed [305].

Limited reports of studies carried out in rodents with non-toxic doses of parenterally administered PPMO have provide data on the tissue distribution pattern of PPMO (*e.g.,* high level in liver, kidney, and spleen, and lower level in lung and heart) and functional activity of PPMO in various tissues. The differing relative amounts of PPMO found in the different tissues may be attributed to differences between the routes of administration used. Specifically, PPMO apparently does not cross the blood brain barrier efficiently, and that the rate of blood perfusion to an organ apparently affects the likelihood that systemically administered PPMO will be found there.

Detailed examination of the pharmacokinetics of PPMO has been conducted in two studies, one in mouse [270] and one in rat [338]. Despite the differences in routes of administration, PPMO sequences, methods of detection, and presence or absence of a fluorescein conjugate, both studies concluded that the pharmacokinetic profile of PPMO was characterized by rapid distribution from the vascular space to tissues, with an initial distribution (plasma) half-life that averaged about 1.5 h and an elimination (tissue) half-life averaging about 6 h.

Targeting Host Factors: A New Rationale for Antiviral ASOs Development

All viruses are gene poor relative to their hosts, therefore, most steps in virus infection involve interactions between relatively few different types of viral components and much more complex pools of host factors. The sea of host factors constitutes the essential milieu that viruses must adapt for survival and a tremendous resource that viruses can manipulate. The increasingly revealed close integration between viral and host functions in infection suggests that a more holistic view of the virus-infected cell as a unified entity constitutes the functional unit of infection. Accordingly, identifying such host factors and their contributions has long been recognized as an important frontier.

Amount of evidence have shown that host factors participate in most, if not all, steps of virus infection, including entry, viral gene expression, virion assembly, and release. Moreover, host factors are targeted by viruses to modulate host gene expression and defenses [339]. Each of these virus-host interactions may contribute to the host specificity, tissue specificity, or pathology of infections, and also represents a potential target for virus control or for optimization to improve beneficial uses of viruses and their components. Besides, targeting host cofactors of the virus life cycle is an attractive concept because it imposes a higher genetic barrier for resistance than direct antiviral compounds [340].

A key objective for discovery of new antisense antimicrobial agents is the determination of the genes essential for the survival of pathogenic organisms. In addition to viral targets, inhibitory antisense reagents can be directed against host factors. Host factors are an underappreciated target for antiviral therapy, and in recent years, researches based on host-directed antiviral strategy has been applied successfully to suppress replication of several viruses *in vitro* and *in vivo*, including HBV, HCV, HIV-1 and HSV-1 (Table **10**). Plenty types of host factors (*i.e.,* genes, proteins, miRNAs, cytokines, *etc.*) have been studied to yield future targets of inhibitory ASOs. Notably, anti-miRNA inhibitors Miravirsen, is the first miRNA-targeted drug to receive investigational new drug approval by the FDA, which will enable phase II studies for the treatment of HCV.

Table 10: Examples of effective host targets for ASO-based antiviral strategy.

Effective Host Targets (Functions)[a]	ASOs (Target Sites) [b]	Experimental Systems[c]	Delivery Method[d]	Validated Efficacy[e]	Ref.
1. Hepatitis B virus (HBV)					
Hsp60 (a molecular chaperone that activates HBV pol)	PS-ODN (5 to -10 nt)	HepG2 cells	—	down-regulation of Hsp60 severely reduced the level of replication-competent HBV without influencing cell proliferation and capsid assembly	[341]
ASGPR1 (a candidate receptor for HBV attachment to hepatocytes)	PS-ODN (749-768 nt)	HepG2.2.15 cells	—	down-regulation of ASGPR1 reduced HBV-DNA, HBsAg and HBeAg level without influencing cell viability	[342]
miR-199a-3p (target S protein coding region)	2'-OMe	HepG2 2.2.15	—	both miRNAs efficiently reduced HBsAg expression and suppressed viral replication without affecting cell proliferation.	[343]

Table 10: contd…

miR-210 (target the HBV pre-S1 region)					
ABHD2	AS-ODN	HepG2 2.2.15	—	antisense inhibition of ABHD2 or EREG significantly blocked HBV propagation	[344]
EREG					
2. Human immunodeficiency virus (HIV)					
gamma interferon, interleukin-6, interleukin-1, and tissue necrosis factor-alpha	PNA	H9 and peripheral blood mononuclear cells (PBMCs)	— / Electropor-ation	PNA significantly inhibited replication of HIV-1 as measured by synthesis of viral mRNA and p24 protein, reverse transcriptase activity, and syncitial cell formation	[345]
miR-28, miR-125b, miR-150, miR-223 and miR-382 (targeting the 3' ends of HIV-1 messenger RNAs that contribute to HIV-1 latency)	2'-OMe	HIV-1 protein translation in resting CD4+ T cells transfected with HIV-1 infectious clones, or HIV-1 virus production from resting CD4+ T cells isolated from HIV-1-infected individuals on suppressive HAART	—	the combined miRNA inhibitors resulted in increased viral production and increased HIV-1 infectious particles at least 10-fold, which reverse HIV latency to be exposed to immune surveillance	[346]
Interleukin-4 (IL-4 enhances the replication of HIV stranis that use CXCR4 (X4) coreceptor)	AS DNA	SHIV X4, SHIV$_{KU-2}$ and SHIV$_{89.6}$P in cultures of CD4$^+$ T cells and macrophages	—	AS IL-4 oligomer caused inhibition of virus replication in cultures of CD4 T cells and macrophages derived from macaques	[347]
		SHIV89.6P-infected macaques		4 of 6 macaques inoculated with the virus and treated with AS IL-4 DNA resulted in a significant decrease in viral RNA concentrations in the liver, lungs, and spleen tissues that are all sites of virus replication in macrophages	
3. Hepatitis C virus (HCV)					
miR-122 (a microRNA specifically expressed in	2'-OMe	HCV infected Huh -7. 5 cells	transfection method	2'-OMe AS-ON depletion of miR-122 inhibited HCV genotype 2a replication and infectious virus production	[348]
	16 nt LNA	non-human	—	long-lasting and reversible decrease in	[349]

the liver, a host factor that binds HCV RNA and upregulates viral RNA levels)	modified ON (5' end of miR-122, also called Miravirsen or SPC3649)	primates (African green monkeys)		total plasma cholesterol without any evidence for LNA-associated toxicities or histopathological changes in the study animals	
		chronically HCV genotype 1a and 1b infected chimpanzees		long-lasting suppression of HCV viremia, with no evidence of viral resistance or side effects in the treated animals, down-regulation of interferon-regulated genes, and improvement of HCV-induced liver pathology	[350]
		5'UTR-NS2 recombinants of HCV genotypes 1a, 1b, 2a, 2b, 3a, 4a,5a, and 6a with efficient growth in Huh7.5 cells		SPC3649-induced miR-122 antagonism had a potent antiviral effect against HCV genotypes 1–6 5' UTR-NS2 viruses	[351]
		HCV patients		Phase I: miravirsen administered subcutaneously weekly or biweekly is well tolerated	[352]
				Phase IIa: miravirsen shows dose-dependent, prolonged viral reduction of 2-3 logs HCV RNA after four-week treatment	http://www.santaris.com/news
4. Herpes simplex virus 1 (HSV-1)					
murine TNF-α-mRNA (proinflamma tory cytokine that participates in intraocular inflammatory disease, and most up-regulated mouse model of HSV-1 induced acute retinal necrosis	AS-ON	von Szily model: right anterior chamber in BALB/c mice inoculated with HSV-1 virus (KOS strain)	— (subconjunctival injection)	TNF-α expression and the incidence and severity of retinitis were reduced on day 8 postinfection in the TNF-α ASON-treated eyes	[353]
			— (intravitreal injection)	decreased expression of TNF-α in the eye, and reduced incidence and severity of retinitis on day 10 after infection could be found after treatment of eyes with ASON	[354]

[a]Hsp60, heat sock protein 60; HBV pol, hepatitis B virus polymerase gene; ASGPR1, asialoglycoprotein receptor 1; miR, micro RNA; IL-4, Interleukin-4; TNF, tumor necrosis factor.

[b]ASO, antisense oligonucleotide; AS-ODN, antisense oligodeoxynucleotide; PS-ODN, phosphotioates oligodeoxynucleotide; 2'-OMe, 2'-O-methylnucleosides; PNA, peptide nucleic acid; LNA, locked nucleic acid.

[c]DHBV, duck hepatitis B virus. DHBV shares a common genomic size, organization, and mode of replication to the human HBV, and chronic infection can be readily induced by neonatal infection; DHBV infection is a useful *in vivo* model for the screening of antiviral drugs for HBV; The 2.2.15 cell is the hepatoblastoma cell line HepG2 transfected with cloned HBV DNA; Huh-7.5 cells, a subline derived from Huh-7 hepatoma cells that are highly permissive for the initiation of HCV replication; SHIV, siman human immunodeficiency virus; HSV-1, herpes simplex virus 1; HAART, highly active antiretroviral therapy.

[d]"—", no delivery method used.

CONCLUDING REMARKS

The human race has been susceptible to a number of disease-causing bacteria and viruses since times immemorial. Over the years many technologies have been proposed and successfully implemented to contain them. The wining antibiotic pipeline combined with the severe situation of inefficient responding to newly-emerging and re-emerging pathogenic bacteria and viruses that may be resistant or otherwise insensitive to available drugs in recent years, has prompted the call for enhancing industrial ability to rapidly develop and produce therapeutics. Highly specific ASOs can be quickly designed, synthesized and delivered in response to such a need in medical crisis or biodefense preparations. In 2002 when a lethal outbreak of WNV in Humboldt penguins occurred at the Milwaukee County Zoo [355], PMOs were designed, synthesized and delivered for treatment within 7 days. All 3 PMO-treated penguins survived, while 7 of 8 untreated penguins died. In 2004, anti-EBOV morpholino-based oligomers were designed, synthesized, and shipped to a biocontainment medical facility as a contingency therapeutic within 7 days of a suspected accidental laboratory exposure event [356]. These situations required only a relatively small-scale synthesis of ASOs; obstacles associated with large-scale production technologies will need to be overcome to produce the large quantities of therapeutics needed for emergency mass treatment during outbreaks involving large numbers of patients or potentially infected individuals.

Based on preclinical studies, antisense-based antibacterial and antiviral compounds appear to hold a great deal of promise and will increasingly find their way through the FDA approval process and into the clinic. However, we expect the efficacy of these drug candidates to improve as ASO design rules become even better understood and entire bacterial or viral genomes are systematically screened for appropriate targets. Further, delivery remains a significant challenge, as mentioned that the only FDA-approved ASO drug is administered locally. CPP-mediated delivery of ASOs after systemic administration is an important advance, because such delivery methods are adaptable to overcome both bacterial and viral cellular uptake problems and have the potential to increase the efficacy of many ASO compounds. The development of more simplified mimics of PASO, and establishment of systemic and postexposure treatment regimens will be critical for the successful and widespread use of these compounds.

The importance of ASO-based therapeutics has been underscored in eukaryotic cell biology. Antibacterial and antiviral ASO strategies co-opt existing mechanisms and may therefore be more efficient than other nucleic acid–based therapeutics (*e.g.,* RNAi, aptamers, and DNAzymeS). However, we should take into account that bacteria as well as viruses and their host cells have evolved for millions of years, thereby bacteria and viruses have developed unique mechanisms (*i.e.,* more strigent cell walls or genetic alterations) to escape from ASOs. Thus, a better understanding of natural pathogen-ASO interactions is key for the development of an effective ASO-based anti-pathogen drug. Improvement of the efficacy, specificity and delivery of ASOs is a critical first step, followed by a detailed screen for unwanted side effects. It is possible that more potent antibacterials and antivirals can be constructed by designing steric-blocking ASOs [25]. Not all bacterial targets are equally suitable or effectively inhibited by ASOs, and it will be important to identify the bacterial Achilles heel for ASO attack with broad antibacterial spectrum. Besides, given the immense interest in ASO as an antiviral therapeutic modality, the coming years are likely to see an increasing range of clinical applications of ASO as an advanced and simplified system with integrated functionality. The realization of the potential of ASO-based therapies to address human pathogens suggests that this field has a very promising future.

ACKNOWLEDGEMENTS

This work was supported by grants from National Fund for Natural Science, China (Nos. 30973666, 81102419, 30600531, 30625041 and 81301477) and a grant from the Important National Science and Technology Specific Projects for Infectious Diseases (2008ZX10002-011).

CONFLICT OF INTEREST

The authors declare no conflict of interest.

REFERENCES

[1]   Paterson BM, Roberts BE, Kuff EL. Structural gene identification and mapping by DNA-mRNA hybrid-arrested cell-free translation. Proc Natl Acad Sci U S A 1977; 74(10): 4370-4.
[2]   Van AA. Oligonucleotides as antivirals: dream or realistic perspective? Antiviral Res 2006; 71(2-3): 307-16.

[3]   Mishra S, Kim S, Lee DK. Recent patents on nucleic acid-based antiviral therapeutics. Recent Pat Antiinfect Drug Discov 2010; 5(3): 255-71.

[4]   Zamecnik PC, Stephenson ML. Inhibition of Rous sarcoma virus replication and cell transformation by a specific oligodeoxynucleotide. Proc Natl Acad Sci U S A 1978; 75(1): 280-4.

[5]   Rahman MA, Summerton J, Foster E *et al.* Antibacterial activity and inhibition of protein synthesis in Escherichia coli by antisense DNA analogs. Antisense Res Dev 1991; 1(4): 319-27.

[6]   Bai H, Xue X, Hou Z, Zhou Y, Meng J, Luo X. Antisense antibiotics: a brief review of novel target discovery and delivery. Curr Drug Discov Technol 2010; 7(2): 76-85.

[7]   Haasnoot J, Westerhout EM, Berkhout B. RNA interference against viruses: strike and counterstrike. Nat Biotechnol 2007; 25(12): 1435-43.

[8]   Lopez-Fraga M, Wright N, Jimenez A. RNA interference-based therapeutics: new strategies to fight infectious disease. Infect Disord Drug Targets 2008; 8(4): 262-73.

[9]   Haasnoot J, Berkhout B. Nucleic acids-based therapeutics in the battle against pathogenic viruses. Handb Exp Pharmacol 2009; 189): 243-63.

[10]   Chan JH, Lim S, Wong WS. Antisense oligonucleotides: from design to therapeutic application. Clin Exp Pharmacol Physiol 2006; 33(5-6): 533-40.

[11]   Bennett CF, Swayze EE. RNA targeting therapeutics: molecular mechanisms of antisense oligonucleotides as a therapeutic platform. Annu Rev Pharmacol Toxicol 2010; 50:259-93.

[12]   Manoharan M. 2'-carbohydrate modifications in antisense oligonucleotide therapy: importance of conformation, configuration and conjugation. Biochim Biophys Acta 1999; 1489(1): 117-30.

[13]   Wilds CJ, Damha MJ. 2'-Deoxy-2'-fluoro-beta-D-arabinonucleosides and oligonucleotides (2'F-ANA): synthesis and physicochemical studies. Nucleic Acids Res 2000; 28(18): 3625-35.

[14]   Verbeure B, Lescrinier E, Wang J, Herdewijn P. RNase H mediated cleavage of RNA by cyclohexene nucleic acid (CeNA). Nucleic Acids Res 2001; 29(24): 4941-7.

[15]   Monia BP, Lesnik EA, Gonzalez C *et al.* Evaluation of 2'-modified oligonucleotides containing 2'-deoxy gaps as antisense inhibitors of gene expression. J Biol Chem 1993; 268(19): 14514-22.

[16]   Deere J, Iversen P, Geller BL. Antisense phosphorodiamidate morpholino oligomer length and target position effects on gene-specific inhibition in Escherichia coli. Antimicrob Agents Chemother 2005; 49(1): 249-55.

[17]   Nielsen PE. Peptide nucleic acids as antibacterial agents *via* the antisense principle. Expert Opin Investig Drugs 2001; 10(2): 331-41.

[18]   Pooga M, Land T, Bartfai T, Langel U. PNA oligomers as tools for specific modulation of gene expression. Biomol Eng 2001; 17(6): 183-92.

[19]   Nielsen PE. PNA Technology. Mol Biotechnol 2004; 26(3): 233-48.

[20]   Alter J, Lou F, Rabinowitz A *et al.* Systemic delivery of morpholino oligonucleotide restores dystrophin expression bodywide and improves dystrophic pathology. Nat Med 2006; 12(2): 175-77.

[21]   Koshkin AA, Wengel J. Synthesis of Novel 2',3'-Linked Bicyclic Thymine Ribonucleosides. J Org Chem 1998; 63(8): 2778-81.

[22]   Kurreck J, Wyszko E, Gillen C, Erdmann VA. Design of antisense oligonucleotides stabilized by locked nucleic acids. Nucleic Acids Res 2002; 30(9): 1911-8.

[23]   Wahlestedt C, Salmi P, Good L *et al.* Potent and nontoxic antisense oligonucleotides containing locked nucleic acids. Proc Natl Acad Sci U S A 2000; 97(10): 5633-8.

[24]   Swayze EE, Siwkowski AM, Wancewicz EV *et al*. Antisense oligonucleotides containing locked nucleic acid improve potency but cause significant hepatotoxicity in animals. Nucleic Acids Res 2007; 35(2): 687-700.

[25]   Kole R, Krainer AR, Altman S. RNA therapeutics: beyond RNA interference and antisense oligonucleotides. Nat Rev Drug Discov 2012; 11(2): 125-40.

[26]   Geller BL. Antibacterial antisense. Curr Opin Mol Ther 2005; 7(2): 109-13.

[27]   Rasmussen LC, Sperling-Petersen HU, Mortensen KK. Hitting bacteria at the heart of the central dogma: sequence-specific inhibition. Microb Cell Fact 2007; 6:24.

[28]   Wright GD. Making sense of antisense in antibiotic drug discovery. Cell Host Microbe 2009; 6(3): 197-8.

[29]   Nielsen PE. Peptide nucleic acid targeting of double-stranded DNA. Methods Enzymol 2001; 340:329-40.

[30]   Shao Y, Wu Y, Chan CY, McDonough K, Ding Y. Rational design and rapid screening of antisense oligonucleotides for prokaryotic gene modulation. Nucleic Acids Res 2006; 34(19): 5660-9.

[31]   Ding Y, Lawrence CE. A statistical sampling algorithm for RNA secondary structure prediction. Nucleic Acids Res 2003; 31(24): 7280-301.

[32]   Dryselius R, Aswasti SK, Rajarao GK, Nielsen PE, Good L. The translation start codon region is sensitive to antisense PNA inhibition in Escherichia coli. Oligonucleotides 2003; 13(6): 427-33.

[33]   Deere J, Iversen P, Geller BL. Antisense phosphorodiamidate morpholino oligomer length and target position effects on gene-specific inhibition in Escherichia coli. Antimicrob Agents Chemother 2005; 49(1): 249-55.

[34]   Good L, Awasthi SK, Dryselius R, Larsson O, Nielsen PE. Bactericidal antisense effects of peptide-PNA conjugates. Nat Biotechnol 2001; 19(4): 360-4.

[35]   Xue-Wen H, Jie P, Xian-Yuan A, Hong-Xiang Z. Inhibition of bacterial translation and growth by peptide nucleic acids targeted to domain II of 23S rRNA. J Pept Sci 2007; 13(4): 220-6.

[36]   Good L, Nielsen PE. Antisense inhibition of gene expression in bacteria by PNA targeted to mRNA. Nat Biotechnol 1998; 16(4): 355-8.

[37]   Shen N, Ko JH, Xiao G *et al*. Inactivation of expression of several genes in a variety of bacterial species by EGS technology. Proc Natl Acad Sci U S A 2009; 106(20): 8163-8.

[38]   Harth G, Horwitz MA, Tabatadze D, Zamecnik PC. Targeting the Mycobacterium tuberculosis 30/32-kDa mycolyl transferase complex as a therapeutic strategy against tuberculosis: Proof of principle by using antisense technology. Proc Natl Acad Sci U S A 2002; 99(24): 15614-19.

[39]   Harth G, Zamecnik PC, Tabatadze D, Pierson K, Horwitz MA. Hairpin extensions enhance the efficacy of mycolyl transferase-specific antisense oligonucleotides targeting Mycobacterium tuberculosis. Proc Natl Acad Sci U S A 2007; 104(17): 7199-204.

[40]   Harth G, Zamecnik PC, Tang JY, Tabatadze D, Horwitz MA. Treatment of Mycobacterium tuberculosis with antisense oligonucleotides to glutamine synthetase mRNA inhibits glutamine synthetase activity, formation of the poly-L-glutamate/glutamine cell wall structure, and bacterial replication. Proc Natl Acad Sci U S A 2000; 97(1): 418-23.

[41]   Kulyte A, Nekhotiaeva N, Awasthi SK, Good L. Inhibition of Mycobacterium smegmatis gene expression and growth using antisense peptide nucleic acids. J Mol Microbiol Biotechnol 2005; 9(2): 101-9.

[42] Shimizu T, Sato K, Sano C, Sano K, Tomioka H. [Effects of antisense oligo DNA on the antimicrobial activity of reactive oxygen intermediates and antimycobacterial agents against Mycobacterium avium complex]. Kekkaku 2003; 78(1): 33-5.

[43] Kurupati P, Tan KS, Kumarasinghe G, Poh CL. Inhibition of gene expression and growth by antisense peptide nucleic acids in a multiresistant beta-lactamase-producing Klebsiella pneumoniae strain. Antimicrob Agents Chemother 2007; 51(3): 805-11.

[44] Stock RP, Olvera A, Scarfi S *et al.* Inhibition of neomycin phosphorotransferase expression in Entamoeba histolytica with antisense peptide nucleic acid (PNA) oligomers. Arch Med Res 2000; 31(4 Suppl): S271-2.

[45] Stock RP, Olvera A, Sanchez R *et al.* Inhibition of gene expression in Entamoeba histolytica with antisense peptide nucleic acid oligomers. Nat Biotechnol 2001; 19(3): 231-4.

[46] Guo QY, Xiao G, Li R, Guan SM, Zhu XL, Wu JZ. Treatment of Streptococcus mutans with antisense oligodeoxyribonucleotides to gtfB mRNA inhibits GtfB expression and function. FEMS Microbiol Lett 2006; 264(1): 8-14.

[47] Nekhotiaeva N, Awasthi SK, Nielsen PE, Good L. Inhibition of Staphylococcus aureus gene expression and growth using antisense peptide nucleic acids. Mol Ther 2004; 10(4): 652-9.

[48] Ji Y, Yin D, Fox B, Holmes DJ, Payne D, Rosenberg M. Validation of antibacterial mechanism of action using regulated antisense RNA expression in Staphylococcus aureus. FEMS Microbiol Lett 2004; 231(2): 177-84.

[49] Deere J, Iversen P, Geller BL. Antisense phosphorodiamidate morpholino oligomer length and target position effects on gene-specific inhibition in Escherichia coli. Antimicrob Agents Chemother 2005; 49(1): 249-255.

[50] Tilley LD, Mellbye BL, Puckett SE, Iversen PL, Geller BL. Antisense peptide-phosphorodiamidate morpholino oligomer conjugate: dose-response in mice infected with Escherichia coli. J Antimicrob Chemother 2007; 59(1): 66-73.

[51] Mellbye BL, Puckett SE, Tilley LD, Iversen PL, Geller BL. Variations in amino acid composition of antisense peptide-phosphorodiamidate morpholino oligomer affect potency against Escherichia coli *in vitro* and *in vivo*. Antimicrob Agents Chemother 2009; 53(2): 525-30.

[52] Mellbye BL, Weller DD, Hassinger JN *et al.* Cationic phosphorodiamidate morpholino oligomers efficiently prevent growth of Escherichia coli *in vitro* and *in vivo*. J Antimicrob Chemother 2010; 65(1): 98-106.

[53] Geller BL, Deere JD, Stein DA, Kroeker AD, Moulton HM, Iversen PL. Inhibition of gene expression in Escherichia coli by antisense phosphorodiamidate morpholino oligomers. Antimicrob Agents Chemother 2003; 47(10): 3233-9.

[54] Tan XX, Actor JK, Chen Y. Peptide nucleic acid antisense oligomer as a therapeutic strategy against bacterial infection: proof of principle using mouse intraperitoneal infection. Antimicrob Agents Chemother 2005; 49(8): 3203-7.

[55] Greenberg DE, Marshall-Batty KR, Brinster LR *et al.* Antisense phosphorodiamidate morpholino oligomers targeted to an essential gene inhibit Burkholderia cepacia complex. J Infect Dis 2010; 201(12): 1822-30.

[56] Tilley LD, Hine OS, Kellogg JA *et al.* Gene-specific effects of antisense phosphorodiamidate morpholino oligomer-peptide conjugates on Escherichia coli and Salmonella enterica serovar typhimurium in pure culture and in tissue culture. Antimicrob Agents Chemother 2006; 50(8): 2789-96.

[57]   Mitev GM, Mellbye BL, Iversen PL, Geller BL. Inhibition of intracellular growth of Salmonella enterica serovar Typhimurium in tissue culture by antisense peptide-phosphorodiamidate morpholino oligomer. Antimicrob Agents Chemother 2009; 53(9): 3700-4.

[58]   McKinney J, Guerrier-Takada C, Wesolowski D, Altman S. Inhibition of Escherichia coli viability by external guide sequences complementary to two essential genes. Proc Natl Acad Sci U S A 2001; 98(12): 6605-10.

[59]   Wesolowski D, Tae HS, Gandotra N, Llopis P, Shen N, Altman S. Basic peptide-morpholino oligomer conjugate that is very effective in killing bacteria by gene-specific and nonspecific modes. Proc Natl Acad Sci U S A 2011; 108(40): 16582-7.

[60]   Bai H, Sang G, You Y *et al.* Targeting RNA Polymerase Primary sigma as a Therapeutic Strategy against Methicillin-Resistant Staphylococcus aureus by Antisense Peptide Nucleic Acid. PLoS One 2012; 7(1): e29886.

[61]   Bai H, You Y, Yan H *et al.* Antisense inhibition of gene expression and growth in gram-negative bacteria by cell-penetrating peptide conjugates of peptide nucleic acids targeted to rpoD gene. Biomaterials 2012; 33(2): 659-67.

[62]   Good L, Nielsen PE. Inhibition of translation and bacterial growth by peptide nucleic acid targeted to ribosomal RNA. Proc Natl Acad Sci U S A 1998; 95(5): 2073-6.

[63]   Good L, Awasthi SK, Dryselius R, Larsson O, Nielsen PE. Bactericidal antisense effects of peptide-PNA conjugates. Nat Biotechnol 2001; 19(4): 360-4.

[64]   Hatamoto M, Nakai K, Ohashi A, Imachi H. Sequence-specific bacterial growth inhibition by peptide nucleic acid targeted to the mRNA binding site of 16S rRNA. Appl Microbiol Biotechnol 2009; 84(6): 1161-8.

[65]   Gruegelsiepe H, Brandt O, Hartmann RK. Antisense inhibition of RNase P: mechanistic aspects and application to live bacteria. J Biol Chem 2006; 281(41): 30613-20.

[66]   Nikravesh A, Dryselius R, Faridani OR *et al.* Antisense PNA accumulates in Escherichia coli and mediates a long post-antibiotic effect. Mol Ther 2007; 15(8): 1537-42.

[67]   Geller BL, Deere J, Tilley L, Iversen PL. Antisense phosphorodiamidate morpholino oligomer inhibits viability of Escherichia coli in pure culture and in mouse peritonitis. J Antimicrob Chemother 2005; 55(6): 983-8.

[68]   Tilley LD, Mellbye BL, Puckett SE, Iversen PL, Geller BL. Antisense peptide-phosphorodiamidate morpholino oligomer conjugate: dose-response in mice infected with Escherichia coli. J Antimicrob Chemother 2007; 59(1): 66-73.

[69]   Mellbye BL, Puckett SE, Tilley LD, Iversen PL, Geller BL. Variations in amino acid composition of antisense peptide-phosphorodiamidate morpholino oligomer affect potency against Escherichia coli *in vitro* and *in vivo*. Antimicrob Agents Chemother 2009; 53(2): 525-30.

[70]   Mellbye BL, Weller DD, Hassinger JN *et al.* Cationic phosphorodiamidate morpholino oligomers efficiently prevent growth of Escherichia coli *in vitro* and *in vivo*. J Antimicrob Chemother 2010; 65(1): 98-106.

[71]   Greenberg DE, Marshall-Batty KR, Brinster LR *et al.* Antisense phosphorodiamidate morpholino oligomers targeted to an essential gene inhibit Burkholderia cepacia complex. J Infect Dis 2010; 201(12): 1822-30.

[72]   Hatamoto M, Ohashi A, Imachi H. Peptide nucleic acids (PNAs) antisense effect to bacterial growth and their application potentiality in biotechnology. Appl Microbiol Biotechnol 2010; 86(2): 397-402.

[73] Kulyte A, Nekhotiaeva N, Awasthi SK, Good L. Inhibition of Mycobacterium smegmatis gene expression and growth using antisense peptide nucleic acids. J Mol Microbiol Biotechnol 2005; 9(2): 101-9.

[74] Guerrier-Takada C, Salavati R, Altman S. Phenotypic conversion of drug-resistant bacteria to drug sensitivity. Proc Natl Acad Sci U S A 1997; 94(16): 8468-72.

[75] White DG, Maneewannakul K, von HE *et al.* Inhibition of the multiple antibiotic resistance (mar) operon in Escherichia coli by antisense DNA analogs. Antimicrob Agents Chemother 1997; 41(12): 2699-704.

[76] Torres VC, Tsiodras S, Gold HS *et al.* Restoration of vancomycin susceptibility in Enterococcus faecalis by antiresistance determinant gene transfer. Antimicrob Agents Chemother 2001; 45(3): 973-5.

[77] Chen H, Ferbeyre G, Cedergren R. Efficient hammerhead ribozyme and antisense RNA targeting in a slow ribosome Escherichia coli mutant. Nat Biotechnol 1997; 15(5): 432-5.

[78] Gao MY, Xu CR, Chen R, Liu SG, Feng JN. Chloromycetin resistance of clinically isolated E coli is conversed by using EGS technique to repress the chloromycetin acetyl transferase. World J Gastroenterol 2005; 11(46): 7368-73.

[79] Sarno R, Ha H, Weinsetel N, Tolmasky ME. Inhibition of aminoglycoside 6'-N-acetyltransferase type Ib-mediated amikacin resistance by antisense oligodeoxynucleotides. Antimicrob Agents Chemother 2003; 47(10): 3296-3304.

[80] Soler Bistue AJ, Ha H, Sarno R, Don M, Zorreguieta A, Tolmasky ME. External guide sequences targeting the aac(6')-Ib mRNA induce inhibition of amikacin resistance. Antimicrob Agents Chemother 2007; 51(6): 1918-25.

[81] Kedar GC, Brown-Driver V, Reyes DR *et al.* Evaluation of the metS and murB loci for antibiotic discovery using targeted antisense RNA expression analysis in Bacillus anthracis. Antimicrob Agents Chemother 2007; 51(5): 1708-18.

[82] Meng J, Hu B, Liu J *et al.* Restoration of oxacillin susceptibility in methicillin-resistant Staphylococcus aureus by blocking the MecR1-mediated signaling pathway. J Chemother 2006; 18(4): 360-5.

[83] Meng J, Wang H, Hou Z *et al.* Novel anion liposome-encapsulated antisense oligonucleotide restores susceptibility of methicillin-resistant Staphylococcus aureus and rescues mice from lethal sepsis by targeting mecA. Antimicrob Agents Chemother 2009; 53(7): 2871-8.

[84] Jeon B, Zhang Q. Sensitization of Campylobacter jejuni to fluoroquinolone and macrolide antibiotics by antisense inhibition of the CmeABC multidrug efflux transporter. J Antimicrob Chemother 2009; 63(5): 946-8.

[85] Wang H, Meng J, Jia M *et al.* oprM as a new target for reversion of multidrug resistance in Pseudomonas aeruginosa by antisense phosphorothioate oligodeoxynucleotides. FEMS Immunol Med Microbiol 2010; 60(3): 275-82.

[86] Garcia-Chaumont C, Seksek O, Grzybowska J, Borowski E, Bolard J. Delivery systems for antisense oligonucleotides. Pharmacology & Therapeutics 2008; 87(2-3): 255-77.

[87] Good L, Sandberg R, Larsson O, Nielsen PE, Wahlestedt C. Antisense PNA effects in Escherichia coli are limited by the outer-membrane LPS layer. Microbiology 2000; 146 (Pt 10):2665-70.

[88] Nekhotiaeva N, Elmquist A, Rajarao GK, Hallbrink M, Langel U, Good L. Cell entry and antimicrobial properties of eukaryotic cell-penetrating peptides. FASEB J 2004; 18(2): 394-6.

[89]   Lebleu B, Moulton HM, Abes R *et al*. Cell penetrating peptide conjugates of steric block oligonucleotides. Adv Drug Deliv Rev 2008; 60(4-5): 517-29.

[90]   Henke E, Perk J, Vider J *et al*. Peptide-conjugated antisense oligonucleotides for targeted inhibition of a transcriptional regulator *in vivo*. Nat Biotechnol 2008; 26(1): 91-100.

[91]   Heitz F, Morris MC, Divita G. Twenty years of cell-penetrating peptides: from molecular mechanisms to therapeutics. Br J Pharmacol 2009; 157(2): 195-206.

[92]   Thompson AJV, Patel K. Antisense Inhibitors, Ribozymes, and siRNAs. Clinics in Liver Disease 2009; 13(3): 375-90.

[93]   Venkatesan N, Kim BH. Peptide conjugates of oligonucleotides: synthesis and applications. Chem Rev 2006; 106(9): 3712-61.

[94]   Alksne LE, Projan SJ. Bacterial virulence as a target for antimicrobial chemotherapy. Current Opinion in Biotechnology 2000; 11(6): 625-36.

[95]   Amantana A, Moulton HM, Cate ML *et al*. Pharmacokinetics, biodistribution, stability and toxicity of a cell-penetrating peptide-morpholino oligomer conjugate. Bioconjug Chem 2007; 18(4): 1325-31.

[96]   Wu RP, Youngblood DS, Hassinger JN *et al*. Cell-penetrating peptides as transporters for morpholino oligomers: effects of amino acid composition on intracellular delivery and cytotoxicity. Nucleic Acids Res 2007; 35(15): 5182-91.

[97]   Youngblood DS, Hatlevig SA, Hassinger JN, Iversen PL, Moulton HM. Stability of cell-penetrating peptide-morpholino oligomer conjugates in human serum and in cells. Bioconjug Chem 2007; 18(1): 50-60.

[98]   Amantana A, Moulton HM, Cate ML *et al*. Pharmacokinetics, biodistribution, stability and toxicity of a cell-penetrating peptide-morpholino oligomer conjugate. Bioconjug Chem 2007; 18(4): 1325-31.

[99]   Zorko M, Langel U. Cell-penetrating peptides: mechanism and kinetics of cargo delivery. Adv Drug Deliv Rev 2005; 57(4): 529-545.

[100]   Boberek JM, Stach J, Good L. Genetic Evidence for Inhibition of Bacterial Division Protein FtsZ by Berberine. PLoS One 2010; 5(10): e13745.

[101]   Bai H, Zhou Y, Hou Z, Xue X, Meng J, Luo X. Targeting bacterial RNA polymerase: promises for future antisense antibiotics development. Infect Disord Drug Targets 2011; 11(2): 175-87.

[102]   Stephenson ML, Zamecnik PC. Inhibition of Rous sarcoma viral RNA translation by a specific oligodeoxyribonucleotide. Proc Natl Acad Sci U S A 1978; 75(1): 285-8.

[103]   Field AK. Viral targets for antisense oligonucleotides: a mini review. Antiviral Res 1998; 37(2): 67-81.

[104]   Kilkuskie RE, Field AK. Antisense inhibition of virus infections. Adv Pharmacol 1997; 40:437-83.

[105]   Crooke ST, Bennett CF. Progress in antisense oligonucleotide therapeutics. Annu Rev Pharmacol Toxicol 1996; 36:107-29.

[106]   Verheijen JC, van der Marel GA, van Boom JH, Bayly SF, Player MR, Torrence PF. 2,5-oligoadenylate-peptide nucleic acids (2-5A-PNAs) activate RNase L. Bioorg Med Chem 1999; 7(3): 449-55.

[107]   Zhou L, Civitello ER, Gupta N *et al*. Endowing RNase H-inactive antisense with catalytic activity: 2-5A-morphants. Bioconjug Chem 2005; 16(2): 383-90.

[108]   Stein DA. Inhibition of RNA virus infections with peptide-conjugated morpholino oligomers. Curr Pharm Des 2008; 14(25): 2619-34.

[109] Arora V, Devi GR, Iversen PL. Neutrally charged phosphorodiamidate morpholino antisense oligomers: uptake, efficacy and pharmacokinetics. Curr Pharm Biotechnol 2004; 5(5): 431-9.

[110] Stein DA, Skilling DE, Iversen PL, Smith AW. Inhibition of Vesivirus infections in mammalian tissue culture with antisense morpholino oligomers. Antisense Nucleic Acid Drug Dev 2001; 11(5): 317-25.

[111] Neuman BW, Stein DA, Kroeker AD *et al.* Antisense morpholino-oligomers directed against the 5' end of the genome inhibit coronavirus proliferation and growth. J Virol 2004; 78(11): 5891-9.

[112] Deas TS, Binduga-Gajewska I, Tilgner M *et al.* Inhibition of flavivirus infections by antisense oligomers specifically suppressing viral translation and RNA replication. J Virol 2005; 79(8): 4599-609.

[113] Swenson DL, Warfield KL, Warren TK *et al.* Chemical modifications of antisense morpholino oligomers enhance their efficacy against Ebola virus infection. Antimicrob Agents Chemother 2009; 53(5): 2089-99.

[114] Warren TK, Warfield KL, Wells J *et al.* Advanced antisense therapies for postexposure protection against lethal filovirus infections. Nat Med 2010; 16(9): 991-4.

[115] Goodarzi G, Gross SC, Tewari A, Watabe K. Antisense oligodeoxyribonucleotides inhibit the expression of the gene for hepatitis B virus surface antigen. J Gen Virol 1990; 71 (Pt 12):3021-5.

[116] Offensperger WB, Offensperger S, Walter E *et al. In vivo* inhibition of duck hepatitis B virus replication and gene expression by phosphorothioate modified antisense oligodeoxynucleotides. EMBO J 1993; 12(3): 1257-62.

[117] Soni PN, Brown D, Saffie R *et al.* Biodistribution, stability, and antiviral efficacy of liposome-entrapped phosphorothioate antisense oligodeoxynucleotides in ducks for the treatment of chronic duck hepatitis B virus infection. Hepatology 1998; 28(5): 1402-10.

[118] Korba BE, Gerin JL. Antisense oligonucleotides are effective inhibitors of hepatitis B virus replication *in vitro*. Antiviral Res 1995; 28(3): 225-42.

[119] Nash KL, Alexander GJ, Lever AM. Inhibition of hepatitis B virus by lentiviral vector delivered antisense RNA and hammerhead ribozymes. J Viral Hepat 2005; 12(4): 346-56.

[120] Deng YB, Nong LG, Huang W, Pang GG, Wang YF. [Inhibition of hepatitis B virus (HBV) replication using antisense LNA targeting to both S and C genes in HBV]. Zhonghua Gan Zang Bing Za Zhi 2009; 17(12): 900-4.

[121] He L, Wu X, Chen C. [Antisense oligodeoxynucleotide targeted at the pre-s region of DHBV could inhibit virus replication *in vivo*]. Zhonghua Shi Yan He Lin Chuang Bing Du Xue Za Zhi 1998; 12(1): 1-4.

[122] Madon J, Blum HE. Receptor-mediated delivery of hepatitis B virus DNA and antisense oligodeoxynucleotides to avian liver cells. Hepatology 1996; 24(3): 474-81.

[123] Zhoug S, Wen SM, Zhang DF, Wang QL, Wang SQ, Ren H. Sequencing of PCR amplified HBV DNA pre-c and c regions in the 2.2.15 cells and antiviral action by targeted antisense oligonucleotide directed against sequence. World J Gastroenterol 1998; 4(5): 434-6.

[124] Zhong S, Zheng SJ, Chen F *et al.* [*In vivo* inhibition of hepatitis B virus replication and gene expression by targeted phosphorothioate modified antisense oligodeoxynucleotides]. Zhonghua Gan Zang Bing Za Zhi 2002; 10(4): 283-6.

[125] Xin W, Wang JH. Treatment of duck hepatitis B virus by antisense poly-2'-O-(2,4-dinitrophenyl)-oligoribonucleotides. Antisense Nucleic Acid Drug Dev 1998; 8(6): 459-68.

[126] zu PJ, Wieland S, Blum HE, Wands JR. Antisense RNA complementary to hepatitis B virus specifically inhibits viral replication. Gastroenterology 1998; 115(3): 702-13.

[127] Yang L, Yao J, Deng L. [Inhibitory effect of phosphorothioae oligodeoxynucleotides on HBV replication and synthesis of antigen *in vitro*]. Zhonghua Yi Xue Za Zhi 1999; 79(11): 857-9.

[128] Liu S, Sun W, Cao Y. Study on anti-HBV effects by antisense oligodeoxynucleotides *in vitro*. Zhonghua Yu Fang Yi Xue Za Zhi 2001; 35(5): 338-40.

[129] Matsukura M, Koike K, Zon G. Antisense phosphorothioates as antivirals against human immunodeficiency virus (HIV) and hepatitis B virus (HBV). Toxicol Lett 1995; 82-83:435-8.

[130] Zhao W, Chen H, Peng ZY, Li WG, Xi HL, Xu XY. [Antiviral effects of dual-target antisense rna: an experimental study with hepatitis B virus transgenic mice]. Zhonghua Yi Xue Za Zhi 2005; 85(49): 3486-90.

[131] Robaczewska M, Narayan R, Seigneres B *et al*. Sequence-specific inhibition of duck hepatitis B virus reverse transcription by peptide nucleic acids (PNA). J Hepatol 2005; 42(2): 180-7.

[132] Smith CC, Aurelian L, Reddy MP, Miller PS, Ts'o PO. Antiviral effect of an oligo(nucleoside methylphosphonate) complementary to the splice junction of herpes simplex virus type 1 immediate early pre-mRNAs 4 and 5. Proc Natl Acad Sci U S A 1986; 83(9): 2787-91.

[133] Kulka M, Smith CC, Aurelian L *et al*. Site specificity of the inhibitory effects of oligo(nucleoside methylphosphonate)s complementary to the acceptor splice junction of herpes simplex virus type 1 immediate early mRNA 4. Proc Natl Acad Sci U S A 1989; 86(18): 6868-72.

[134] Kulka M, Wachsman M, Miura S *et al*. Antiviral effect of oligo(nucleoside methylphosphonates) complementary to the herpes simplex virus type 1 immediate early mRNAs 4 and 5. Antiviral Res 1993; 20(2): 115-30.

[135] Kean JM, Kipp SA, Miller PS, Kulka M, Aurelian L. Inhibition of herpes simplex virus replication by antisense oligo-2'-O-methylribonucleoside methylphosphonates. Biochemistry 1995; 34(45): 14617-20.

[136] Poddevin B, Meguenni S, Elias I, Vasseur M, Blumenfeld M. Improved anti-herpes simplex virus type 1 activity of a phosphodiester antisense oligonucleotide containing a 3'-terminal hairpin-like structure. Antisense Res Dev 1994; 4(3): 147-54.

[137] Peyman A, Helsberg M, Kretzschmar G, Mag M, Grabley S, Uhlmann E. Inhibition of viral growth by antisense oligonucleotides directed against the IE110 and the UL30 mRNA of herpes simplex virus type-1. Biol Chem Hoppe Seyler 1995; 376(3): 195-8.

[138] Peyman A, Helsberg M, Kretzschmar G, Mag M, Ryte A, Uhlmann E. Nuclease stability as dominant factor in the antiviral activity of oligonucleotides directed against HSV-1 IE110. Antiviral Res 1997; 33(2): 135-9.

[139] Cantin EM, Podsakoff G, Willey DE, Openshaw H. Antiviral effects of herpes simplex virus specific anti-sense nucleic acids. Adv Exp Med Biol 1992; 312:139-49.

[140] Feng CP, Kulka M, Smith C, Aurelian L. Herpes simplex virus-mediated activation of human immunodeficiency virus is inhibited by oligonucleoside methylphosphonates that target immediate-early mRNAs 1 and 3. Antisense Nucleic Acid Drug Dev 1996; 6(1): 25-35.

[141] Shoji Y, Shimada J, Mizushima Y *et al*. Cellular uptake and biological effects of antisense oligodeoxynucleotide analogs targeted to herpes simplex virus. Antimicrob Agents Chemother 1996; 40(7): 1670-75.

[142] Shoji Y, Ishige H, Tamura N *et al.* Enhancement of anti-herpetic activity of antisense phosphorothioate oligonucleotides 5' end modified with geraniol. J Drug Target 1998; 5(4): 261-73.

[143] Zhu J, Trang P, Kim K, Zhou T, Deng H, Liu F. Effective inhibition of Rta expression and lytic replication of Kaposi's sarcoma-associated herpesvirus by human RNase P. Proc Natl Acad Sci U S A 2004; 101(24): 9073-8.

[144] Hoke GD, Draper K, Freier SM *et al.* Effects of phosphorothioate capping on antisense oligonucleotide stability, hybridization and antiviral efficacy *versus* herpes simplex virus infection. Nucleic Acids Res 1991; 19(20): 5743-8.

[145] Flores-Aguilar M, Besen G, Vuong C *et al.* Evaluation of retinal toxicity and efficacy of anti-cytomegalovirus and anti-herpes simplex virus antiviral phosphorothioate oligonucleotides ISIS 2922 and ISIS 4015. J Infect Dis 1997; 175(6): 1308-16.

[146] Moerdyk-Schauwecker M, Stein DA, Eide K *et al.* Inhibition of HSV-1 ocular infection with morpholino oligomers targeting ICP0 and ICP27. Antiviral Res 2009; 84(2): 131-41.

[147] Eide K, Moerdyk-Schauwecker M, Stein DA, Bildfell R, Koelle DM, Jin L. Reduction of herpes simplex virus type-2 replication in cell cultures and in rodent models with peptide-conjugated morpholino oligomers. Antivir Ther 2010; 15(8): 1141-9.

[148] Draper KG, Ceruzzi M, Kmetz ME, Sturzenbecker LJ. Complementary oligonucleotide sequence inhibits both Vmw65 gene expression and replication of herpes simplex virus. Antiviral Res 1990; 13(4): 151-64.

[149] Kmetz ME, Ceruzzi M, Schwartz J. Vmw65 phosphorothioate oligonucleotides inhibit HSV KOS replication and Vmw65 protein synthesis. Antiviral Res 1991; 16(2): 173-84.

[150] Aurelian L, Smith CC. Herpes simplex virus type 2 growth and latency reactivation by cocultivation are inhibited with antisense oligonucleotides complementary to the translation initiation site of the large subunit of ribonucleotide reductase (RR1). Antisense Nucleic Acid Drug Dev 2000; 10(2): 77-85.

[151] Zhang YJ, Bonaparte RS, Patel D, Stein DA, Iversen PL. Blockade of viral interleukin-6 expression of Kaposi's sarcoma-associated herpesvirus. Mol Cancer Ther 2008; 7(3): 712-20.

[152] Gao WY, Stein CA, Cohen JS, Dutschman GE, Cheng YC. Effect of phosphorothioate homo-oligodeoxynucleotides on herpes simplex virus type 2-induced DNA polymerase. J Biol Chem 1989; 264(19): 11521-6.

[153] Gao WY, Hanes RN, Vazquez-Padua MA, Stein CA, Cohen JS, Cheng YC. Inhibition of herpes simplex virus type 2 growth by phosphorothioate oligodeoxynucleotides. Antimicrob Agents Chemother 1990; 34(5): 808-12.

[154] Fennewald SM, Mustain S, Ojwang J, Rando RF. Inhibition of herpes simplex virus in culture by oligonucleotides composed entirely of deoxyguanosine and thymidine. Antiviral Res 1995; 26(1): 37-54.

[155] Pagano JS, Jimenez G, Sung NS, Raab-Traub N, Lin JC. Epstein-Barr viral latency and cell immortalization as targets for antisense oligomers. Ann N Y Acad Sci 1992; 660:107-16.

[156] Roth G, Curiel T, Lacy J. Epstein-Barr viral nuclear antigen 1 antisense oligodeoxynucleotide inhibits proliferation of Epstein-Barr virus-immortalized B cells. Blood 1994; 84(2): 582-7.

[157] Diabata M, Enzinger EM, Monroe JE, Kilkuskie RE, Field AK, Mulder C. Antisense oligodeoxynucleotides against the BZLF1 transcript inhibit induction of productive Epstein-Barr virus replication. Antiviral Res 1996; 29(2-3): 243-60.

[158] Mattia E, Chichiarelli S, Hickish T *et al*. Inhibition of *in vitro* proliferation of Epstein Barr Virus infected B cells by an antisense oligodeoxynucleotide targeted against EBV latent membrane protein LMP1. Oncogene 1997; 15(4): 489-93.

[159] Galletti R, Masciarelli S, Conti C *et al*. Inhibition of Epstein Barr Virus LMP1 gene expression in B lymphocytes by antisense oligonucleotides: uptake and efficacy of lipid-based and receptor-mediated delivery systems. Antiviral Res 2007; 74(2): 102-10.

[160] Azad RF, Driver VB, Tanaka K, Crooke RM, Anderson KP. Antiviral activity of a phosphorothioate oligonucleotide complementary to RNA of the human cytomegalovirus major immediate-early region. Antimicrob Agents Chemother 1993; 37(9): 1945-1954.

[161] Detrick B, Nagineni CN, Grillone LR, Anderson KP, Henry SP, Hooks JJ. Inhibition of human cytomegalovirus replication in a human retinal epithelial cell model by antisense oligonucleotides. Invest Ophthalmol Vis Sci 2001; 42(1): 163-9.

[162] Pari GS, Field AK, Smith JA. Potent antiviral activity of an antisense oligonucleotide complementary to the intron-exon boundary of human cytomegalovirus genes UL36 and UL37. Antimicrob Agents Chemother 1995; 39(5): 1157-61.

[163] Smith JA, Pari GS. Expression of human cytomegalovirus UL36 and UL37 genes is required for viral DNA replication. J Virol 1995; 69(3): 1925-31.

[164] Zhang Y, Katakura Y, Seto P, Shirahata S. Evidence that phosphatidylcholine-specific phospholipase C is a key molecule mediating insulin-induced enhancement of gene expression from human cytomegalovirus promoter in CHO cells. Cytotechnology 1997; 23(1-3): 193-6.

[165] Dal MP, Bessia C, Ripalti A *et al*. Stably expressed antisense RNA to cytomegalovirus UL83 inhibits viral replication. J Virol 1996; 70(4): 2086-94.

[166] Yao JX, Cui GH, Xia LH, Song SJ. [*In vitro* infection of human megakaryocyte precursors by human cytomegalovirus (HCMV) and the antiviral effect of HCMV antisense oligonucleotides]. Zhonghua Xue Ye Xue Za Zhi 2004; 25(12): 720-3.

[167] Kawamura M, Hayashi M, Furuichi T, Nonoyama M, Isogai E, Namioka S. The inhibitory effects of oligonucleotides, complementary to Marek's disease virus mRNA transcribed from the BamHI-H region, on the proliferation of transformed lymphoblastoid cells, MDCC-MSB1. J Gen Virol 1991; 72 (Pt 5):1105-11.

[168] Cowsert LM, Fox MC, Zon G, Mirabelli CK. *In vitro* evaluation of phosphorothioate oligonucleotides targeted to the E2 mRNA of papillomavirus: potential treatment for genital warts. Antimicrob Agents Chemother 1993; 37(2): 171-7.

[169] Lewis EJ, Agrawal S, Bishop J *et al*. Non-specific antiviral activity of antisense molecules targeted to the E1 region of human papillomavirus. Antiviral Res 2000; 48(3): 187-96.

[170] Roberts S, Ashmole I, Rookes SM, Gallimore PH. Mutational analysis of the human papillomavirus type 16 E1--E4 protein shows that the C terminus is dispensable for keratin cytoskeleton association but is involved in inducing disruption of the keratin filaments. J Virol 1997; 71(5): 3554-62.

[171] Cho CW, Poo H, Cho YS *et al*. HPV E6 antisense induces apoptosis in CaSki cells *via* suppression of E6 splicing. Exp Mol Med 2002; 34(2): 159-66.

[172] Clawson GA, Miranda GQ, Sivarajah A *et al*. Inhibition of papilloma progression by antisense oligonucleotides targeted to HPV11 E6/E7 RNA. Gene Ther 2004; 11(17): 1331-41.

[173] Abdulkarim B, Bourhis J. Antiviral approaches for cancers related to Epstein-Barr virus and human papillomavirus. Lancet Oncol 2001; 2(10): 622-30.

[174] Alvarez-Salas LM, Benitez-Hess ML, DiPaolo JA. Advances in the development of ribozymes and antisense oligodeoxynucleotides as antiviral agents for human papillomaviruses. Antivir Ther 2003; 8(4): 265-78.

[175] Zamecnik PC, Goodchild J, Taguchi Y, Sarin PS. Inhibition of replication and expression of human T-cell lymphotropic virus type III in cultured cells by exogenous synthetic oligonucleotides complementary to viral RNA. Proc Natl Acad Sci U S A 1986; 83(12): 4143-46.

[176] Agrawal S, Tang JY. GEM 91--an antisense oligonucleotide phosphorothioate as a therapeutic agent for AIDS. Antisense Res Dev 1992; 2(4): 261-6.

[177] Turner JJ, Fabani M, Arzumanov AA, Ivanova G, Gait MJ. Targeting the HIV-1 RNA leader sequence with synthetic oligonucleotides and siRNA: chemistry and cell delivery. Biochim Biophys Acta 2006; 1758(3): 290-300.

[178] Wei X, Gotte M, Wainberg MA. Human immunodeficiency virus type-1 reverse transcription can be inhibited *in vitro* by oligonucleotides that target both natural and synthetic tRNA primers. Nucleic Acids Res 2000; 28(16): 3065-74.

[179] Freund F, Boulme F, Michel J, Ventura M, Moreau S, Litvak S. Inhibition of HIV-1 replication *in vitro* and in human infected cells by modified antisense oligonucleotides targeting the tRNALys3/RNA initiation complex. Antisense Nucleic Acid Drug Dev 2001; 11(5): 301-15.

[180] Matsukura M, Shinozuka K, Zon G *et al.* Phosphorothioate analogs of oligodeoxynucleotides: inhibitors of replication and cytopathic effects of human immunodeficiency virus. Proc Natl Acad Sci U S A 1987; 84(21): 7706-10.

[181] Mori K, Boiziau C, Cazenave C *et al.* Phosphoroselenoate oligodeoxynucleotides: synthesis, physico-chemical characterization, anti-sense inhibitory properties and anti-HIV activity. Nucleic Acids Res 1989; 17(20): 8207-19.

[182] Agrawal S, Goodchild J, Civeira MP, Thornton AH, Sarin PS, Zamecnik PC. Oligodeoxynucleoside phosphoramidates and phosphorothioates as inhibitors of human immunodeficiency virus. Proc Natl Acad Sci U S A 1988; 85(19): 7079-83.

[183] Letsinger RL, Zhang GR, Sun DK, Ikeuchi T, Sarin PS. Cholesteryl-conjugated oligonucleotides: synthesis, properties, and activity as inhibitors of replication of human immunodeficiency virus in cell culture. Proc Natl Acad Sci U S A 1989; 86(17): 6553-6.

[184] Zaia JA, Rossi JJ, Murakawa GJ *et al.* Inhibition of human immunodeficiency virus by using an oligonucleoside methylphosphonate targeted to the tat-3 gene. J Virol 1988; 62(10): 3914-7.

[185] Sarin PS, Agrawal S, Civeira MP, Goodchild J, Ikeuchi T, Zamecnik PC. Inhibition of acquired immunodeficiency syndrome virus by oligodeoxynucleoside methylphosphonates. Proc Natl Acad Sci U S A 1988; 85(20): 7448-51.

[186] Shibahara S, Mukai S, Morisawa H, Nakashima H, Kobayashi S, Yamamoto N. Inhibition of human immunodeficiency virus (HIV-1) replication by synthetic oligo-RNA derivatives. Nucleic Acids Res 1989; 17(1): 239-52.

[187] Renneisen K, Leserman L, Matthes E, Schroder HC, Muller WE. Inhibition of expression of human immunodeficiency virus-1 *in vitro* by antibody-targeted liposomes containing antisense RNA to the env region. J Biol Chem 1990; 265(27): 16337-42.

[188] Laurence J, Sikder SK, Kulkosky J, Miller P, Ts'o PO. Induction of chronic human immunodeficiency virus infection is blocked *in vitro* by a methylphosphonate oligodeoxynucleoside targeted to a U3 enhancer element. J Virol 1991; 65(1): 213-9.

[189] Daum T, Engels JW, Mag M *et al.* Antisense oligodeoxynucleotide: inhibitor of splicing of mRNA of human immunodeficiency virus. Intervirology 1992; 33(2): 65-75.

[190] Degols G, Leonetti JP, Benkirane M, Devaux C, Lebleu B. Poly(L-lysine)-conjugated oligonucleotides promote sequence-specific inhibition of acute HIV-1 infection. Antisense Res Dev 1992; 2(4): 293-301.

[191] Kinchington D, Galpin S, Jaroszewski JW, Ghosh K, Subasinghe C, Cohen JS. A comparison of gag, pol and rev antisense oligodeoxynucleotides as inhibitors of HIV-1. Antiviral Res 1992; 17(1): 53-62.

[192] Svinarchuk FP, Konevetz DA, Pliasunova OA, Pokrovsky AG, Vlassov VV. Inhibition of HIV proliferation in MT-4 cells by antisense oligonucleotide conjugated to lipophilic groups. Biochimie 1993; 75(1-2): 49-54.

[193] Li G, Lisziewicz J, Sun D *et al.* Inhibition of Rev activity and human immunodeficiency virus type 1 replication by antisense oligodeoxynucleotide phosphorothioate analogs directed against the Rev-responsive element. J Virol 1993; 67(11): 6882-8.

[194] Tang JY, Temsamani J, Agrawal S. Self-stabilized antisense oligodeoxynucleotide phosphorothioates: properties and anti-HIV activity. Nucleic Acids Res 1993; 21(11): 2729-35.

[195] Lisziewicz J, Sun D, Metelev V, Zamecnik P, Gallo RC, Agrawal S. Long-term treatment of human immunodeficiency virus-infected cells with antisense oligonucleotide phosphorothioates. Proc Natl Acad Sci U S A 1993; 90(9): 3860-4.

[196] Weichold FF, Lisziewicz J, Zeman RA *et al.* Antisense phosphorothioate oligodeoxynucleotides alter HIV type 1 replication in cultured human macrophages and peripheral blood mononuclear cells. AIDS Res Hum Retroviruses 1995; 11(7): 863-7.

[197] Zelphati O, Zon G, Leserman L. Inhibition of HIV-1 replication in cultured cells with antisense oligonucleotides encapsulated in immunoliposomes. Antisense Res Dev 1993; 3(4): 323-38.

[198] Zelphati O, Imbach JL, Signoret N, Zon G, Rayner B, Leserman L. Antisense oligonucleotides in solution or encapsulated in immunoliposomes inhibit replication of HIV-1 by several different mechanisms. Nucleic Acids Res 1994; 22(20): 4307-14.

[199] Bongartz JP, Aubertin AM, Milhaud PG, Lebleu B. Improved biological activity of antisense oligonucleotides conjugated to a fusogenic peptide. Nucleic Acids Res 1994; 22(22): 4681-8.

[200] Boiziau C, Debart F, Rayner B, Imbach JL, Toulme JJ. Chimeric alpha-beta oligonucleotides as antisense inhibitors of reverse transcription. FEBS Lett 1995; 361(1): 41-45.

[201] Matsukura M, Koike K, Zon G. Antisense phosphorothioates as antivirals against human immunodeficiency virus (HIV) and hepatitis B virus (HBV). Toxicol Lett 1995; 82-83:435-8.

[202] Morelli D, Pozzi B, Maier JA, Menard S, Colnaghi MI, Balsari A. A monoclonal antibody extends the half-life of an anti-HIV oligodeoxynucleotide and targets it to CD4+ cells. Nucleic Acids Res 1995; 23(22): 4603-7.

[203] Anazodo MI, Salomon H, Friesen AD, Wainberg MA, Wright JA. Antiviral activity and protection of cells against human immunodeficiency virus type-1 using an antisense oligodeoxyribonucleotide phosphorothioate complementary to the 5'-LTR region of the viral genome. Gene 1995; 166(2): 227-32.

[204] Bordier B, Helene C, Barr PJ, Litvak S, Sarih-Cottin L. *In vitro* effect of antisense oligonucleotides on human immunodeficiency virus type 1 reverse transcription. Nucleic Acids Res 1992; 20(22): 5999-6006.

[205]   Bordier B, Perala-Heape M, Degols G *et al.* Sequence-specific inhibition of human immunodeficiency virus (HIV) reverse transcription by antisense oligonucleotides: comparative study in cell-free assays and in HIV-infected cells. Proc Natl Acad Sci U S A 1995; 92(20): 9383-7.

[206]   Anazodo MI, Wainberg MA, Friesen AD, Wright JA. Inhibition of human immunodeficiency virus type 1 (HIV) gag gene expression by an antisense oligodeoxynucleotide phosphorothioate. Leukemia 1995; 9 Suppl 1:S86-8.

[207]   Anazodo MI, Salomon H, Friesen AD, Wainberg MA, Wright JA. Antiviral activity and protection of cells against human immunodeficiency virus type-1 using an antisense oligodeoxyribonucleotide phosphorothioate complementary to the 5'-LTR region of the viral genome. Gene 1995; 166(2): 227-32.

[208]   Anazodo MI, Wainberg MA, Friesen AD, Wright JA. Sequence-specific inhibition of gene expression by a novel antisense oligodeoxynucleotide phosphorothioate directed against a nonregulatory region of the human immunodeficiency virus type 1 genome. J Virol 1995; 69(3): 1794-801.

[209]   Skripkin E, Paillart JC, Marquet R, Blumenfeld M, Ehresmann B, Ehresmann C. Mechanisms of inhibition of *in vitro* dimerization of HIV type I RNA by sense and antisense oligonucleotides. J Biol Chem 1996; 271(46): 28812-7.

[210]   Elmen J, Zhang HY, Zuber B *et al.* Locked nucleic acid containing antisense oligonucleotides enhance inhibition of HIV-1 genome dimerization and inhibit virus replication. FEBS Lett 2004; 578(3): 285-90.

[211]   Selvam MP, Buck SM, Blay RA, Mayner RE, Mied PA, Epstein JS. Inhibition of HIV replication by immunoliposomal antisense oligonucleotide. Antiviral Res 1996; 33(1): 11-20.

[212]   Lavigne C, Thierry AR. Enhanced antisense inhibition of human immunodeficiency virus type 1 in cell cultures by DLS delivery system. Biochem Biophys Res Commun 1997; 237(3): 566-71.

[213]   Boulme F, Perala-Heape M, Sarih-Cottin L, Litvak S. Specific inhibition of *in vitro* reverse transcription using antisense oligonucleotides targeted to the TAR regions of HIV-1 and HIV-2. Biochim Biophys Acta 1997; 1351(3): 249-55.

[214]   Lee R, Kaushik N, Modak MJ, Vinayak R, Pandey VN. Polyamide nucleic acid targeted to the primer binding site of the HIV-1 RNA genome blocks *in vitro* HIV-1 reverse transcription. Biochemistry 1998; 37(3): 900-10.

[215]   Duzgunes N, Pretzer E, Simoes S *et al.* Liposome-mediated delivery of antiviral agents to human immunodeficiency virus-infected cells. Mol Membr Biol 1999; 16(1): 111-8.

[216]   Sei S, Yang QE, O'Neill D, Yoshimura K, Nagashima K, Mitsuya H. Identification of a key target sequence to block human immunodeficiency virus type 1 replication within the gag-pol transframe domain. J Virol 2000; 74(10): 4621-33.

[217]   Fraternale A, Paoletti MF, Casabianca A *et al.* Erythrocytes as carriers of antisense PNA addressed against HIV-1 gag-pol transframe domain. J Drug Target 2009; 17(4): 278-85.

[218]   Lavigne C, Yelle J, Sauve G, Thierry AG. Lipid-based delivery of combinations of antisense oligodeoxynucleotides for the *in vitro* inhibition of HIV-1 replication. AAPS PharmSci 2001; 3(1): E7.

[219]   Vickers T, Baker BF, Cook PD *et al.* Inhibition of HIV-LTR gene expression by oligonucleotides targeted to the TAR element. Nucleic Acids Res 1991; 19(12): 3359-68.

[220] Boulme F, Freund F, Moreau S *et al.* Modified (PNA, 2'-O-methyl and phosphoramidate) anti-TAR antisense oligonucleotides as strong and specific inhibitors of *in vitro* HIV-1 reverse transcription. Nucleic Acids Res 1998; 26(23): 5492-500.

[221] Arzumanov A, Stetsenko DA, Malakhov AD *et al.* A structure-activity study of the inhibition of HIV-1 Tat-dependent trans-activation by mixmer 2'-O-methyl oligoribonucleotides containing locked nucleic acid (LNA), alpha-L-LNA, or 2'-thio-LNA residues. Oligonucleotides 2003; 13(6): 435-53.

[222] Arzumanov A, Walsh AP, Rajwanshi VK, Kumar R, Wengel J, Gait MJ. Inhibition of HIV-1 Tat-dependent trans activation by steric block chimeric 2'-O-methyl/LNA oligoribonucleotides. Biochemistry 2001; 40(48): 14645-54.

[223] Hamma T, Saleh A, Huq I, Rana TM, Miller PS. Inhibition of HIV tat-TAR interactions by an antisense oligo-2'-O-methylribonucleoside methylphosphonate. Bioorg Med Chem Lett 2003; 13(11): 1845-8.

[224] Mayhood T, Kaushik N, Pandey PK, Kashanchi F, Deng L, Pandey VN. Inhibition of Tat-mediated transactivation of HIV-1 LTR transcription by polyamide nucleic acid targeted to TAR hairpin element. Biochemistry 2000; 39(38): 11532-9.

[225] Chaubey B, Tripathi S, Ganguly S, Harris D, Casale RA, Pandey VN. A PNA-transportan conjugate targeted to the TAR region of the HIV-1 genome exhibits both antiviral and virucidal properties. Virology 2005; 331(2): 418-28.

[226] Kaushik N, Basu A, Pandey VN. Inhibition of HIV-1 replication by anti-trans-activation responsive polyamide nucleotide analog. Antiviral Res 2002; 56(1): 13-27.

[227] Kaushik N, Basu A, Palumbo P, Myers RL, Pandey VN. Anti-TAR polyamide nucleotide analog conjugated with a membrane-permeating peptide inhibits human immunodeficiency virus type 1 production. J Virol 2002; 76(8): 3881-91.

[228] Tripathi S, Chaubey B, Ganguly S, Harris D, Casale RA, Pandey VN. Anti-HIV-1 activity of anti-TAR polyamide nucleic acid conjugated with various membrane transducing peptides. Nucleic Acids Res 2005; 33(13): 4345-56.

[229] Brown D, Arzumanov AA, Turner JJ, Stetsenko DA, Lever AM, Gait MJ. Antiviral activity of steric-block oligonucleotides targeting the HIV-1 trans-activation response and packaging signal stem-loop RNAs. Nucleosides Nucleotides Nucleic Acids 2005; 24(5-7): 393-6.

[230] Matskevich AA, Ziogas A, Heinrich J, Quast SA, Moelling K. Short partially double-stranded oligodeoxynucleotide induces reverse transcriptase/RNase H-mediated cleavage of HIV RNA and contributes to abrogation of infectivity of virions. AIDS Res Hum Retroviruses 2006; 22(12): 1220-30.

[231] Heinrich J, Mathur S, Matskevich AA, Moelling K. Oligonucleotide-mediated retroviral RNase H activation leads to reduced HIV-1 titer in patient-derived plasma. AIDS 2009; 23(2): 213-21.

[232] Jakobsen MR, Haasnoot J, Wengel J, Berkhout B, Kjems J. Efficient inhibition of HIV-1 expression by LNA modified antisense oligonucleotides and DNAzymes targeted to functionally selected binding sites. Retrovirology 2007; 4:29.

[233] Lim TW, Yuan J, Liu Z, Qiu D, Sall A, Yang D. Nucleic-acid-based antiviral agents against positive single-stranded RNA viruses. Curr Opin Mol Ther 2006; 8(2): 104-7.

[234] Samuel MA, Diamond MS. Pathogenesis of West Nile Virus infection: a balance between virulence, innate and adaptive immunity, and viral evasion. J Virol 2006; 80(19): 9349-60.

[235] Fusco DN, Chung RT. Novel therapies for hepatitis C: insights from the structure of the virus. Annu Rev Med 2012; 63:373-87.

[236] Schinazi RF, Bassit L, Gavegnano C. HCV drug discovery aimed at viral eradication. J Viral Hepat 2010; 17(2): 77-90.

[237] Martinand-Mari C, Lebleu B, Robbins I. Oligonucleotide-based strategies to inhibit human hepatitis C virus. Oligonucleotides 2003; 13(6): 539-48.

[238] Romero-Lopez C, Sanchez-Luque FJ, Berzal-Herranz A. Targets and tools: recent advances in the development of anti-HCV nucleic acids. Infect Disord Drug Targets 2006; 6(2): 121-45.

[239] Jubin R. Hepatitis C IRES: translating translation into a therapeutic target. Curr Opin Mol Ther 2001; 3(3): 278-87.

[240] Wakita T, Wands JR. Specific inhibition of hepatitis C virus expression by antisense oligodeoxynucleotides. *In vitro* model for selection of target sequence. J Biol Chem 1994; 269(19): 14205-10.

[241] Mizutani T, Kato N, Hirota M, Sugiyama K, Murakami A, Shimotohno K. Inhibition of hepatitis C virus replication by antisense oligonucleotide in culture cells. Biochem Biophys Res Commun 1995; 212(3): 906-11.

[242] Lima WF, Brown-Driver V, Fox M, Hanecak R, Bruice TW. Combinatorial screening and rational optimization for hybridization to folded hepatitis C virus RNA of oligonucleotides with biological antisense activity. J Biol Chem 1997; 272(1): 626-38.

[243] Alt M, Renz R, Hofschneider PH, Caselmann WH. Core specific antisense phosphorothioate oligodeoxynucleotides as potent and specific inhibitors of hepatitis C viral translation. Arch Virol 1997; 142(3): 589-99.

[244] Alt M, Renz R, Hofschneider PH, Paumgartner G, Caselmann WH. Specific inhibition of hepatitis C viral gene expression by antisense phosphorothioate oligodeoxynucleotides. Hepatology 1995; 22(3): 707-17.

[245] Caselmann WH, Eisenhardt S, Alt M. Synthetic antisense oligodeoxynucleotides as potential drugs against hepatitis C. Intervirology 1997; 40(5-6): 394-9.

[246] Alt M, Eisenhardt S, Serwe M, Renz R, Engels JW, Caselmann WH. Comparative inhibitory potential of differently modified antisense oligodeoxynucleotides on hepatitis C virus translation. Eur J Clin Invest 1999; 29(10): 868-76.

[247] Caselmann WH, Serwe M, Lehmann T, Ludwig J, Sproat BS, Engels JW. Design, delivery and efficacy testing of therapeutic nucleic acidsused to inhibit hepatitis C virus gene expression *in vitro* and *in vivo*. World J Gastroenterol 2000; 6(5): 626-29.

[248] Lehmann TJ, Serwe M, Caselmann WH, Engels JW. Design and properties of hepatitis C virus antisense oligonucleotides for liver specific drug targeting. Nucleosides Nucleotides Nucleic Acids 2001; 20(4-7): 1343-6.

[249] Lehmann TJ, Engels JW. Synthesis and properties of bile acid phosphoramidites 5'-tethered to antisense oligodeoxynucleotides against HCV. Bioorg Med Chem 2001; 9(7): 1827-35.

[250] Heintges T, Encke J, zu PJ, Wands JR. Inhibition of hepatitis C virus NS3 function by antisense oligodeoxynucleotides and protease inhibitor. J Med Virol 2001; 65(4): 671-80.

[251] Hanecak R, Brown-Driver V, Fox MC *et al*. Antisense oligonucleotide inhibition of hepatitis C virus gene expression in transformed hepatocytes. J Virol 1996; 70(8): 5203-12.

[252] Zhang H, Hanecak R, Brown-Driver V *et al*. Antisense oligonucleotide inhibition of hepatitis C virus (HCV) gene expression in livers of mice infected with an HCV-vaccinia virus recombinant. Antimicrob Agents Chemother 1999; 43(2): 347-53.

[253] Amin MA, Awadein MR, Gabr H. Evaluation of the inhibitory effect of antisense oligodeoxynucleotides on the growth of hepatitis C-associated hepatocellular carcinoma cells *in vitro*. Chin J Dig Dis 2005; 6(3): 142-8.

[254] McCaffrey AP, Meuse L, Karimi M, Contag CH, Kay MA. A potent and specific morpholino antisense inhibitor of hepatitis C translation in mice. Hepatology 2003; 38(2): 503-8.

[255] Soler M, McHutchison JG, Kwoh TJ, Dorr FA, Pawlotsky JM. Virological effects of ISIS 14803, an antisense oligonucleotide inhibitor of hepatitis C virus (HCV) internal ribosome entry site (IRES), on HCV IRES in chronic hepatitis C patients and examination of the potential role of primary and secondary HCV resistance in the outcome of treatment. Antivir Ther 2004; 9(6): 953-68.

[256] Witherell GW. ISIS-14803 (Isis Pharmaceuticals). Curr Opin Investig Drugs 2001; 2(11): 1523-9.

[257] McHutchison JG, Patel K, Pockros P *et al*. A phase I trial of an antisense inhibitor of hepatitis C virus (ISIS 14803), administered to chronic hepatitis C patients. J Hepatol 2006; 44(1): 88-96.

[258] Tallet-Lopez B, Aldaz-Carroll L, Chabas S, Dausse E, Staedel C, Toulme JJ. Antisense oligonucleotides targeted to the domain IIId of the hepatitis C virus IRES compete with 40S ribosomal subunit binding and prevent *in vitro* translation. Nucleic Acids Res 2003; 31(2): 734-42.

[259] Nulf CJ, Corey D. Intracellular inhibition of hepatitis C virus (HCV) internal ribosomal entry site (IRES)-dependent translation by peptide nucleic acids (PNAs) and locked nucleic acids (LNAs). Nucleic Acids Res 2004; 32(13): 3792-8.

[260] el-Awady MK, el-Din NG, el-Garf WT *et al*. Antisense oligonucleotide inhibition of hepatitis C virus genotype 4 replication in HepG2 cells. Cancer Cell Int 2006; 6:18.

[261] Alotte C, Martin A, Caldarelli SA *et al*. Short peptide nucleic acids (PNA) inhibit hepatitis C virus internal ribosome entry site (IRES) dependent translation *in vitro*. Antiviral Res 2008; 80(3): 280-7.

[262] Ahn DG, Shim SB, Moon JE, Kim JH, Kim SJ, Oh JW. Interference of hepatitis C virus replication in cell culture by antisense peptide nucleic acids targeting the X-RNA. J Viral Hepat 2011; 18(7): e298-e306.

[263] Laxton C, Brady K, Moschos S *et al*. Selection, optimization, and pharmacokinetic properties of a novel, potent antiviral locked nucleic acid-based antisense oligomer targeting hepatitis C virus internal ribosome entry site. Antimicrob Agents Chemother 2011; 55(7): 3105-14.

[264] Estrada-Pena A, Ayllon N, de la Fuente J. Impact of climate trends on tick-borne pathogen transmission. Front Physiol 2012; 3:64.

[265] Deas TS, Binduga-Gajewska I, Tilgner M *et al*. Inhibition of flavivirus infections by antisense oligomers specifically suppressing viral translation and RNA replication. J Virol 2005; 79(8): 4599-609.

[266] Deas TS, Bennett CJ, Jones SA *et al*. *In vitro* resistance selection and *in vivo* efficacy of morpholino oligomers against West Nile virus. Antimicrob Agents Chemother 2007; 51(7): 2470-82.

[267] Torrence PF, Gupta N, Whitney C, Morrey JD. Evaluation of synthetic oligonucleotides as inhibitors of West Nile virus replication. Antiviral Res 2006; 70(2): 60-5.

[268] Kinney RM, Huang CY, Rose BC *et al.* Inhibition of dengue virus serotypes 1 to 4 in vero cell cultures with morpholino oligomers. J Virol 2005; 79(8): 5116-28.

[269] Holden KL, Stein DA, Pierson TC *et al.* Inhibition of dengue virus translation and RNA synthesis by a morpholino oligomer targeted to the top of the terminal 3' stem-loop structure. Virology 2006; 344(2): 439-52.

[270] Stein DA, Huang CY, Silengo S *et al.* Treatment of AG129 mice with antisense morpholino oligomers increases survival time following challenge with dengue 2 virus. J Antimicrob Chemother 2008; 62(3): 555-65.

[271] Yoo JS, Kim CM, Kim JH, Kim JY, Oh JW. Inhibition of Japanese encephalitis virus replication by peptide nucleic acids targeting cis-acting elements on the plus- and minus-strands of viral RNA. Antiviral Res 2009; 82(3): 122-33.

[272] Deas TS, Binduga-Gajewska I, Tilgner M *et al.* Inhibition of flavivirus infections by antisense oligomers specifically suppressing viral translation and RNA replication. J Virol 2005; 79(8): 4599-609.

[273] Frolova TV, Nomokonova NI, Roikhel' VM, Fokina GI, Pogodkina VV, Vlasov VV. Antiviral effect of various oligonucleotide derivatives, complementary to tick-born encephalitis virus RNA. Vopr Virusol 1994; 39(5): 232-5.

[274] Karmysheva VI, Roikhel' VM, Fokina GI, Frolova MP, Pogodina VV. [Some mechanisms of the action of oligonucleotides: stimulation of the immune system and decreased infection of the brain in tickborne encephalitis]. Vopr Virusol 1998; 43(1): 39-42.

[275] Lim T, Yuan J, Zhang HM *et al.* Antisense DNA and RNA agents against picornaviruses. Front Biosci 2008; 13:4707-25.

[276] Chen TC, Weng KF, Chang SC, Lin JY, Huang PN, Shih SR. Development of antiviral agents for enteroviruses. J Antimicrob Chemother 2008; 62(6): 1169-73.

[277] Sun H, Liu Z, Zhang T. [Antisense oligonucleotides resistance to coxsackievirus B3 infection in HeLa cells]. Zhonghua Liu Xing Bing Xue Za Zhi 2000; 21(4): 295-7.

[278] Wang A, Cheung PK, Zhang H *et al.* Specific inhibition of coxsackievirus B3 translation and replication by phosphorothioate antisense oligodeoxynucleotides. Antimicrob Agents Chemother 2001; 45(4): 1043-52.

[279] Yuan J, Cheung PK, Zhang H *et al.* A phosphorothioate antisense oligodeoxynucleotide specifically inhibits coxsackievirus B3 replication in cardiomyocytes and mouse hearts. Lab Invest 2004; 84(6): 703-14.

[280] Yuan J, Stein DA, Lim T *et al.* Inhibition of coxsackievirus B3 in cell cultures and in mice by peptide-conjugated morpholino oligomers targeting the internal ribosome entry site. J Virol 2006; 80(23): 11510-9.

[281] Gutierrez A, Rodriguez A, Pintado B, Sobrino F. Transient inhibition of foot-and-mouth disease virus infection of BHK-21 cells by antisense oligonucleotides directed against the second functional initiator AUG. Antiviral Res 1993; 22(1): 1-13.

[282] Vagnozzi A, Stein DA, Iversen PL, Rieder E. Inhibition of foot-and-mouth disease virus infections in cell cultures with antisense morpholino oligomers. J Virol 2007; 81(21): 11669-80.

[283] Stone JK, Rijnbrand R, Stein DA *et al.* A morpholino oligomer targeting highly conserved internal ribosome entry site sequence is able to inhibit multiple species of picornavirus. Antimicrob Agents Chemother 2008; 52(6): 1970-81.

[284] Peiris JS, Lai ST, Poon LL *et al*. Coronavirus as a possible cause of severe acute respiratory syndrome. Lancet 2003; 361(9366): 1319-25.

[285] Rota PA, Oberste MS, Monroe SS *et al*. Characterization of a novel coronavirus associated with severe acute respiratory syndrome. Science 2003; 300(5624): 1394-9.

[286] Neuman BW, Stein DA, Kroeker AD *et al*. Inhibition and escape of SARS-CoV treated with antisense morpholino oligomers. Adv Exp Med Biol 2006; 581:567-71.

[287] Neuman BW, Stein DA, Kroeker AD *et al*. Inhibition, escape, and attenuated growth of severe acute respiratory syndrome coronavirus treated with antisense morpholino oligomers. J Virol 2005; 79(15): 9665-76.

[288] Ahn DG, Lee W, Choi JK *et al*. Interference of ribosomal frameshifting by antisense peptide nucleic acids suppresses SARS coronavirus replication. Antiviral Res 2011; 91(1): 1-10.

[289] van den Born E, Stein DA, Iversen PL, Snijder EJ. Antiviral activity of morpholino oligomers designed to block various aspects of Equine arteritis virus amplification in cell culture. J Gen Virol 2005; 86(Pt 11): 3081-90.

[290] Zhang J, Stein DA, Timoney PJ, Balasuriya UB. Curing of HeLa cells persistently infected with equine arteritis virus by a peptide-conjugated morpholino oligomer. Virus Res 2010; 150(1-2): 138-42.

[291] Patel D, Opriessnig T, Stein DA *et al*. Peptide-conjugated morpholino oligomers inhibit porcine reproductive and respiratory syndrome virus replication. Antiviral Res 2008; 77(2): 95-107.

[292] Han X, Fan S, Patel D, Zhang YJ. Enhanced inhibition of porcine reproductive and respiratory syndrome virus replication by combination of morpholino oligomers. Antiviral Res 2009; 82(1): 59-66.

[293] Patel D, Stein DA, Zhang YJ. Morpholino oligomer-mediated protection of porcine pulmonary alveolar macrophages from arterivirus-induced cell death. Antivir Ther 2009; 14(7): 899-909.

[294] Opriessnig T, Patel D, Wang R *et al*. Inhibition of porcine reproductive and respiratory syndrome virus infection in piglets by a peptide-conjugated morpholino oligomer. Antiviral Res 2011; 91(1): 36-42.

[295] Neuman BW, Stein DA, Kroeker AD *et al*. Antisense morpholino-oligomers directed against the 5' end of the genome inhibit coronavirus proliferation and growth. J Virol 2004; 78(11): 5891-9.

[296] Burrer R, Neuman BW, Ting JP *et al*. Antiviral effects of antisense morpholino oligomers in murine coronavirus infection models. J Virol 2007; 81(11): 5637-48.

[297] Paessler S, Rijnbrand R, Stein DA *et al*. Inhibition of alphavirus infection in cell culture and in mice with antisense morpholino oligomers. Virology 2008; 376(2): 357-70.

[298] Stein DA, Skilling DE, Iversen PL, Smith AW. Inhibition of Vesivirus infections in mammalian tissue culture with antisense morpholino oligomers. Antisense Nucleic Acid Drug Dev 2001; 11(5): 317-25.

[299] Smith AW, Iversen PL, O'Hanley PD *et al*. Virus-specific antiviral treatment for controlling severe and fatal outbreaks of feline calicivirus infection. Am J Vet Res 2008; 69(1): 23-32.

[300] Smith AW, Matson DO, Stein DA *et al*. Antisense treatment of caliciviridae: an emerging disease agent of animals and humans. Curr Opin Mol Ther 2002; 4(2): 177-84.

[301] Hoenen T, Groseth A, Falzarano D, Feldmann H. Ebola virus: unravelling pathogenesis to combat a deadly disease. Trends Mol Med 2006; 12(5): 206-15.

[302] de WE, Feldmann H, Munster VJ. Tackling Ebola: new insights into prophylactic and therapeutic intervention strategies. Genome Med 2011; 3(1): 5.

[303] Wong JP, Christopher ME, Salazar AM *et al.* Broad-spectrum and virus-specific nucleic acid-based antivirals against influenza. Front Biosci (Schol Ed) 2010; 2:791-800.

[304] Warfield KL, Swenson DL, Olinger GG *et al.* Gene-specific countermeasures against Ebola virus based on antisense phosphorodiamidate morpholino oligomers. PLoS Pathog 2006; 2(1): e1.

[305] Enterlein S, Warfield KL, Swenson DL *et al.* VP35 knockdown inhibits Ebola virus amplification and protects against lethal infection in mice. Antimicrob Agents Chemother 2006; 50(3): 984-93.

[306] Warren TK, Warfield KL, Wells J *et al.* Advanced antisense therapies for postexposure protection against lethal filovirus infections. Nat Med 2010; 16(9): 991-4.

[307] Jairath S, Vargas PB, Hamlin HA, Field AK, Kilkuskie RE. Inhibition of respiratory syncytial virus replication by antisense oligodeoxyribonucleotides. Antiviral Res 1997; 33(3): 201-13.

[308] Lai SH, Stein DA, Guerrero-Plata A *et al.* Inhibition of respiratory syncytial virus infections with morpholino oligomers in cell cultures and in mice. Mol Ther 2008; 16(6): 1120-8.

[309] Kabanov AV, Vinogradov SV, Ovcharenko AV *et al.* A new class of antivirals: antisense oligonucleotides combined with a hydrophobic substituent effectively inhibit influenza virus reproduction and synthesis of virus-specific proteins in MDCK cells. FEBS Lett 1990; 259(2): 327-30.

[310] Vlasov VV, Gorn VV, Nomokonova NI, Fokina TN, Iurchenko LV. Suppression of translation *in vitro* of the mRNA of the M1 protein of influenza virus using antisense oligonucleotides. Mol Biol (Mosk) 1991; 25(5): 1332-7.

[311] Hatta T, Nakagawa Y, Takai K, Nakada S, Yokota T, Takaku H. Inhibition of influenza virus RNA polymerase and nucleoprotein of gene expression by antisense oligonucleotides. Nucleic Acids Symp Ser 1995; 34): 129-30.

[312] Hatta T, Takai K, Nakada S, Yokota T, Takaku H. Specific inhibition of influenza virus RNA polymerase and nucleoprotein genes expression by liposomally endocapsulated antisense phosphorothioate oligonucleotides: penetration and localization of oligonucleotides in clone 76 cells. Biochem Biophys Res Commun 1997; 232(2): 545-9.

[313] Abe T, Suzuki S, Hatta T, Takai K, Yokota T, Takaku H. Specific inhibition of influenza virus RNA polymerase and nucleoprotein gene expression by liposomally encapsulated antisense phosphorothioate oligonucleotides in MDCK cells. Antivir Chem Chemother 1998; 9(3): 253-62.

[314] Mizuta T, Fujiwara M, Hatta T *et al.* Antisense oligonucleotides directed against the viral RNA polymerase gene enhance survival of mice infected with influenza A. Nat Biotechnol 1999; 17(6): 583-7.

[315] Abe T, Mizuta T, Suzuki S *et al. In vitro* and *in vivo* anti-influenza A virus activity of antisense oligonucleotides. Nucleosides Nucleotides 1999; 18(6-7): 1685-8.

[316] Chen Z, Wang S, Guan W, Yang B, Sun Z. [*In vitro* antiviral activity of antisense oligonucleotides against influenza virus]. Wei Sheng Wu Xue Bao 2000; 40(5): 482-7.

[317] Mizuta T, Fujiwara M, Abe T *et al*. Inhibitory effects of an antisense oligonucleotide in an experimentally infected mouse model of influenza A virus. Biochem Biophys Res Commun 2000; 279(1): 158-61.

[318] Abe T, Mizuta T, Hatta T *et al*. Antisense therapy of influenza. Eur J Pharm Sci 2001; 13(1): 61-69.

[319] Ge Q, Pastey M, Kobasa D *et al*. Inhibition of multiple subtypes of influenza A virus in cell cultures with morpholino oligomers. Antimicrob Agents Chemother 2006; 50(11): 3724-33.

[320] Wong JP, Christopher ME, Salazar AM, Dale RM, Sun LQ, Wang M. Nucleic acid-based antiviral drugs against seasonal and avian influenza viruses. Vaccine 2007; 25(16): 3175-3178.

[321] Gabriel G, Nordmann A, Stein DA, Iversen PL, Klenk HD. Morpholino oligomers targeting the PB1 and NP genes enhance the survival of mice infected with highly pathogenic influenza A H7N7 virus. J Gen Virol 2008; 89(Pt 4): 939-48.

[322] Duan M, Zhou Z, Lin RX, Yang J, Xia XZ, Wang SQ. *In vitro* and *in vivo* protection against the highly pathogenic H5N1 influenza virus by an antisense phosphorothioate oligonucleotide. Antivir Ther 2008; 13(1): 109-14.

[323] Lupfer C, Stein DA, Mourich DV, Tepper SE, Iversen PL, Pastey M. Inhibition of influenza A H3N8 virus infections in mice by morpholino oligomers. Arch Virol 2008; 153(5): 929-37.

[324] Giannecchini S, Clausi V, Nosi D, Azzi A. Oligonucleotides derived from the packaging signal at the 5' end of the viral PB2 segment specifically inhibit influenza virus *in vitro*. Arch Virol 2009; 154(5): 821-32.

[325] Giannecchini S, Wise HM, Digard P *et al*. Packaging signals in the 5'-ends of influenza virus PA, PB1, and PB2 genes as potential targets to develop nucleic-acid based antiviral molecules. Antiviral Res 2011; 92(1): 64-72.

[326] Neuman BW, Bederka LH, Stein DA, Ting JP, Moulton HM, Buchmeier MJ. Development of peptide-conjugated morpholino oligomers as pan-arenavirus inhibitors. Antimicrob Agents Chemother 2011; 55(10): 4631-8.

[327] Bell AF, Whitton JL, Fujinami RS. Antisense-mediated resistance to measles virus infection in HeLa cells. J Infect Dis 1997; 176(1): 258-61.

[328] Sleeman K, Stein DA, Tamin A, Reddish M, Iversen PL, Rota PA. Inhibition of measles virus infections in cell cultures by peptide-conjugated morpholino oligomers. Virus Res 2009; 140(1-2): 49-56.

[329] Lemaitre M, Bayard B, Lebleu B. Specific antiviral activity of a poly(L-lysine)-conjugated oligodeoxyribonucleotide sequence complementary to vesicular stomatitis virus N protein mRNA initiation site. Proc Natl Acad Sci U S A 1987; 84(3): 648-52.

[330] Prasad V, Hashim S, Mukhopadhyay A, Basu SK, Roy RP. Oligonucleotides tethered to a short polyguanylic acid stretch are targeted to macrophages: enhanced antiviral activity of a vesicular stomatitis virus-specific antisense oligonucleotide. Antimicrob Agents Chemother 1999; 43(11): 2689-96.

[331] Bajramovic JJ, Munter S, Syan S, Nehrbass U, Brahic M, Gonzalez-Dunia D. Borna disease virus glycoprotein is required for viral dissemination in neurons. J Virol 2003; 77(22): 12222-31.

[332] Amantana A, Moulton HM, Cate ML *et al*. Pharmacokinetics, biodistribution, stability and toxicity of a cell-penetrating peptide-morpholino oligomer conjugate. Bioconjug Chem 2007; 18(4): 1325-31.

[333] Youngblood DS, Hatlevig SA, Hassinger JN, Iversen PL, Moulton HM. Stability of cell-penetrating peptide-morpholino oligomer conjugates in human serum and in cells. Bioconjug Chem 2007; 18(1): 50-60.

[334] Abes R, Arzumanov A, Moulton H *et al.* Arginine-rich cell penetrating peptides: design, structure-activity, and applications to alter pre-mRNA splicing by steric-block oligonucleotides. J Pept Sci 2008; 14(4): 455-60.

[335] Warren TK, Shurtleff AC, Bavari S. Advanced morpholino oligomers: a novel approach to antiviral therapy. Antiviral Res 2012; 94(1): 80-88.

[336] Mulamba GB, Hu A, Azad RF, Anderson KP, Coen DM. Human cytomegalovirus mutant with sequence-dependent resistance to the phosphorothioate oligonucleotide fomivirsen (ISIS 2922). Antimicrob Agents Chemother 1998; 42(4): 971-3.

[337] Stone JK, Rijnbrand R, Stein DA *et al.* A morpholino oligomer targeting highly conserved internal ribosome entry site sequence is able to inhibit multiple species of picornavirus. Antimicrob Agents Chemother 2008; 52(6): 1970-81.

[338] Amantana A, Moulton HM, Cate ML *et al.* Pharmacokinetics, biodistribution, stability and toxicity of a cell-penetrating peptide-morpholino oligomer conjugate. Bioconjug Chem 2007; 18(4): 1325-31.

[339] Ahlquist P, Noueiry AO, Lee WM, Kushner DB, Dye BT. Host factors in positive-strand RNA virus genome replication. J Virol 2003; 77(15): 8181-6.

[340] Khattab MA. Targeting host factors: a novel rationale for the management of hepatitis C virus. World J Gastroenterol 2009; 15(28): 3472-9.

[341] Park SG, Lee SM, Jung G. Antisense oligodeoxynucleotides targeted against molecular chaperonin Hsp60 block human hepatitis B virus replication. J Biol Chem 2003; 278(41): 39851-7.

[342] Yang J, Bo XC, Ding XR *et al.* Antisense oligonucleotides targeted against asialoglycoprotein receptor 1 block human hepatitis B virus replication. J Viral Hepat 2006; 13(3): 158-165.

[343] Zhang GL, Li YX, Zheng SQ, Liu M, Li X, Tang H. Suppression of hepatitis B virus replication by microRNA-199a-3p and microRNA-210. Antiviral Res 2010; 88(2): 169-75.

[344] Ding XR, Yang J, Sun DC, Lou SK, Wang SQ. Whole genome expression profiling of hepatitis B virus-transfected cell line reveals the potential targets of anti-HBV drugs. Pharmacogenomics J 2008; 8(1): 61-70.

[345] Hirschman SZ, Chen CW. Peptide nucleic acids stimulate gamma interferon and inhibit the replication of the human immunodeficiency virus. J Investig Med 1996; 44(6): 347-51.

[346] Huang J, Wang F, Argyris E *et al.* Cellular microRNAs contribute to HIV-1 latency in resting primary CD4+ T lymphocytes. Nat Med 2007; 13(10): 1241-7.

[347] Dhillon NK, Sui Y, Potula R *et al.* Inhibition of pathogenic SHIV replication in macaques treated with antisense DNA of interleukin-4. Blood 2005; 105(8): 3094-9.

[348] Randall G, Panis M, Cooper JD *et al.* Cellular cofactors affecting hepatitis C virus infection and replication. Proc Natl Acad Sci U S A 2007; 104(31): 12884-12889.

[349] Elmen J, Lindow M, Schutz S *et al.* LNA-mediated microRNA silencing in non-human primates. Nature 2008; 452(7189): 896-9.

[350] Lanford RE, Hildebrandt-Eriksen ES, Petri A *et al.* Therapeutic silencing of microRNA-122 in primates with chronic hepatitis C virus infection. Science 2010; 327(5962): 198-201.

[351] Li YP, Gottwein JM, Scheel TK, Jensen TB, Bukh J. MicroRNA-122 antagonism against hepatitis C virus genotypes 1-6 and reduced efficacy by host RNA insertion or mutations in the HCV 5' UTR. Proc Natl Acad Sci U S A 2011; 108(12): 4991-6.

[352] Fusco DN, Chung RT. Novel therapies for hepatitis C: insights from the structure of the virus. Annu Rev Med 2012; 63:373-87.

[353] Li J, Wasmuth S, Bauer D, Baehler H, Hennig M, Heiligenhaus A. Subconjunctival antisense oligonucleotides targeting TNF-alpha influence immunopathology and viral replication in murine HSV-1 retinitis. Graefes Arch Clin Exp Ophthalmol 2008; 246(9): 1265-73.

[354] Grajewski RS, Li J, Wasmuth S, Hennig M, Bauer D, Heiligenhaus A. Intravitreal treatment with antisense oligonucleotides targeting tumor necrosis factor-alpha in murine herpes simplex virus type 1 retinitis. Graefes Arch Clin Exp Ophthalmol 2012; 250(2): 231-8.

[355] Potera C. Antisense--down, but not out. Nat Biotechnol 2007; 25(5): 497-9.

[356] Kortepeter MG, Martin JW, Rusnak JM *et al.* Managing potential laboratory exposure to ebola virus by using a patient biocontainment care unit. Emerg Infect Dis 2008; 14(6): 881-7.

Send Orders for Reprints to reprints@benthamscience.net

CHAPTER 4

Anti-Infective Agents Against Flavivirus

S. John Vennison[*] and S. Gowri Sankar

Department of Biotechnology, Anna University-BIT Campus, Tiruchirappalli-620 024, Tamil Nadu, India

Abstract: Flavivirus become a re-emerging problem to humans and animals. Through the availability of efficient whole inactivated viral vaccines against Yellow Fever (YF), Japanese Encephalitis (JEV), the burden of these two diseases has been minimized to some extent. Restrictions in the use of these vaccines, accessibility and cost for these usage in places where the disease is endemic and failures in attempt to produce vaccines against Dengue virus (DENV), West Nile virus (WNV), has prompted to go for other effective anti-viral strategies. Incomplete understanding in the host immune mechanism, lack of suitable animal models is the main reasons for the unavailability of an effective antiviral treatment. Hence supportive measures and symptomatic treatment are the only available choice. In this chapter, the major vaccines currently available against flavivirus and new approaches utilized for designing novel vaccine candidates and antiviral agents targeting various pathways in the flavivirus life-cycle are discussed.

Keywords: Anti-infective agents, arbo-viruses, dengue virus, flavivirus, Japanese encephalitis virus, mosquito pathogens, non-structural genes, structural genes, vaccines, yellow fever virus, West Nile virus.

INTRODUCTION

The Flaviviridae, (from Latin flavus, yellow), is a virus family consisting of three genus: Flavivirus, Pestivirus (from the Latin pestis, plague), and Hepacivirus (from the Greek hepar, hepatos, liver) [1]. The family distribution is based on the evolutionary relatedness of the RNA-dependent RNA polymerases (RdRP). Members of the family share similar in virion morphology, genome organization and replication strategy (Figs. **1** and **3**) but exhibit diverse biological properties and serology. Pestiviruses and hepaciviruses are also having their genome

---

*Address correspondence to S. John Vennison: Department of Biotechnology, Anna University-BIT Campus, Tiruchirappalli-620 024, Tamil Nadu, India; Tel: +00 91 9486230714; Fax: +00 91 431 2407333; E-mail: johnvennison36@gmail.com

Atta-ur-Rahman (Ed)

replication strategies similar to flaviviruses but are antigenically distinct and are not arthropod–borne. An insect virus known as Cell-Fusing Agent also been placed in the genus on the basis of similar genome strategy and partial homology with flaviviruses [2].

Unlike the cellular mRNA, genome of flavivirus lacks poly-A tail. However a type I cap is present on the 5' end while type II caps have been found in the RNA of infected cells. The flaviviridae genome composed of one single open reading frame (ORF) and on both ends of this ORF there exists non-structural regions of approximately 100 nucleotides in length. 5' non-coding regions are not highly conserved among different flaviviridae species while 3' non-coding regions tend to be highly variable between the viruses that are transmitted by mosquitoes and viruses transmitted by ticks. In flaviviruses structural genes are found at the 5' of the genome and non-structural genes are located at the 3' end of the genome. This organization allows the virus to maximize production of structural genes, since viral assembly requires more structural proteins than non-structural proteins.

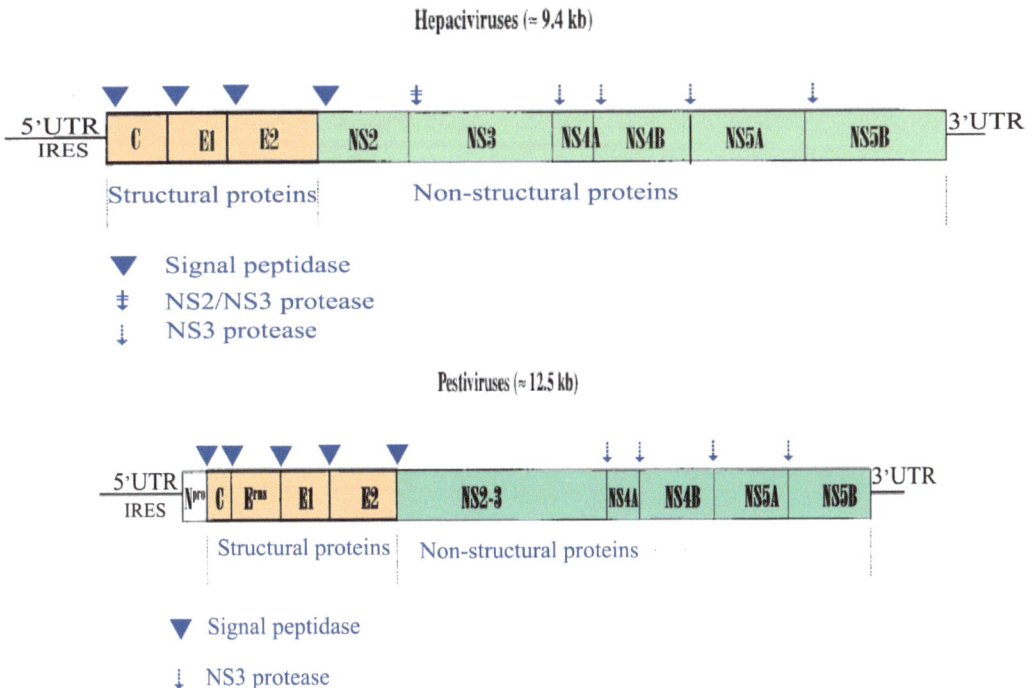

Figure 1: Genome organization of Hepacivirus and Pestivirus.

FLAVIVIRUS

Flavivirus in flaviviridae family are usually arthropod-borne. The genus consists of more than 70 viruses, many of which are arthropod-borne human pathogens. The first human virus discovered over a century ago was a flavivirus when Walter Reed demonstrated yellow fever from an infected individual [3]. Flaviviruses cause a variety of diseases, including fevers, encephalitis, and hemorrhagic fevers. Entities of major global concern include dengue virus (DENV) with its associated dengue hemorrhagic fever (DHF) and dengue shock syndrome (DSS), Japanese encephalitis virus (JEV), West Nile virus (WNV), and Yellow fever Virus (YFV) (Fig. **2**). They consist of single stranded enveloped positive sense RNA, which has 11,000 nucleotides in length [4]. Members of the family share similarities in virion morphology, genome organization and replication strategy. The genome encodes a poly-protein which is co-and post-translationally modified by host and cellular proteases in to three structural (C, M and E) proteins and seven non-structural proteins (NS1, NS2A, NS2B, NS3, NS4A, NS4B and NS5) [5] (Fig. **3**).

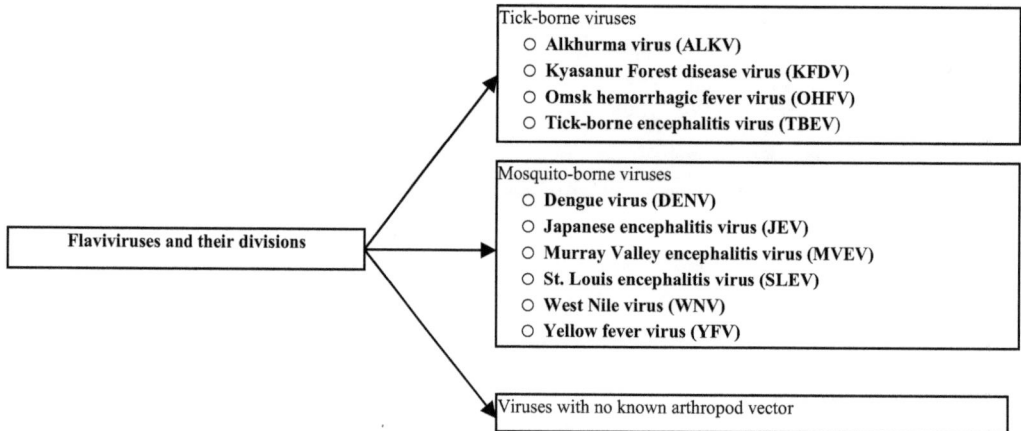

Figure 2: Flavivirus classification and important pathogens to humans.

Flaviviruses are becoming more concerned now a days because of increase in number of dengue infections, increasing mortality and morbidity associated with DHF and DSS, spread of JEV across East Asian countries, persistence of YFV in Africa and emergence of WNV infection in western countries [4]. Recent outbreaks of TBEV in North Europe and Japan, JEV infection in Australia and

WNV infection in north and South America highlights the geographical expansion of flavivirus infections [6].

Virus Structure

The outer surface of virus particles contains two viral proteins, E (envelope) and M (membrane). E protein is the major antigenic determinant of the mature dengue virus and elicits neutralizing antibodies. E protein enables receptor mediated attachment of virions to host cell and fusion with host cell membrane [7]. M protein produced during maturation of nascent virus particles within the secretory pathway, is a small proteolytic fragment of the precursor prM protein. Capsid (C) is a highly basic protein consisting of charged residues [8] mediating RNA interaction [9].

The NS1 glycoprotein is translocated into the ER during synthesis and cleaved from E protein by host signal peptidase, whereas an unknown ER-resident host enzyme cleaves NS1/2A junction [10, 11]. NS1 protein contains two conserved N-glycosylation sites and 12 invariant cysteine residues forming six disulfide bonds [12, 13]. The role of nonstructural glycoprotein-1 (NS1) of dengue virus has not been clearly outlined sofar. Besides its confirmed role in viral RNA replication [14], NS1 involvement in assembly and maturation of virus are not conclusively studied. The protein exists in different forms like monomer, dimer and tetramer in different regions of the infected mosquito and mammalian cells [15]. It is expressed as a soluble monomer in infected mammalian cells forms dimer in the lumen of endoplasmic reticulum and subsequently transported to surface where it remains as membrane associated protein or released extra cellularly [16, 17]. NS1 protein is secreted out in blood during acute stage of DENV infection. Detection of NS1 antigen [15] and specific IgM response against NS1 could help in detecting dengue at early stage [18].

NS2B is a small membrane-associated protein [19] and forms a stable complex along with NS3 and act as a cofactor for NS2B-NS3 serine protease [20]. NS3 is a large multifunctional protein, containing several activities required for polyprotein processing and RNA replication. The N-terminal end is the catalytic domain of the NS2B-NS3 serine protease complex [21, 22, 23] cleaving the NS2A/NS2B,

NS2B/NS3, NS3/NS4A and NS4B/NS5 junctions and generating C-terminal region of mature capsid [24, 25] and NS4A proteins [26]. Besides this, it has RNA helicase [22], RNA-stimulated nucleoside triphosphatase (NTPase) activity [27] and RNA unwinding property [28]. NS4A and NS4B are small hydrophobic proteins. NS4A role in RNA replication is evident from the presence of protein in replication complex. NS4B co-localizes along with NS3 and viral double stranded RNA (dsRNA) in ER-derived membrane structures presumed to be site of RNA replication [29, 26]. NS4A and NS4B can also block type I IFN signaling [30]. NS5 is another highly conserved, multifunctional protein with methyltransferase (MTase) and RdRP activities. The N-terminal region of NS5 was found to exhibit homology with S-adenosyl-methionine (SAM)-dependent MTase, and the protein is involved in the modification of the 5' cap [31]. The C-terminal contains RdRPs [32, 33] and NS5 form complex with NS3 [34, 35] and stimulate both NTPase and RTPase activities of NS3 [36, 37].

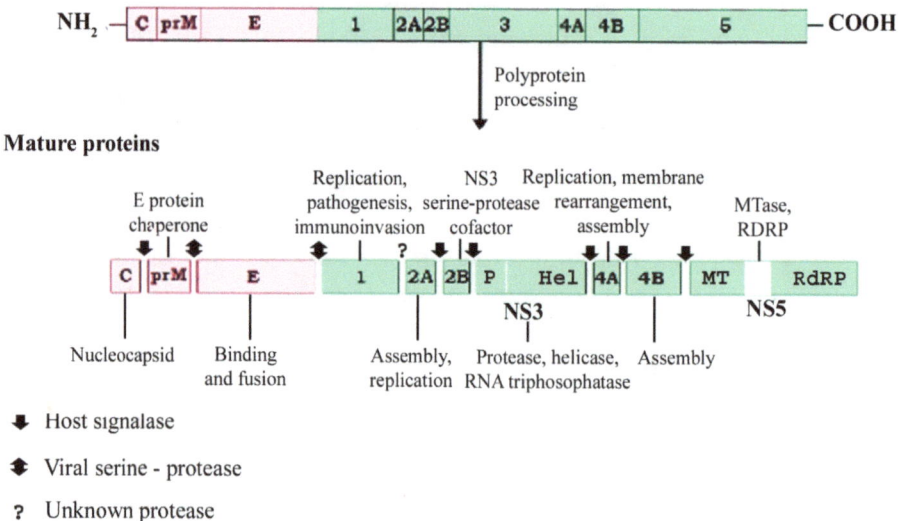

Figure 3: Flavivirus genome organization and poly-protein processing.

Viral Replication

Flaviviruses are considered to be cytoplasmic as most of its life cycle (translation, replication) takes place inside. They avoid host immune response by their virally encoded nucleoproteins which alter host gene response and expression [38]. The viral RNA dependent RNA polymerase (RdRP) is the product of viral NS5 gene [39, 40]. The protein also codes for 2'-*O*- Methyl transferase essential for methylation [39]. RdRp forms a replication complex consists of viral and host cell proteins. The replication complex uses genomic RNA to generate double stranded replication form which act as a template to produce single stranded viral RNA [41].

Cell surface receptors responsible for internalization of virus in to host cell are not clearly defined so far. Several ligands and surface molecules are thought to be receptors for flavivirus (Table **1**).

Table 1: List of identified flavivirus receptors in humans

Receptor	Expression Cells	References
DC-SIGN/L-SIGN/CD-209	Dendritic cell lectin	[42, 43, 44]
Mannose receptor	Macrophages	[45]
HSP70/HSP90	Macrophages	[46]
GRP78	Plasma membrane	[47, 48, 49]
Laminin receptor	Laminin	[50, 51]
DC-SIGNR (WNV)	Dendritic cell lectin	[52]
Heparan sulfate	Cell surface and Matrix proteins	[53, 54]
Fc receptor	Immunoglobulins	[55, 56]
C-type lectin domain family 5, member A (CLEC5, MDL-1) (DENV)	Macrophages	[57]
α_vB3 integrin (WNV, JEV)	Endothelial cells	[58]
Rab GTPase (DENV, WNV)	Peripheral membranes	[59]

Most of the identified receptors were discovered while studying DENV infection. Except a few unusual receptors for some specific flavivirus, all the flaviviruses share common receptors for cell entry. Besides the identification of putative cellular receptors for human and mammalian cells, several receptors were identified for insect cells (reviewed by Hidari and Suzuki [60], Alen and Schols [61]). YFV does not have any glycan modifications on protein E and it attaches to cells in lectin-independent manner [62].

In host cell, virus entry and internalization *via* clathrin- mediated endocytosis is mediated by E protein [63, 64]. Low pH of endosome disrupt virus bounded by lipid bi-layer, and releases the virus in to cytoplasm [65]. The RNA is translated in to polyprotein which is cleaved by host and cellular proteases to structural and nonstructural proteins [66]. Virus assembly takes place in the endoplasmic reticulum (ER) [67]. RNA and capsid proteins are enveloped leading to the formation of non-infectious immature virions. These immature particles are transported to golgi apparatus and at the acidic pH of golgi network, furin mediated cleavage results in the conversion of prM (pre-membrane) to M (membrane) leading to the maturation of virus [68, 69]. Maturation is accompanied by structural rearrangements of glycoprotein shell [70, 71].One of the characteristic feature of the flavivirus life cycle is the production of capsidless, membrane-containing particles usually found in the supernatants of flavivirus-infected cells [72]. These capsidless particles can be generated recombinantly (recombinant sub viral particles) by co-expressing prM and E proteins, which indicates the non-requirement of capsid for viral particle formation in ER [73]. These particles also lack genomic RNA, and are non-infectious [74]. Mature virus and sub-viral particles are released from the host cell by exocytosis (Fig. **4**).

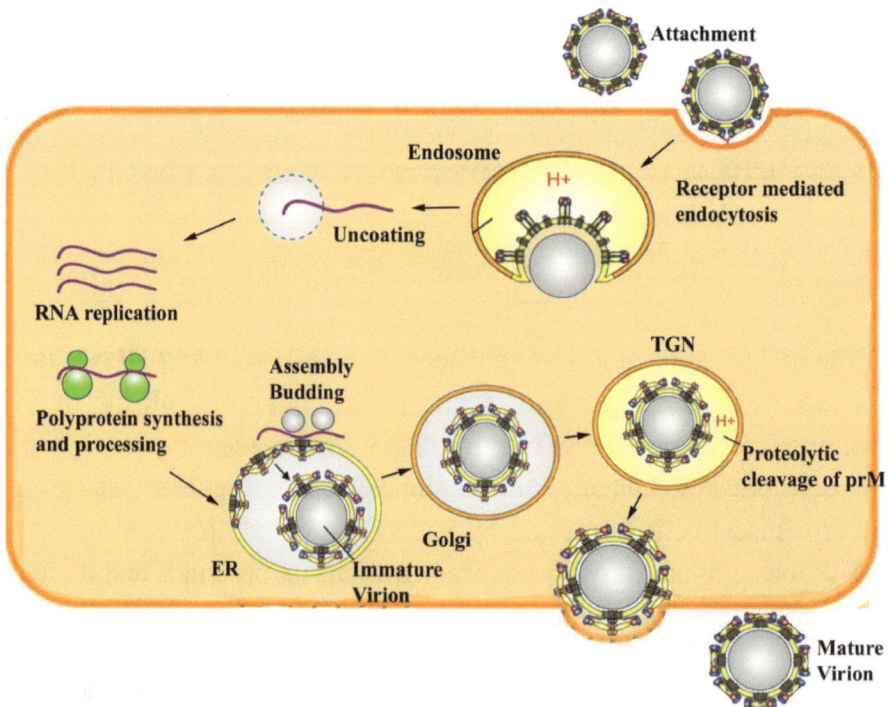

Figure 4: Life cycle of Flavivirus.

Medical Importance of Flavivirus

Despite the long history of flaviviruses and their medical importance, the availability of effective vaccines and better antiviral treatments are rare. Re-infestation of vector *Aedes aegypti* (DENV, YFV), failure in vaccination programs in endemic sub-Saharan Africa countries (YFV) [75, 76], increase in travel (YFV, DENV) and commerce from endemic regions causing epidemics in urban areas (*e.g.* YFV in southern US parts) [77, 78, 79], unplanned urbanization and breakdown of vector control efforts (DENV) are the main reasons for the continued circulation and burden of the arthropod borne flavivirus. Dengue is a major worldwide public health problem with an estimated 100 million annual cases of dengue fever (DF) and 500,000 annual cases of dengue hemorrhagic fever (DHF), the severe form of the disease, resulting in about 25,000 fatal cases, mainly in children under the age of 15. Primary dengue infection is usually subclinical or manifests in to an acute, self-limited febrile illness (DF) and results in production of long-term immunity to the infecting serotype [80]. A secondary infection by a heterologous serotype may results in dengue hemorrhagic fever (DHF) or dengue shock syndrome (DSS) [81, 82]. The pathogenesis is mediated by pre-existing antibodies produced against the previous infection enhancing the clinical severity through antibody dependent enhancement (ADE) [83-87]. DHF/DSS has high mortality rate of 5% [88].

YF is a another major public health problem with an estimated 200,000 cases and 30,000 deaths annually, Even though, the safe and efficacy of the vaccine 17D-204 has been documented, use of this vaccine is restricted to children below 6 months of age, persons hypersensitivity to eggs, immuno-suppressed/ compromised individuals either by HIV/Leukemia, Lymphoma, radiation therapy, drugs. Studies have also shown that persons aged ≥65 years are susceptible to systemic adverse events following vaccination [23, 70, 89, 90] and there is an increase in post-vaccine adverse effects during mass vaccination [91-95]. Once the clinical symptoms sets in there is no specific treatment available for YF, and the disease has a case-fatality rate of approximately 20% [96]. WNV is a part of Japanese encephalitis (JE) antigenic complex, found in both tropical and temperate regions. It mainly infects birds, but also known to infect humans, horses, dogs, cats, bats. The disease can cause cognitive dysfunction and flaccid paralysis [97, 98]. West Nile control is achieved through mosquito control, by the elimination of mosquito breeding sites, breeding areas and by personal use of mosquito repellents. Though an effective WNV vaccine for horse (West Nile-INNOVATOR) is available, vaccine for human WNV has not been found out sofar.

Japanese encephalitis (JE) is the most common cause of viral encephalitis in the Asian Pacific region. JE carries with a heavy burden of permanent neuropsychiatric sequelae. Mental retardation developed from this disease usually leads to coma. Mortality of this disease varies but is generally much higher in children. Some 50,000 cases of JE occur annually, with 25%-35% case fatality rates, and more than 30% with severe long-term disabilities in survivors [99]. There is no specific treatment for Japanese encephalitis and treatment is supportive; with assistance given for feeding, breathing or seizure control as required. Infection with JEV confers lifelong immunity. All current vaccines are based on the genotype III virus. A formalin-inactivated mouse-brain derived vaccine was first produced in Japan during 1930s and widely used. The widespread use of vaccine and urbanisation has led to control of the disease in Japan, Korea, Taiwan and Singapore. The high cost of the vaccine, (which has to be grown in live mice), renders the use of vaccine in under developed countries for routine immunization programme [100].

Vaccines Against Flavivirus Infection

Vaccination is the most effective means of disease prevention for the flavi-viral infections. A live attenuated vaccine for YF and inactivated vaccine for JE and TBE have reduced the incidence of the disease drastically, while licensed vaccines for DEN and WN are still under development. The first flavivirus vaccine was developed over 60 years and it's for YF [101]. The vaccine made from attenuated YF 17D strains and still remains one of the safest and most effective vaccine ever developed. Major strategies of vaccine development against DENV, WNV are discussed below elaborately.

In Activated Whole Virus Vaccines

Inactivated virus vaccines require small animal or cell culture, for producing virus in sufficient quantity. The crude antigens/ viruses are then purified, inactivated, and formulated with adjuvants. If the inactivated virus retains its antigenic structure, the resulting vaccine will provide adequate protection. Most of the successful vaccines available now are inactivated vaccines.

Dengue

Attempts to develop inactivated vaccines for DEN began over 70 years ago, when phenol, formalin, or ox bile were used to inactivate DEN viruses in infectious mosquito pools or viremic plasma. Initial studies were conducted using DEN-2 virus established in fetal rhesus lung cell cultures. Supernatants were separated, filtered and inactivated at 22°C with 0.05% formalin. This unpurified vaccine induced anti-DEN neutralizing antibodies in mice and protects vaccinated animals from lethal virus challenge. Similarly, Putnak and his colleagues [102, 103], used low-passage strains of DEN virus adapted to grow at high titers and prepare prototype DEN-2 purified inactivated vaccine (PIV). Virus harvested from serum-free supernatant culture fluid, was concentrated, purified on sucrose gradients and inactivated with 0.05% formalin for 10 days at 22°C [101]. This vaccine contained viral structural proteins E, prM and trace amounts of NS1. Mice immunized with two 1.5-µg doses of vaccine adjuvanted with 0.01% alum given for one month produce high-titered virus neutralizing and hemagglutination inhibiting (HAI) antibodies. Rhesus monkeys that received vaccines demonstrated significant protection against virus challenge after 2 months of vaccination.

Japanese Encephalitis

JE infection usually subclinical has the potential to cause fatal encephalitis and life-long neurologic problems in survivors. Before the widespread use of inactivated JE vaccine in Japan and inactivated and live-attenuated vaccine in China, thousands of cases of JE occurred each year in those countries [104, 105].

JE remains a serious health threat in many other countries where vaccination is not routinely practiced. A purified inactivated vaccine developed at the Research Foundation for Microbial Diseases of Osaka University (BIKEN) in Japan, made from wild-type strain of the JE virus (Nakayama or Beijing) propagated in the brains of suckling mice, formalin-inactivated, purified, and formulated with gelatin stabilizer [106], has been in continuous use in Japan for over 30 years, where it was proven to be relatively safe and effective.

A large field efficacy trial sponsored by the United States Army was performed in Thailand to compare monovalent BIKEN vaccine with a bivalent version made from Nakayama and Beijing-1 strain viruses [107] and the efficacy of both monovalent and bivalent vaccines was found to be 91%. There were no major side effects or serious adverse reactions. Minor side effects and reactions such as headache, sore arm, rash, and swelling were observed. This vaccine continues to be produced in Japan and now commercially distributed outside of Japan by Aventis-Pasteur under the trade name JE-VAX. An inactivated vaccine made from the P3 strain of JE virus grown in primary hamster kidney cells is used in China, although lately it has been largely supplanted by the SA14-14-2 live-attenuated vaccine, which is cheaper to produce and more efficacious [108]. The SA14-14-2 virus, after additional passages in primary dog kidney (PDK) and Vero cell cultures, was used to make second-generation, inactivated purified JE vaccine [109].

Tick-Borne Encephalitis Virus

Tick-Borne Encephalitis virus (TBE) is of great concern to Central and Eastern Europe before vaccination program in 1973. Inactivated mouse brain–derived virus was the first vaccine to made, followed by inactivated vaccine made from virus grown in chick embryo cells [110].

These first-generation vaccines were highly, effective but also reactogenic [111]. Second generation TBE vaccine developed in 1980 used continuous-flow zonal ultracentrifugation for purification of formalin-inactivated virus. The resultant vaccine, was 100-fold effective and pure than previous preparations and highly immunogenic and less reactogenic [112]. Further improvements with inactivated vaccine from TBE strain K23, purified by zonal centrifugation stabilized with depolygeline and 0.2% aluminum hydroxide as an adjuvant was reported in 1990 [113]. Clinical trials conducted with the vaccine (**Encepur**) testing at 0 and 28-day with doses ranging from 0.03 to 3 mg, demonstrated that the vaccine was well tolerated with a total of 10 adverse vaccination events including local reactions, headache, flu-like symptoms, and nausea. The vaccine was immunogenic in a dose-dependent manner.

The frequency of adverse events (both vaccine-related and unrelated) following primary vaccination was 37% for recipients of the conventional schedule and 46% for recipients of the abbreviated schedule. These incidences decreased to 9% and 21% respectively following the second dose, and 5% and 15% following the third dose. Side effects included were headache, fever, and local reactions; however, these were reported to be generally mild and no serious adverse events were reported. Later large scale trials showed that the antibody titers in children who received lower doses have approximately same titers of antibody like adults who received the full dose. However, this effect diminished with age and was not seen in children older than 12 years of age.

West Nile Virus

West Nile virus causes disease varying from asymptomatic to febrile influenza-like illness and to lethal meningo-encephalitis [114]. The name "West Nile" derives from the location from where it was first isolated (West Nile district in Uganda). WN disease used to be considered a minor risk for human and horses. Before 1999, West Nile (WN) virus disease was primarily a problem confined to the Middle East and Africa. In 1999, the first outbreak in the United States occurred in New York and now the virus has spread across United States.

A DNA vaccine encoding prM and E proteins, an inactivated cell culture derived whole virus and a canarypox-vector recombinant vaccine have all been approved for veterinary use in the USA [114, 115]. Formalin-inactivated WN vaccine produced in Israel from infected mouse brain was found to protect geese up to 94% from lethal WN virus challenge following immunization with two doses [116]. These results suggest that similar approaches can be used for developing an inactivated WN vaccine for human use. Several potential WNV vaccine candidates are in the course of development for human use. The live attenuated Schwarz strain of measles virus expressing the secreted form of E protein from a virulent strain of WNV induced high level of NAb and showed protection against a lethal challenge of WNV in mice [117]. Preclinical trials of a live attenuated WNV/ Dengue 4 chimeric virus showed moderate to high titers of NAb in a non-human primate model [118]. The most recent vaccine candidate for WNV is ChimeriVax-WN02, a live attenuated recombinant chimeric YF-17D backbone with WNV prM and a mutated E. Other vaccine approaches that have shown promise in mouse models includes a live attenuated WNV isolate and RepliVax WN, a defective pseudo infectious WNV, lacking functional C gene [118, 119, 120] (explained under 'Chimeric vaccines').

Yellow Fever Virus

YFV causes approximately 200,000 cases annually of which 90% occur in Africa alone. Besides this it also poses threat to millions of people who travel to endemic areas [121]. It causes a broad spectrum disease including fever, hemorrhage, jaundice and myocardial injury.

The first effective YF vaccine the French neurotropic vaccine (FNV) was developed by passaging the French viscerotropic virus (Dakar) in mouse brain [122]. Though it had high efficiency, FNV was discontinued due to high incidence of adverse effects [123]. The second vaccine YF-17D, derived from clinical sample was repeatedly passaged on mouse and chicken embryonic tissues for 176 passages, resulting in the original YF-17D strain [100]. To eliminate the inconsistency amongst vaccine lots, the World Health Organization (WHO) developed the seed lot system for the production of YF vaccine in 1945. The primary seed lot with a determined number of virus passages was established. Currently, two sub-strains 17DD and 17D- 204 are used as YF vaccines. YF-17DD, which at passages 287 - 289, is being used in Brazil, and YF-17D-204, which at passages 235 - 240, is used in other countries [124]. Both vaccine strains differ from the parental strain by 48 nucleotides, which results in 22 amino acid substitutions [125]. A single dose of vaccine provides a protective level of NAb's in 90% of the recipients within 10 days of immunization and increases up to 99% in 30 days [126]. Though the immunity can lasts for 30 years, WHO recommend revaccination after 10 years [124, 127]. Common adverse effects are generally mild and severe adverse effects are extremely rare. YF vaccine associated neurotropic disease (YEL-AND) [123] and more severe form of adverse event called vaccine associated viscerotropic disease (YEL-AVD) has been described. All YF-17D vaccine produced around the world is propagated in embryonated chicken eggs; individuals who are allergic to egg should not receive the vaccine [123]. (Review by Frierson [128] includes a detailed view of historical background on the progress and development of Yellow fever vaccine).

Live Attenuated Vaccines

DENV

There are no animal models that reproduce human dengue fever or dengue hemorrhagic fever. Mice are sensitive to intracerebral inoculation and generally only after neuroadaptation.

Initial attempts to develop live attenuated vaccines was tried in early 1940s and 50s [79, 129]. Sabin and Schlesinger (1945) propagated Dengue virus 1 (Hawaii strain) in mice by intracerebral inoculation. Test of the virus after seventh, ninth, and 10th mouse passage indicated changes in pathogenicity for humans. Successfully after this pilot attempt, several researchers around the world tried to develop a live attenuated vaccine using different dengue types and strains [130, 131]. The genetic basis of attenuation of the Sabin-Wisseman vaccine has been discovered. Ten amino acid differences separate the parental and mouse-adapted DEN2 vaccine strain developed by Sabin. Dissection of mutations implicated a change from negatively charged to positively charged amino acid (Glu $\longrightarrow$ Lys) at E126 within the hinge region of the E glycoprotein [132].

Interestingly, another experiment involving neuroadaptation of DEN 1 virus by passage in mouse brain was associated with increased neuro virulence for mice and apoptotic cell death in neural cells but reduced apoptosis in human hepatocytes in culture [133]. These divergent effects on the induction of apoptosis in neural vs hepatic cells were associated with mutations in the virus replication and assembly regions: the hinge region at E196 (Met $\longrightarrow$ Val), the interface between domains I and III at E365 (Val $\longrightarrow$ Ile) as well as two mutations in the proximal stem-anchor (E405, Thr $\longrightarrow$ Ile) and the NS3 helicase region. Researchers from Mahidol University, Thailand have developed live-attenuated vaccine for dengue 1, 2 and 4 from primary dog kidney cells and dengue 3 from green monkey kidney cells. These candidate vaccines were found to be safe and immunogenic [134]. While monovalent dengue vaccines found to produce type-specific antibody response, tetravalent vaccine failed protect flavivirus naive-North American volunteers [135].

JEV

Besides the whole virus inactivated vaccine, a live-attenuated JE vaccine has been licensed in China. The parental strain (SA14) was isolated from *Culex pipiens* larvae in 1954. The vaccine was obtained by serial passage on primary hamster kidney (PHK) cells, plaque selection and cloning in chick embryo cells. Subpassage in subcutaneous tissue of suckling mice followed by two cloning passages in PHK cells were also done to get a stable non-neurovirulent virus (SA14-14-2).

Phenotypic characteristics of the SA14-14-2 strain include small plaque size and reduced mouse neurovirulence. Liquid vaccine is lyophilized in the presence of a stabilizer consisting of gelatin (1%) and sucrose (5%). The vaccine titer must exceed 106.7 TCID50 per milliliter. The genome sequence has been determined for the parental SA14 virus (from several different laboratory passages) and the attenuated vaccine SA14-14-2 grown in primary hamster kidney cells [136] and primary dog kidney cells (PDK) [118] (Table **2**).

Table 2: Amino acid changes observed in live attenuated vaccine candidates

Amino Acid (in envelope protein)	SA14 [137]	SA14-14-2 PHK [138]	SA14-14-2-PDK [136]
107	Leu	Phe	Phe
138	Glu	Lys	Lys
176	Ile	Val	Val
177	Thr	Ala	Thr
243	Glu	Glu	Lys
244	Glu	Gly	Gly
264	Gln	His	Gln
279	Lys	Met	Met
315	Ala	Val	Val
439	Lys	Arg	Arg

The numbers given in the superscript are the respective references.

Subunit Vaccine Candidates

Recombinant subunit vaccines may provide better efficiency than other methods. Subunit vaccines are safe as they eliminate the chance for exposure to viruses and can be easily manipulated. Although, subunit vaccines promised to be a good agent they are poor immunogens thereby requires several immunizations and must be formulated with a strong adjuvant [127].

WNV

A bacterial vector expressing domain III of WNV E protein has been proposed as a subunit vaccine candidate [133, 139]. Despite the high titer of Abs and protection in murine model, the use of unlicensed adjuvant and long vaccination schedule appear to be significant obstacles [80]. A more promising subunit vaccine is the recombinant truncated E (trE) and NS1 proteins expressed in Drosophila S2 cells. The high yield of the system and the excellent efficacy profile of the vaccine make it one of the best flavivirus subunit vaccines [131, 132].

DENV

Initially *E. coli* was used for producing DENV immunogens and later replaced by mammalian and eukaryotic systems since the antigenicity and immunogenicity of the proteins were poor [122]. Recombinant baculovirus has been used extensively for expressing viral genes in Sf9 cells (reviewed in [140]). Baculovirus-expressed DEN-1, DEN-2, and DEN-3 E proteins, truncated by varying amounts at the C-terminal region elicits low-titered neutralizing antibodies and conferred partial protection in mice [141,128,117,118]. Recently, lipidated consensus DENV E domain III has shown to be capable of activating antigen-presenting cells and enhancing the cellular and humoral immune response. They were also able to elicit neutralizing antibodies against DENV 1-4. The study also shows, a strong memory response in response to challenge and offers protection [127].

DNA Vaccines

DNA vaccines have been explored for development of flavivirus vaccines [142,143]. Injected plasmid DNA containing viral antigens induces both arms of immune system and confers protection against flavivirus infection in animals like mice [144] and monkeys [145, 146]. Usually plasmids containing virus like particles (VLPs), are used for DNA vaccines. These VLPs contain prM and E proteins and form capsid free virus like particles. They are secreted from cells and induce neutralizing antibodies and strong immune response [147, 148-150, 151].

Recent researches are aiming at increasing the immunogenicity of the DNA vaccine. Some of the methods include,

Optimization of vector delivery, increasing transfection efficiency in to cells: *in vivo* electroporation, micro-particle injection, use of gene gun [152,153]

Co-Expression of foreign genes and sequences: administration of cytokine along plasmid addition,

With viral antigen to modulate immune response: human immuno-stimulatory sequence [154].

Japanese Encephalitis Virus

Amongst the JEV viral proteins, E protein seems to be the most suitable for plasmid DNA based vaccines, since antibodies against E protein are capable of neutralizing JEV activity. Plasmid constructs containing E protein with prM protein have been shown to provide a protective immune response to lethal JEV challenge in adult mice [155,156]. There were some DNA vaccines utilizing non-structural proteins of JEV. Immunization of mice with plasmid DNA constructs containing JEV NS1 provided 90% protection against lethal challenge with JEV while plasmids bearing longer constructs failed to provide protection [157]. However, vaccines utilizing JEV NS5 or NS3 failed to raise an effective immune response [158-160].

DEN Virus

DNA vaccines based on the E and prM protein have been reported by several researchers [161], as prM act as a chaperone molecule for protein folding during virus exit [145, 162]. Several DNA vaccines constructed yields low levels of neutralizing antibodies and confers partial or short-term protection in various animal models ([160] and reference therein). Recently, Azevedo and his colleagues (2011) have shown that neutralizing antibodies elicited against the whole E protein confers more protection than domain III of E protein alone in mice [161].

RNA Vaccines

Application of *in vitro* synthesized infectious RNA or DNA, from which infectious RNA is transcribed, has been studied for its ability to induce protective immunity without any adverse effects. As there is no spread of virus in the host, the method is safe as like inactivated or subunit vaccines. *In vivo* replication of the subgenomic RNA at the inoculation site and the expression of nonstructural proteins induce better immune response than inactivated or subunit vaccine and this is similar to one elicited by a live virus vaccine [163]. Two experiments with TBE have shown the efficiency of RNA vaccines.

Direct inoculation of *in vitro* transcribed flavivirus RNA from attenuated infectious (470 nt deletion in the 3′ UTR) TBE [164] and non-infectious [61] amino acid deletion in the C protein) TBEV [139] protects immunized mice against lethal dose of wild-type TBEV. Though both tests provide highly titer and protective immune responses in mice, cytotoxic T cell responses are similar and antibody titers were 3–10 times lower than those observed when mice were immunized with a widely used commercial TBEV vaccine [139]. Single immunization with 1 µg of the replicon induced high neutralizing antibody titers lasting for 1 year [165].

Chimeric Vaccines

The yellow fever vaccine 17D is an attenuated form of Asibi strain and it is one of the highly efficient vaccine used for human vaccination (over 70 years) [166]. The significant property of the virus includes, limited viral replication with greater dissemination of the viral mass producing a robust and long-lived immune response [167, 168]. It also induces T cell response [169-172]. YF 17D virus has become very attractive as an expression vector for the development of new live attenuated vaccines [173, 174].

YFV 17D genome cloned as cDNA

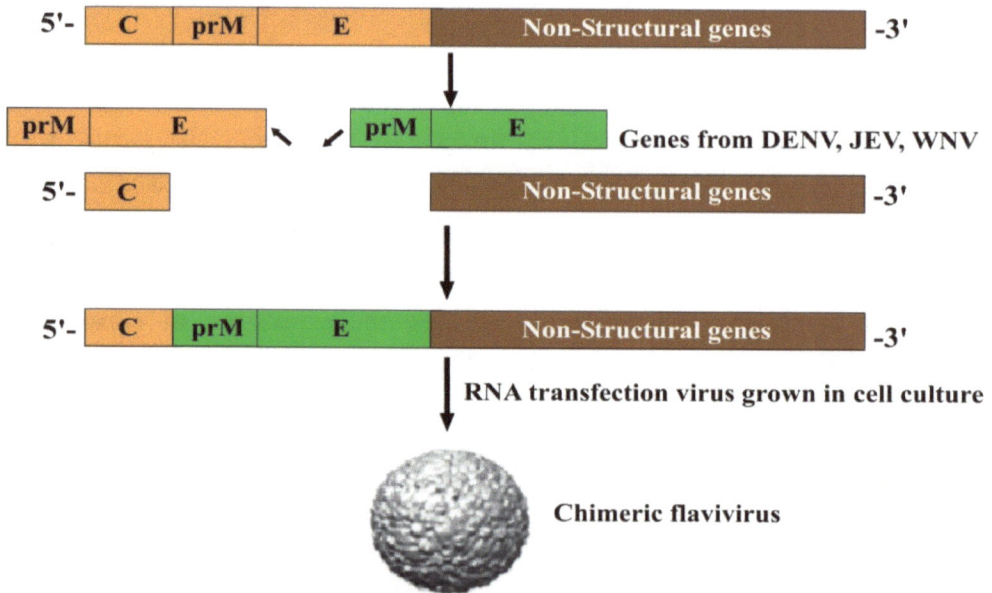

Figure 5: An example for chimeric flavivirus vaccine designed using YFV 17D genome.

Several chimeric flavivirus vaccines use YFV 17D viral vector (which expresses YFV C and NS protein) to deliver heterologous flavivirus prM and E proteins (Fig. **5**). Some of this vaccines produced by Acambis are already in clinical trials (ChimeriVax-WNV, ChimeriVax-JE, and ChimeriVax-DEN). A second-generation WNV vaccine (ChimeriVax-WNV02) engineered by introducing three point mutations, two in E protein to attenuate the virus and another one to reduce the neuro-invasiveness in mice has been tested in monkeys. This vaccine protects against lethal wild-type WNV challenge [175]. Similarly, prM/E genes from the attenuated SA14-14-2 strain were used to replace corresponding genes in YFV 17D and point mutations introduced in the ChimeriVax-WNV02 have also been introduced in ChimeriVax-JE. The modified vaccine candidate induces high titers of neutralizing antibodies in humans [176]. For dengue, a tetravalent dengue vaccine ChimeriVax-DEN1-4 has been tested in monkey models. Monkeys were protected from DENV challenge (100% from DENV2 and DENV3; 83% for DENV1 and DENV4) 6 months after a single immunization [177, 178]. Another vaccine candidate D2/WN replacing DENV prM/E genes with WNV, is highly immunogenic and induces protective immunity against WNV in mice. The DENV PDK-53 backbone has also shown to highly effective and immunogenic [178]. The chimeric vaccines contain the structural proteins of each of the serotypes (C, prM, and E) together with the DEN-2 PDK-53 nonstructural protein and ancillary sequences [179]. Recently sequential immunization with heterologous chimeric flavivirus antigens and its effect on immunity has been studied. Secondary homologous or heterologous immunization induces cross-reactive CD8+ T-cell response depending on both sequence of infecting viruses and variation in their epitopes [180].

Antiviral Agents Against Flaviviruses

Availability of safe and efficient vaccine for controlling JEV, YFV and TBEV has resulted in a more potential, focused approach for targeting DENV, WNV. Anti-flaviviral agent targeting DENV has been in the center stage for a long time. An effective anti-flaviviral agent should be i) administered orally *i.e.*, the oral availability should be high ii) administered not more than few times a day iii) highly rigid in developing resistance iv) very safe, since the target group comprises mostly of children v) thermally stable, as the intended use is mainly on tropical areas vi) production should be simple and be low-cost [5]. Many approaches are being evaluated for designing antiviral agents and some of them includes i) designing based on crystal structure of viral proteins/ based on secondary and tertiary structures ii) identification of small molecule inhibitors iii) screening chemical libraries iv) testing known antiviral agents of other viruses v) modifying existing inhibitors to enhance antiviral effect.

Inhibition of Flavivirus Assembly

Compounds like N-nonyl deoxynojirimycin, N-butyl deoxynojirimycin and 6-*O*-butanoyl castanospermine inhibit DEN-2 and WNV infection (Fig. **6**) [181, 182]. N-nonyl deoxynojirimycin competitively inhibit ER α-glucosidase, disturb maturation, secretion and function of viral glycoproteins [183]. Derivatives of deoxynojirimycin with an oxygenated side chain and terminal restricted ring structure shows higher antiviral activity with low cytotoxicity. These modified iminosugar derivatives inhibit DENV infection *in vitro* with 90% effective at sub micro molar concentration [183].

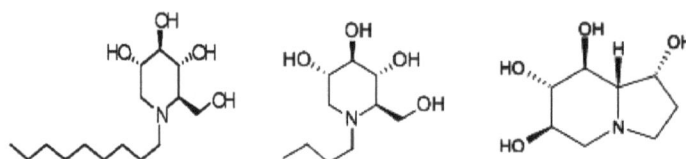

N-nonyl deoxynojirimycin N-butyl deoxynojirimycin Castanospermine

Figure 6: Flavivirus assembly inhibitors.

Castanospermine is a natural alkaloid derived from black bean (*Castanospermum australae*) [184] which has shown to block DENV-1 infection in BHK cells [185]. Later experiments by Whitby and his colleagues have shown castanospermine able to inhibit all DENV serotypes but not YFV and WNV. They also shown *in vivo* that, it prevents mortality associated with DENV infection in mice pre-infected with DENV [186]. Inhibitors of the c-Src protein kinase exhibit a potent inhibitory effect on dengue virus (serotypes 1–4) and murine flavivirus Modoc. Studies on the mechanism of action demonstrated that the c-Src protein kinase inhibitor dasatinib prevents the assembly of dengue virions within the virus-induced membranous replication complex [187]. Recently Anwar *et al.* [188] has identified SFV785 a kinase inhibitor, has selective effects on NTRK1 and MAPKAPK5 kinase activity and has anti-viral activity on DENV and yellow fever viruses. SFV785 inhibited DENV propagation without inhibiting DENV RNA synthesis or translation. Precisely, SFV785 inhibited the recruitment and assembly of the nucleocapsid in specific ER compartments during the DENV assembly process and ultimately the production of infectious DENV particles.

Viral Entry Inhibitors

Inhibition of viral entry in to host cell would be an advantage since it is the first step in RNA replication cycle. Some of the antiviral compounds like sulfated polysaccharides and polyoxotungstates [189] inhibit DENV and WNV infections. Sulfated polysaccharides inhibit DEN-2 (IC$_{50}$ 1.0>g/ml) adsorption and internalization in human and monkey cells not in mosquito cells [190, 191]. Peptide inhibitors targeting E protein have been studied. DN59 peptide (MAILGDTAWDFGSLGGVFTSIGKALHQVFGAIY) corresponding to stem region of dengue virus E protein inhibit plaque formation (>99%) at <25µM concentration and surprisingly it inhibits WNV also. WNV peptide inhibitor

WN83 (TFLVHREWFMDLNLPWSSA) corresponding to domain IIb of E protein inhibit WNV infection but not DENV infection [192]. Peptides corresponding from position 442 to 444 derived from stem region of E protein of DENV2 have shown to inhibit virus entry and affect late stage fusion intermediates [193]. High throughput docking (HTD) enabled searching on hydrophobic pocket region of dengue E protein, yielded compound 6 as one of the inhibitors. Cell culture studies have shown that with an average EC_{50} of 119 nanomolar against dengue virus serotype 2 and type 3. Mechanism of action studies demonstrate that compound 6 acts at early stage during dengue virus infection. It inhibits NS3 expression [194].

Protease and Helicase Inhibitors

Inhibition of helicase and protease activities of NS3 will be a useful target since they are responsible for attaining a productive flavivirus infection. Studies with WNV have shown that, HMC-H04 a nucleoside analog inhibits WNV replication *in vitro* [195]. Similarly, other small flavivirus-specific helicase inhibitors like halogenated benzimidazoles, benzotriazoles have shown to inhibit WNV NS3 mediated RNA unwinding. One of the advantages of these compounds is they specifically act against viral helicases [196-198].

Flavivirus proteases recognize sites containing dibasic aminoacids residues (position P1 and P2) and small amino acid side chain (at position P1). Hence conventional cellular protease inhibitors like PMSF and benzamidine are inactive against DENV and WNV NS3 proteases. Several short peptides mimicking protease cleavage site and binding to catalytic site have been discovered. Ivermectin, a commonly used anti-helminthic drug is a highly potent inhibitor of YFV replication (EC_{50} values in the sub-nanomolar range). Though Ivermectin can inhibit the replication of dengue fever, Japanese encephalitis and tick-borne encephalitis viruses it is very less efficient [199].

Intravenous Immunoglobulin Based Therapy

Intravenous administration of passive immunoglobulin has been studied for anti-flaviviral treatment. The efficiency of human immune γ-globulin in treating mice infected with WNV has been well documented [200, 201]. Studies conducted in

Israel, revealed that administration of pooled human plasma from healthy donors who have WNV antibodies, protect mice from WNV infection. It also confirmed that the recovery depends on dose and time of IVIG administration [200]. Recently, Eagle and Diamond [201] have shown that WNV CNS infection can be treated by administration of IVIG.

Nucleic Acid Based Approaches

Microinjection of modified phosphorothiolate antisense agents to DENV-2 reduces virus titer by 50 to 75% after post infection administration [202]. Reports of RNAi silencing in suppressing DEN-2 [203] and WNV infection in cell culture [204] has been studied. The inability of RNA interference to target flavivirus RNA synthesis unless provided intracellularly prior to viral infection [205, 206] is related to the lack of unavailability and accessibility of nucleases and proteases for membrane bound flavivirus replication complex [207].

Phosphoro-diamidate morpholino oligomers (PMO) targeting complementary RNA sequences located at both 5' and 3' West Nile virus genome were studied by Deas and his colleagues [208]. They showed that PMOs targeting 20 nucleotides at 5'terminal results in inhibition of viral RNA translation, whereas PMOs targeting 3' terminal involving in viral genome cyclization results in suppression of viral RNA replication rather than translation inhibition. Virus titers were reduced by 5 to 6 logs at 5µM concentration without any detectable cytotoxicity. Conjugation of Arg-rich peptides to PMOs greatly enhances cellular delivery of oligomers.

Table 3: Nucleoside tri-phosphate synthesis inhibitors and their classifications

OMPDC İnhibitors	6-azauridine, 6-azauridine acetate, pyrozofurin
IMPDH inhibitors	Ribavirin, Mycophenolic acid Ribavirin act as a mutagen and as a analog of RNA cap [209, 210] Mycophenolic acid binds directly to IMPDH, decreases guanosine to a level where RNA synthesis is affected [211]
CTP inhibitors	Cyclopenetyl cytosine The active metabolite cyclopentenyl cytosine 5'-triphosphate competitively inhibits cytidine triphosphate (CTP) synthase, depleting intracellular cytidine pools and inhibiting RNA synthesis

Nucleoside Tri-Phosphate Synthesis Inhibitors

Nucleoside triphosphate synthesis inhibitors inhibit either one of the three pathways: Orotidine monophosphate decarboxylase (OMPDC), inosine monophosphate dehydrogenase (IMPDH) and CTP synthesis (Table **3**). Some of the compounds included in this category are, Mycophenolic acid, 6-azauridine, 6-azauridine acetate, Ribavirin, cyclopenetyl cytosine and pyrozofurin (Fig. **7**).

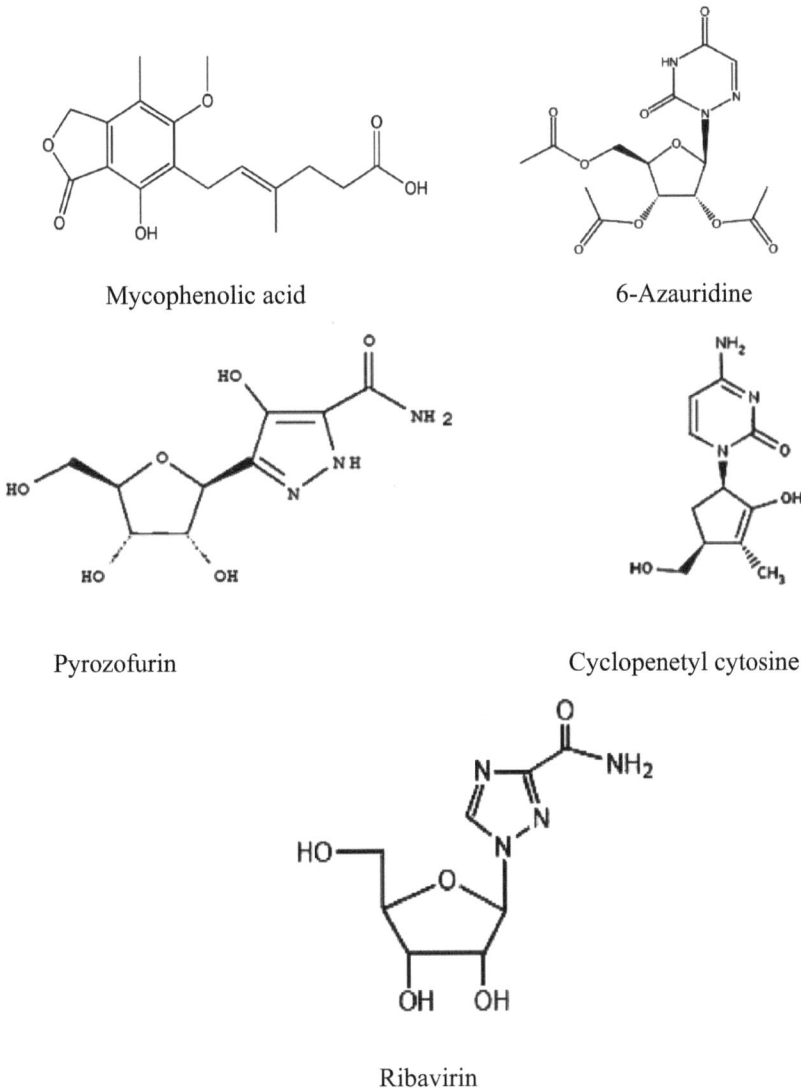

Mycophenolic acid   6-Azauridine

Pyrozofurin   Cyclopenetyl cytosine

Ribavirin

Figure 7: Structure of some nucleoside tri-phosphate synthesis inhibitors.

Nucleoside Analogs

Research conducted at Novartis Institute of Tropical Disease, Singapore identified adenosine analogue NITD008 (Fig. **8**) inhibiting DENV *in vitro* and *in vivo*. The compound act directly on RdRP and cause termination of RNA synthesis. It has broad spectrum of anti-flavivirus activity including WNV, JEV. *In vivo* experiments with DENV infected mice shown that NITD008 able to prevent mortality without any adverse side effects [212].

Another adenosine analog 2-*C*-caetylene-7-deaza-7-carbamoyladenosine (NITD449) has shown promising inhibitory effect on DENV in cell culture but its *in vivo* use is limited as its bioavailability in plasma is low when administered orally. Covalent linking of isobutyric acid to 3' and 5'hydroxyl groups of ribose through ester leads to formation of prodrug 3', 5-*O*-diisobutyryl-2'-*C*- acetylene-7-deaza-7-carbamoyl-adenosin (NITD203) (Fig. **8**) which retained the parental drug NITD449 in plasma for a long time and reduces viremia load by 30 times [213].

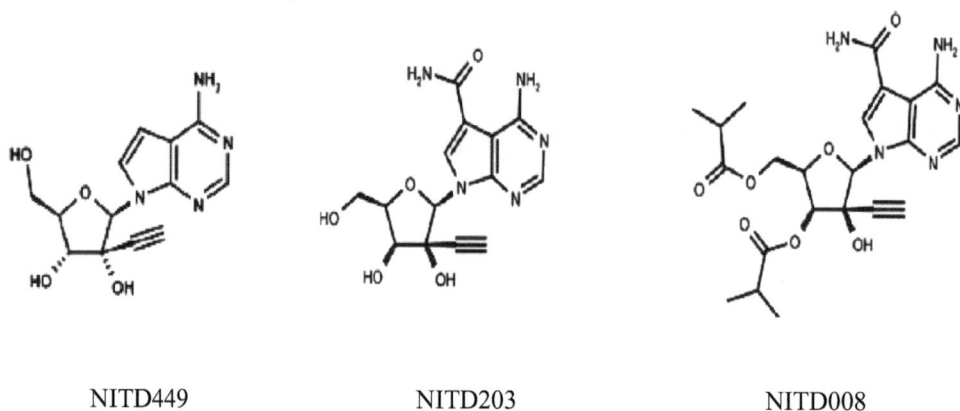

NITD449   NITD203   NITD008

Figure 8: Structure of Nucleoside analogs.

Blocking Methyl Transferase From NS5

Blocking the production of methyl transferase from NS5 could be an attractive target for anti-flaviviral strategy without affecting host cell mechanism as flavivirus MTase recognizes 5' terminus of viral RNA during methylation in a sequence specific manner. This is in contrast to other cellular MTases which mediate capping in a non-sequence specific manner [214]. This paves way for selective blocking of flavivirus MTase. NS5 mutation in both N-7 and 2'-O methylation leads to WNV replication arrest in cell culture [215]. Ribavirin 5-triphosphate, a GTP analog was found to inhibit 2'-O methylation of the DENV-2 MTase by competing with GTP-binding [216]. Using a cell-based assay, Puig-Basagoiti and his colleagues have identified a compound that inhibited the WNV N7 MTase activity with an IC_{50} of 54 µmol/L [217].

Flavivirus methyl transferase (MTase), methylates N7 and 2'-*O* of viral RNA cap (GpppA-RNA→m^7GpppA-RNA→m^7GpppAm-RNA) using *S*- adenosyl-L-Methionine (AdoMet). Sinefugin (SIN) (Fig. **9**) has shown to inhibit WNV infection in cell culture with EC_{50} value of 23µM and CC_{50} value 4.5mM. Similarities between AdoMet and SIN compound pose a problem to host cell since it may invariably affect host's AdoMet utilizing enzymes. Modified SIN analogues interacting specifically with hydrophobic regions of the flavivirus MTases activity has been discovered recently [218].

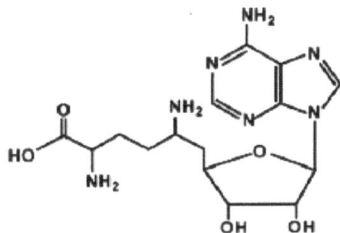

Figure 9: Structure of SIN.

Small Molecule Inhibitors

Most compounds under small molecule inhibitors fall under different categories of flavivirus inhibitors like RNA synthesis inhibitors, viral entry inhibitors and replication inhibitors. As more information's are available in this category, some of the promising small molecule inhibitors are explained under a separate heading.

Sub-genomic cell based replicon screening by Gu and his co-workers [219] resulted in the identification of pyrozolopyrimidine compound with anti-WNV activity. Similarly another study using high throughput screening of compound library reveals a triaryl pyrazoline compound (Fig. **10**) inhibiting WNV epidemic strain at EC_{50} of 28µM concentration. The compound also inhibits DENV, YFV and WNV by suppressing RNA synthesis [220]. Several compounds belonging to secondary sulfonamide family have shown to inhibit WNV, YFV and DENV at sub-micro molar concentration during an evaluation of more than 80,000 chemical compounds. Of the tested ones, compound AP30451 (Fig. **10**) specifically inhibit translation of yellow fever virus [221]. LJ001 (aryl methyldiene rhodanine derivative), another small inhibitor molecule has been found during a high throughput screening analysis. The molecule has shown to inhibit a broad range of enveloped viruses including Ebola, Marburg, Influenza, West Nile and Yellow fever by competitively binding to lipid membrane thus inhibiting the virus-cell fusion leading to the prohibition of virus entry in to cells [150]. High-throughput screening of 235,456 commercially available compounds predict 2-thioxothiazolidin-4-ones family and a molecule within this family (E)-(3-(5-(4-tert-butylbenzylidene)-4-oxo-2-thioxo-1,3-thiazolidin-3-yl) propanoic acid (BG-323)) has able to inhibit GTP binding and guanylyltransferase function of the capping enzyme. Further, the compound also shown to inhibit West Nile and yellow fever viruses in cell culture with low toxicity [222].

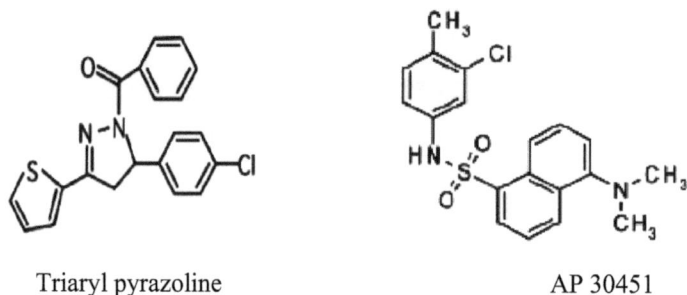

Triaryl pyrazoline   AP 30451

Figure 10: Structure of small molecule inhibitors.

Miscellaneous Anti-Flaviviral Agents

Nonsteroidal anti-inflammatory drugs like aspirin, indomethacin and sodium salicylate were shown to inhibit JEV propagation *in vitro* (Fig. **11**) [223]. The anti-JEV activity of steroid dehydroepiandrosterone (DHEA), a precursor of estrogen and androgen has been found by Chang and his colleagues [223]. DHEA is thought to be involved in the inhibition of glucose-6-phosphate dehydrogenase

(G6PD) activity [224] which lowers the cellular level of ribose 5-phosphate and inhibit ribonucleotide and deoxyribonucleotide synthesis [148]. Two nucleotide analogs, 1-β-D-ribofuranosyl-3-ethynyl-[1,2,4] triazole (ETAR) and 1-β-D-ribofuranosyl-4-ethynyl-[1,3] imidazole (IM18) having ethynyl group and iso-structural relationship with ribavirin exhibits broad range antiviral activity against flaviviruses like DEN 1,3,4, Langat and Mosaic Virus by decreasing cellular GTP level [149]. Alkylated porphyrins and their derivatives especially like chlorine e6 (a metal free chlorophyllide like molecule) inhibits many enveloped viruses, including DENV (EC$_{50}$ of 0.3 nM) [225].

Recently aglycon analogue of antibiotic teichoplanin LCTA-949 has shown to block the binding of virus to host cells. It elicits high antiviral activity against TBEV (EC$_{50}$:0.3 μM) and less to WNV (EC$_{50}$:13μM) [226]. Yang and his co-workers [227] using combinatorial computational approaches targeting envelope protein, identified rolitetracycline and doxycycline derivatives of tetracycline, selectively inhibit the membrane fusion of dengue virus during viral entry (EC$_{50}$ value of 67.1μM and 55.6 μM respectively).

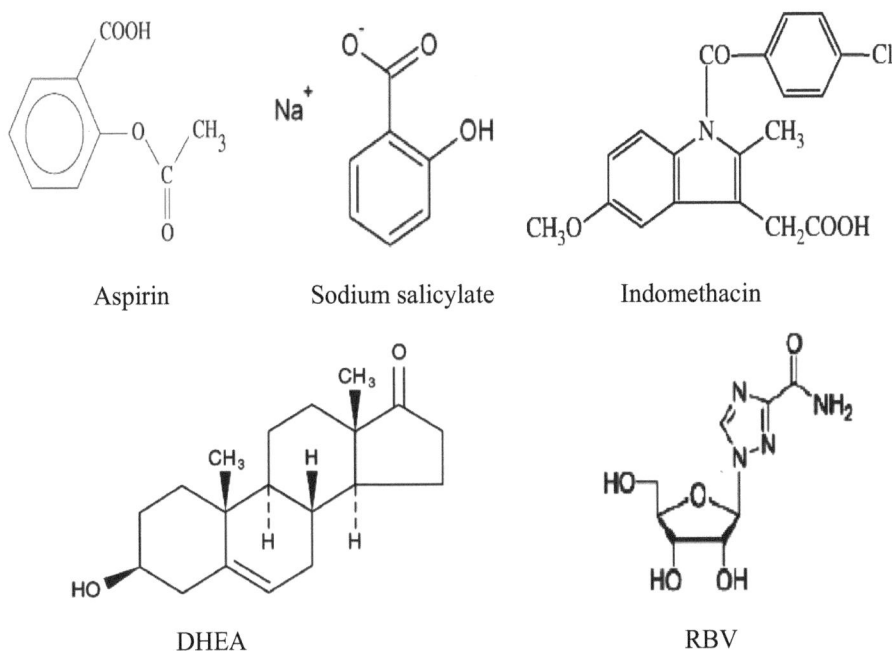

Aspirin   Sodium salicylate   Indomethacin

DHEA   RBV

Fig. 11 contd...

ETAR   IM18

Chlorophyllide   Chlorine e6

Rolitetracycline   Doxycycline

Structure of LCTA-949

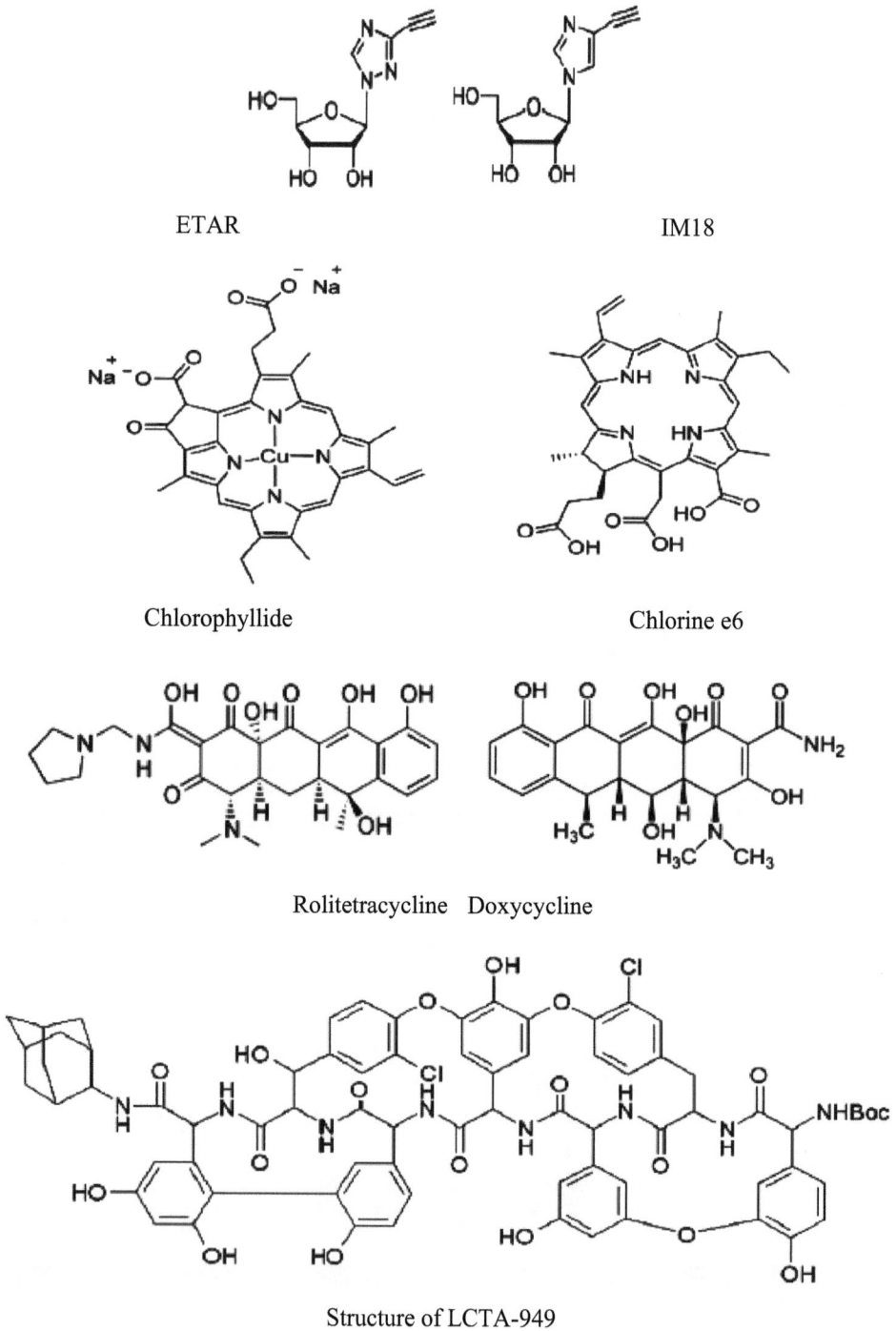

Figure 11: Miscellaneous anti-flavivirus agents.

CONCLUSIONS

The first successful flavivirus vaccine was established for YF as early as 1930's. Since then, only vaccines for JEV and TBEV have been successfully produced. Attempts to produce DENV and WNV vaccines were hindered by the lack of suitable animal models, inability in understanding their pathology, pathogenesis and host immune response against these pathogens. Chimeric vaccines like ChimeriVax- WNV, ChimeriVax-DENV and DNA vaccines containing VLPs are in phase I and phase II clinical trials. Besides these developments, better understanding in virus life cycle and advancement in proteomics and genomics leads to several new anti-viral molecules specifically inhibiting virus translation, replication have been discovered. Several antiviral compounds tested *in vitro* as well as *in vivo* are now in phase I and phase II clinical trials. Both host and viral factors involved in pathogenesis, pathology and immunity needs to be clearly studied, which might help in designing better vaccine and antiviral agents against flavivirus.

ACKNOWLEDGEMENTS

Work on dengue research in author's laboratory is supported by DST-SERC, Government of India and the core facility is supported by DST-FIST. The authors would like to thank D. Immanual Gilwax Prabhu and B. Karpanai Selvan for their technical assistance.

CONFLICT OF INTEREST

The author(s) confirm that this chapter content has no conflict of interest.

REFERENCES

[1]   Thiel H-J, Collett MS, Gould EA, *et al.* Family Flaviviridae. In: Fauquet CM, Mayo MA, Maniloff J, *et al*, Eds. Virus Taxonomy. VIIIth Report of the International Committee on Taxonomy of Viruses. San Diego: Academic Press; 2005 pp: 979–996.

[2]   Cammissa Parks H, Cisar LA, Kane A, Stollar V. The complete nucleotide sequence of cell fusing agent (CFA): Homology between the nonstructural proteins encoded by CFA and the nonstructural proteins encoded by arthropod-borne flaviviruses. Virology 1992; 189:511–524.

[3]   Strode GK, ed. Yellow Fever. New York: McGraw-Hill; 1951.

[4]   Seligman SJ. Constancy and diversity in the flavivirus fusion peptide. Virol J 2008; 5:27.

[5]   Bollati M, Alvarez K, Assenberg R, *et al*. Structure and functionality in flavivirus NS-proteins: Perspectives for drug design. Antiviral Res 2010; 87:125-148.

[6]   Heiss BL, Maximova OA, Pletnev AG. Insertion of microRNA targets in to flavivirus genome alters its highly neurovirulent phenotype. J Virol 2011; 85:1464-1472.

[7]   Kuhn RJ, Zhang W, Rassmann MG, *et al*. Structure of dengue virus: implications for flavivirus organization, maturation and fusion. Cell 2002; 108:717-725.

[8]   Ma L, Jones CT, Groesch TD, *et al*. Solution structure of dengue virus capsid protein reveals another fold. Proc Natl Acad Sci USA 2004; 101:3414–3419.

[9]   Khromykh AA, Westaway EG. RNA binding properties of core protein of the flavivirus Kunjin. Arch Virol 1996; 141:685–699.

[10] Koonin EV. Computer-assisted identification of a putative methyltransferase domain in NS5 protein of flaviviruses and lambda 2 protein of reovirus. J Gen Virol 1993; 74:733–740.

[11] Falgout B, Chanock R, Lai C-J. Proper processing of dengue virus nonstructural glycoprotein NS1 requires the N-terminal hydrophobic signal sequence and the downstream nonstructural protein NS2a. J Virol 1989; 63:1852–1860.

[12] Chambers JJ, Hahn CS, Galler R, Rice CM. Flavivirus genome organization, expression, and expression. Ann Rev Microbiol 1990; 44:649-688.

[13] Lindenbach BD, Rice CM. Trans-complementation of yellow fever virus NS1 reveals a role in early RNA replication. J Virol 1997; 71:9608-9617.

[14] Sampath A, Padmanabhan R. Molecular targets for flavivirus drug discovery. Antiviral Res 2009; 81:6-15.

[15] Young PR, Hilditch PA, Bletchly C, Halloran W. An antigen capture enzyme-linked immunosorbent assay reveals high levels of the dengue virus protein NS1 in the sera of infected patients. J Clin Microbiol 2000; 38:1053-1057.

[16] Flamand M, Megret F, Mathieu M, Lepault J, Rey FA, Deubel V. Dengue virus type 1 nonstructural glycoprotein NS1 is secreted from mammalian cells as a soluble hexamer in a glycosylation-dependent fashion. J Virol 1999; 73:6104-611.

[17] Winkler G, Randolph VB, Cleaves GR, Ryan TE, Stollar V. Evidence that the mature form of the flavivirus nonstructural protein NS1 is a dimer. Virology 1988; 162:187-196.

[18] Sankar SG, Dhananjeyan KJ, Paramasivan R, Thenmozhi V, Tyagi BK, Vennison SJ. Evaluation and use of NS1 IgM antibody detection for acute dengue virus diagnosis: report from an outbreak investigation. Clin Microbiol Infect 2012; 12:E8-E10.

[19] Clum S, Ebner KE, Padmanabhan R. Co-translational membrane insertion of the serine proteinase precursor NS2B-NS3 (Pro) of dengue virus type 2 is required for efficient *in vitro* processing and is mediated through the hydrophobic regions of NS2B. J Biol Chem 1997; 272:30715–30723.

[20] Falgout B, Pethel M, Zhang Y-M, *et al*. Both nonstructural proteins NS2B and NS3 are required for the proteolytic processing of Dengue virus nonstructural proteins. J Virol 1991; 65: 2467–2475.

[21] Chambers TJ, Weir RC, Grakoui A, *et al*. Evidence that the N- terminal domain of nonstructural protein NS3 from yellow fever virus is a serine protease responsible for site-specific cleavages in the viral polyprotein. Proc Natl Acad Sci USA 1990; 87:8898– 8902.

[22] Gorbalenya AE, Donchenko AP, Koonin EV, *et al*. N-terminal domains of putative helicases of flavi- and pestiviruses may be serine proteases. Nucleic Acids Res 1989; 17:3889–3897.

[23]   Martin M, Weld LH, Tsai TF, *et al*. 2001. Advanced age a risk factor for illness temporally associated with yellow fever vaccination. Emerg Infect Dis 2001; 7: 945–951.

[24]   Amberg SM, Nestorowicz A, McCourt DW, *et al*. NS2B-3 proteinase-mediated processing in the yellow fever virus structural region: *in vitro* and *in vivo* studies. J Virol 1994; 68:3794–3802.

[25]   Yamshchikov VF, Compans RW. Processing of the intracellular form of the west Nile virus capsid protein by the viral NS2B-NS3 protease: an *in vitro* study. J Virol 1994; 68:5765–5771.

[26]   Miller S, Sparacio S, Bartenschlager R. Subcellular localization and membrane topology of the dengue virus type 2 nonstructural protein 4B. J Biol Chem 2006; 281:8854–8863.

[27]   Wengler G, Wengler G. The carboxy-terminal part of the NS3 protein of the West Nile flavivirus can be isolated as a soluble protein after proteolytic cleavage and represents an RNA-stimulated NTPase. Virology 1991; 184:707–715.

[28]   Warrener P, Tamura JK, Collett MS. An RNA-stimulated NTPase activity associated with yellow fever virus NS3 protein expressed in bacteria. J Virol 1993; 67:989–996.

[29]   Yamshchikov VF, Compans RW. Processing of the intracellular form of the west Nile virus capsid protein by the viral NS2B-NS3 protease: an *in vitro* study. J Virol 1994; 68:5765–5771.

[30]   Munoz-Jordan JL, Sanchez-Burgos GG, Laurent-Rolle M, *et al*. Inhibition of interferon signaling by dengue virus. Proc Natl Acad Sci USA 2003; 100:14333–14338.

[31]   Koonin EV. Computer-assisted identification of a putative methyltransferase domain in NS5 protein of flaviviruses and lambda 2 protein of reovirus. J Gen Virol 1993; 74:733–740.

[32]   Koonin EV, Dolja VV. Evolution and taxonomy of positive-strand RNA viruses: implications of comparative analysis of amino acid sequences. Crit Rev Biochem Mol Biol 1993; 28:375–430.

[33]   Rice CM, Lenches EM, Eddy SR, *et al*. Nucleotide sequence of yellow fever virus: implications for flavivirus gene expression and evolution. Science 1985; 229:726–733.

[34]   Kapoor M, Zhang L, Ramachandra M, *et al*. Association between NS3 and NS5 proteins of dengue virus type 2 in the putative RNA replicase is linked to differential phosphorylation of NS5. J Biol Chem 1995; 270:19100–19106.

[35]   Johansson M, Brooks AJ, Jans DA, *et al*. A small region of the dengue virus-encoded RNA-dependent RNA polymerase, NS5, confers interaction with both the nuclear transport receptor importin- beta and the viral helicase, NS3. J Gen Virol 2001; 82:735–745.

[36]   Cui T, Sugrue RJ, Xu Q, *et al*. Recombinant dengue virus type1 NS3 protein exhibits specific viral RNA binding and NTPase activity regulated by the NS5 protein. Virology 1998; 246:409– 417.

[37]   Yon C, Teramoto T, Mueller N, *et al*. Modulation of the nucleoside triphosphatase/RNA helicase and 5'-RNA triphosphatase activities of Dengue virus type 2 nonstructural protein 3 (NS3) by interaction with NS5, the RNA-dependent RNA polymerase. J Biol Chem 2005; 280:27412–27419.

[38]   Hiscox JA. The interaction of animal cytoplasmic RNA viruses with the nucleus to facilitate replication. Virus Res 2003; 95:13-22.

[39]   Egloff MP, Benarroch D, Selisko J, Romette L, Canard B. An RNA cap (nucleoside-2'-*O*-) methyl transferases in the flavivirus RNA polymerase NS5: crystal structure and functional characterization. EMBO J 2002; 21:2757-2768.

[40] Tan BH, Fu J, Sugrue RJ, Yap EH, Chan YC, Tan YH. Recombinant dengue type 1 virus NS5 protein expressed in *Escherichia coli* exhibits RNA- dependent RNA polymerase activity. Virology 1999; 216:317-325.

[41] Uchil PD, Kumar AV, Satchidanandam V. Nuclear localization of flavivirus RNA synthesis in infected cells. J Virol 2006; 80:5451-5464.

[42] Navaro-Sanchez E, Altmeyer R, Amara A, *et al*. Dendritic-cell specific ICAM3-grabbing non-integrin is essential for the productive infection of human dendritic cells by mosquito cell derived dengue viruses. EMBO Rep 2003; 4:723-728.

[43] Tassaneetrithep B, Burgess TH, Granelli-Piperno A, *et al*. DC-SIGN (CD209) mediates dengue virus infection of human dendritic cells. J Exp Med 2003; 197:823-829.

[44] Pokidysheva E, Zhang Y, Battisti AJ, Bator-Kelly CM, *et al*. Cryo-EM reconstruction of dengue virus in complex with the carbohydrate recognition domain of DC-SIGN. Cell 2006; 124:485-493.

[45] Miller JL, deWet BJ, Martinez-Pomares L, *et al*. The mannose receptor mediates dengue virus infection of macrophages. PLos Pathog 2008; 4:e17.

[46] Reyes-Delvelle J, Chavez-Salinas S, Medina F, Del Angel RM. Heat shock protein 90 and heat shock protein 70 are components of dengue virus receptor complex in human cells. J Virol 2005; 79:4557-4567.

[47] Jindadasamrongwech S, Thepparit C, Smith DR. Identification of GRP 78 (BiP) as a liver cell expressed receptor element for dengue virus serotype 2. Arch Virol 2004; 149:915-927.

[48] Cabrera-Fernandez A, Thepparit C, Suksanpainan L, Smith DR. Dengue virus entry in to liver (HepG2) cells is independent of hsp90 and hsp70. J Med Virol 2007; 79:386-392.

[49] Upnan S, Kwadkithan A, Smith DR. Identification of dengue virus binding proteins using affinity chromatography. J Virol Methods 2008; 151:325- 328.

[50] Tio P-H, Jong W-W, Cardosa MJ. Two dimensional VOBPA reveals Laminin receptor (LAMR1) interaction with dengue virus serotype 1,2, and 3. Virol J 2005; 2:25.

[51] Thepparit C, Smith DR. Serotype-specific entry of dengue virus into liver cells: identification of the 37-kilodalton/67-kilodalton high-affinity Laminin receptor as a dengue virus serotype 1 receptor. J Virol 2004; 78:12647-12656.

[52] Davis CW, Nguyen HY, Hanna SL, *et al*. West Nile virus discriminates between DC-SIGN and DC-SIGNR for cellular attachment and infection. J Virol 2006; 80:1290–130.

[53] Chen Y, Maguire T, Hileman RE, *et al*. Dengue virus infectivity depends on envelope protein binding to target cell heparan sulfate. Nat Med 1997; 3:866–871.

[54] Kroschewski H, Allison SL, Heinz FX, *et al*. Role of heparan sulfate for attachment and entry of tick-borne encephalitis virus. Virology 2003; 308:92–100.

[55] Peiris JS, Porterfield JS. Antibody-mediated enhancement of Flavivirus replication in macrophage-like cell lines. Nature 1979; 282:509–511.

[56] Schlesinger JJ, Brandriss MW, Monath TP. Monoclonal antibodies distinguish between wild and vaccine strains of yellow fever virus by neutralization, hemagglutination inhibition, and immune precipitation of the virus envelope protein. Virology 1983; 125:8–17.

[57] Chen ST, Lin YL, Huang MT, *et al*. CLEC5A is critical for dengue-virus-induced lethal disease. Nature 2008; 453:672-676.

[58] Chu JJ, Ng ML. Interaction of West Nile virus with alpha v beta 3 integrin mediates virus entry into cells. J Biol Chem 2004; 279:54533–54541.

[59] Krishnan MN, Sukumaran B, Pal U, *et al.* Rab 5 is required for the cellular entry of dengue and West Nile viruses. J. Virol 2007; 81:4881–4885.

[60] Alen MF, Schols D. Dengue virus entry as target for antiviral therapy. J Trop Med 2012; Article ID 787240.

[61] Kidari KI, Suzuki T. Dengue virus receptor. Trop Med Int Health 2011; 39 (Supp 4): 37-43.

[62] Barba-Spaeth G, Longman RS, Albert ML, Rice CM. Live attenuated yellow fever 17D infects human DCs and allows for presentation of endogenous and recombinant T cell epitopes. J. Exp. Med 2005; 202:1179–1184.

[63] Ishak R, Tovey DG, Howard CR. Morphogenesis of yellow fever virus 17 D in infected cell culture. J Gen Virol 1998; 69:325-335.

[64] Van der Schaar HM, Rust MJ, Waarts BZ, *et al.* Characterization of the early events in dengue virus cell entry by biochemical assays and single virus tracking. J Virol 2007; 81:12019-12028.

[65] Modis Y, Ogata S, Clements D, Harrison SC. A ligand-binding pocket in the dengue virus envelope glycoprotein. Proc Natl Acad Sci USA 2003; 100:6986-6991.

[66] Lindenbach BD, Rice CM. Flaviviridae: The viruses and their replication. In: Knipe D, Howley PM, Eds. Fields Virology. 4th ed. Philadelphia: Lippincott Williams and Wilkins 2001; pp. 991-1041.

[67] Mackenzie JM, Westaway EG. Assembly and maturation of the flavivirus Kunjin virus appear to occur in the rough endoplasmic reticulum and along the secretory pathway, respectively. J Virol 2001; 75:10787–10799.

[68] Perera R, Khaliq M, Kuhn RJ. Closing the door on flavivirus: Entry as a target for antiviral drug design. Antiviral Res 2008; 80:11-22.

[69] Elshuber S, Allison SL, Heinz FX, *et al.* Cleavage of protein prM is necessary for infection of BHK-21 cells by tick-borne encephalitis virus. J Gen Virol 2003; 84:183–191.

[70] Massad E, Coutinho FA, Burattini MN, Lopez LF, Struchiner CJ. Yellow fever vaccination: how much is enough? Vaccine 2005; 23:3908–3914.

[71] Modis Y, Ogata S, Clements D, Harrison SC. Variable surface epitopes in the crystal structure of dengue virus type 3 envelope glycoprotein. J Virol 2005; 79:1223-1231.

[72] Smith TJ, Brandt WE, Swanson JL, McCown JM, Buescher EL. Physical and biological properties of dengue-2 virus and associated antigens. J Virol 1970; 5:524–532.

[73] Stiasny K, Heinz FX. Flavivirus membrane fusion. J Gen Virol 2006; 87:2755-2766.

[74] Schalich J, Allison SC, Stiasny K, Mandl CW, Kunz C, Heinz FX. Recombinant sub viral particles from tick-borne encephalitis virus are fusogenic and provide a model system for studying flavivirus envelope glycoprotein functions. J Virol 1996; 70:4549–4557.

[75] Barnett ED. Yellow fever: epidemiology and prevention. Clin Infect Dis 2007; 4:850–856.

[76] Monath TP. Yellow fever as an endemic/epidemic disease and priorities for vaccination. Bull Soc Pathol Exot 2006; 99:341–347.

[77] Gubler DJ. Human arbovirus infections worldwide. Ann NY Acad Sci 2001; 951:13–24.

[78] Gubler DJ. The global emergence/resurgence of arboviral diseases as public health problems. Arch Med Res 2002; 33:330–342.

[79] Gardner CL, Ryman KD. Yellow fever: a reemerging threat. Clin Lab Med 2010; 30:237–260.

[80] Sabin AB. Research on dengue during World War II. Am J Trop Med Hyg 1952; 1:30–50.

[81] Nimmannitya S. Clinical spectrum and management of dengue haemorrhagic fever. Southeast Asian J Trop Med Public Health 1987; 18:392–7.

[82]   Nimmannitya S. Dengue hemorrhagic fever: diagnosis and management. In: Gubler DJ, Kuno G, eds. Dengue and dengue hemorrhagic fever. Cambridge: CAB International, 1997:133–45.

[83]   Rothman AL, Green S, Vaughn DW, *et al*. Dengue hemorrhagic fever. In: Saluzzo JF, Dodet B, eds. Factors in the emergence of arbovirus diseases. Paris: Elsevier, 1997:109–16.

[84]   Halstead SB, O'Rourke EJ. Dengue viruses and mononuclear phagocytes. I. Infection enhancement by non-neutralizing antibody. J Exp Med 1977; 146:201–17.

[85]   Halstead SB. Antibody-dependent enhancement of infection: a mechanism for indirect virus entry into cells. In: Wimmer E, ed. Cellular receptors for animal viruses. Cold Spring Harbor Laboratory Press, 1994:493–515.

[86]   Vaughn DW, Green S, Kalayanarooj S, *et al*. Dengue in the early febrile phase: viremia and antibody responses. J Infect Dis 1997; 176:322–30.

[87]   Libraty DH, Endy TP, Houng HS, *et al*. Differing influences of virus burden and immune activation on disease severity in secondary dengue- 3 virus infections. J Infect Dis 2002; 185:1213–21.

[88]   Ramirez Ronda CH, Garcia CD. Dengue in the Western Hemisphere. Infect. Dis. Clin. North Am 1994; 8:107–128.

[89]   Monath TP, Cetron MS, McCarthy K, *et al*. Yellow fever 17D vaccine safety and immunogenicity in the elderly. Hum. Vaccine 2005; 1:207–214.

[90]   Khromava AY, Eidex RB, Weld LH, *et al*. Yellow fever vaccine: an updated assessment of advanced age as a risk factor for serious adverse events. Vaccine 2005; 23:3256–3263.

[91]   Barrett AD, Niedrig M, Teuwen DE. International laboratory network for yellow fever vaccine-associated adverse events. Vaccine 2008; 26:5441–5442.

[92]   Barrett AD, Teuwen DE. Yellow fever vaccine – how does it work and why do rare cases of serious adverse events take place? Curr Opin Immunol 2009; 21:308–313.

[93]   Doblas A, Domingo C, Bae HG, *et al*. Yellow fever vaccine-associated viscerotropic disease and death in Spain. J Clin Virol 2006; 36:156–158.

[94]   Engel AR, Vasconcelos PF, McArthur MA, Barrett AD, Characterization of a viscerotropic yellow fever vaccine variant from a patient in Brazil. Vaccine 2006; 24:2803–2809.

[95]   Ferguson M, Shin J, Knezevic I, Minor P, Barrett A. WHO Working Group on Technical Specifications for manufacture and evaluation of yellow fever Vaccines, Geneva, Switzerland, 13–14 May 2009. Vaccine 2010; 28:8236–8245.

[96]   Thibodeaux BA, Garbino NC, Liss NM, *et al*. A humanized IgG but not IgM antibody is effective in prophylaxis and therapy of yellow fever infection in an AG129/17D-204 peripheral challenge mouse model. Antiviral Res 2012; 94:1–8.

[97]   Omalu BI, Shakir AA, Wang G, Lipkin WI, Wiley CA. Fatal fulminant pan-meningo-polio encephalitis due to West Nile virus. Brain Pathol 2003; 13:465-472.

[98]   Sejvar JJ, Leis AA, Stokic DS, *et al*. Acute flaccid paralysis and West Nile virus infection. Emerg Infect. Dis 2003; 9:788-793.

[99]   Solomon T, Vaughn DW. Pathogenesis and clinical features of Japanese encephalitis and West Nile virus infections. Curr Top Microbiol Immunol. 2002; 267:171-94.

[100]   Solomon T, Winter PM. Neurovirulence and host factors in flavivirus encephalitis-- evidence from clinical epidemiology. Arch Virol 2004; 18:161-70.

[101]   Theiler M, Smith HH. Use of yellow fever virus modified by *in vitro* cultivation for human immunization. J Exp Med 1937; 65:748–800.

[102] Putnak R, Barvir DA, Burrous JM, *et al*. Development of a purified, inactivated, dengue-2 virus vaccine prototype in Vero cells: Immunogenicity and protection in mice and rhesus monkeys. J Infect Dis 1996; 174:1176–1184.

[103] Putnak R, Cassidy K, Conforti N, *et al*. A purified, inactivated, dengue-2 virus vaccine prototype made in fetal rhesus lung (FRhL-2) cells is immunogenic in mice. Am J Trop Med Hyg 1996; 55:504–510.

[104] Burke DS, Leake CJ. Japanese encephalitis. In: T. P. Monath, Eds. The Arboviruses: Epidemiology and Ecology. Vol III CRC, Boca Raton, FL 1988; pp. 63–92.

[105] Vaughn DW, Hoke CH. The epidemiology of Japanese encephalitis: Prospects for prevention. Epidemiol Rev 1992; 14:197–221.

[106] Takaku K, Yamashita T, Osanai T, *et al*. Japanese encephalitis purified vaccine. Biken J 1968; 11: 25–39.

[107] Hoke CH, Nisalak A, Sangawhipa N, *et al*. Protection against Japanese encephalitis by inactivated vaccines. N Engl J Med 1988; 319:608–614.

[108] Ao J, Yu Y, Tang YS, *et al*. Selection of a better immunogenic and highly attenuated live vaccine virus strain of Japanese encephalitis. II. Safety and immunogenicity of live vaccine SA14-14-2 observed in inoculated children. Chin J Microbiol Immunol 1983; 3:245–248.

[109] Srivastava AK, Putnak JR, Lee SH, *et al*. A purified inactivated Japanese encephalitis virus vaccine made in vero cells. Vaccine 2000; 19:4557–4565.

[110] Kunz C. TBE vaccination and Austrian experience. Vaccine 2003; 21:50–55.

[111] Heinz FX, Kunz C. Concentration and purification of tick-borne encephalitis virus grown in suspensions of chick embryo cells. Acta Virol 1977; 21: 301–307.

[112] Heinz FX, Kunz C, Fauma H. Preparation of a highly purified vaccine against tick-borne encephalitis by continuous flow zonal ultracentrifugation. J Med Virol 1980. 6:213–221.

[113] Bock HL, Klockmann U, Jungst C, *et al*. New vaccine against tick-borne encephalitis: initial trial in man including a dose-response study. Vaccine 1990; 8:22–24.

[114] Dauphin G, Zientara S. West Nile virus: recent trends in diagnosis and vaccine development. Vaccine 2007; 25:5563-5576.

[115] Pugachev KV, Guirakhoo F, Monath TP. New developments in flavivirus vaccines with special attention to yellow fever. Curr Opin Infect Dis 2005; 18:387-394.

[116] Malkinson M, Banet C, Khinich Y, Samina I, Pokamunski S, Weisman Y. Use of live and inactivated vaccines in the control of West Nile fever in domestic geese. Annals N Y Acad Sci 2002; 951:255–261.

[117] Despres P, Combredet C, Frenkiel MP, Lorin C, Brahic M, Tangy F. Live measles vaccine expressing the secreted form of the West Nile virus envelope glycoprotein protects against West Nile virus encephalitis. J Infect Dis 2005; 191:207-214.

[118] Yamshchikov G, Borisevich V, Seregin A, *et al*. An attenuated West Nile prototype virus is highly immunogenic and protects against the deadly NY99 strain: a candidate for live WN vaccine development. Virology 2004; 330:304-312.

[119] Mason PW, Shustov AV, Frolov I. Production and characterization of vaccines based on flaviviruses defective in replication. Virology 2006; 351:432-443.

[120] Widman DG, Frolov I, Mason PW. Third-generation flavivirus vaccines based on single-cycle, encapsidation-defective viruses. Adv Virus Res 2008; 72:77-126.

[121] Putnak JR, Coller BA, Voss G, *et al*. An evaluation of dengue type-2 inactivated, recombinant subunit, and live-attenuated vaccine candidates in the rhesus macaque model. Vaccine 2005; 23:4442–4452.

[122] Sellards A, Langret J. Vaccination de l'homme contre la fievre jaune. C R Acad Sci 1932; 194:1609-1611.

[123] Barnett ED. Yellow fever: epidemiology and prevention. Clin Infect Dis 2007; 44:850-856.

[124] Barrett AD, Teuwen DE. Yellow fever vaccine - how does it work and why do rare cases of serious adverse events take place? Curr Opi Immunol 2009; 21:308-313.

[125] dos Santos CN, Post PR, Carvalho R, Ferreira, Rice CM, Galler R. Complete nucleotide sequence of yellow fever virus vaccine strains 17DD and 17D-213. Virus Res 1995; 35:35-41.

[126] Monath TP. Yellow fever: an update. Lancet Infect Dis 2001; 1:11- 20.

[127] Poland JD, Calisher CH, Monath TP, Downs WG, Murphy K. Persistence of neutralizing antibody 30-35 years after immunization with 17D yellow fever vaccine. Bull World Health Organ 1981; 59:895- 900.

[128] Frierson JG. The Yellow Fever Vaccine: A History. Yale Journal of Biology and Medicine 2010; 83:77-85.

[129] Sabin AB, Schlesinger RW. Production of immunity to dengue virus modified by propagation in mice. Science 1945; 101:640–642.

[130] Hotta S. Experimental studies on dengue. I. Isolation, identification and modification of the virus. J Infect Dis 1952; 90:1–12.

[131] Wisseman CL, Sweet BH, Rosenzweig EC, Eylar OR. Attenuated living type 1 dengue vaccines. Am J Trop Med Hyg 1963; 12:620–623.

[132] Gualano RC, Pryor MI, Cauchi MR, Wright PJ, Davidson AD. Identification of a major determinant of mouse neurovirulence of dengue virus type 2 using stably cloned genomic-length cDNA. J Gen Virol 1998; 79:436–437.

[133] Duarte dos Santos, Frenkiel M-P, Courageot M-P, *et al.* Determinants in the envelope E protein and viral RNA helicase NS3 that influence the induction of apoptosis in response to infection with dengue type 1 virus. Virology 2000; 274: 292–308.

[134] Bhamarapravati N, Sutee Y. Live attenuated tetravalent dengue vaccine. Vaccine 2000; (suppl2):44-47.

[135] Vaughn DW, Hoke CH Jr, Yoksan S, La Chance R, Rice RM, Bhamarapravati N. Testing of a dengue 2 live attenuated vaccine (strain 16681 PDK 53) in ten American volunteers. Vaccine 1996; 14:329-336.

[136] Aihara S, Rao CM, Yu YX. Identification of mutations that occurred on the genome of Japanese encephalitis virus during the attenuation process. Virus Genes 1991; 5:95–109.

[137] Ni H, Burns NJ, Chang C-J, *et al.* Comparison of nucleotide and deduced amino acid sequence of the 50 non-coding region and structural protein genes of the wild-type Japanese encephalitis virus strain SA14 and its attenuated vaccine derivatives. J Gen Virol 1994; 75:1505–1510.

[138] Nitayaphan S, Grant JG, Chang GJ, Trent DW. Nucleotide sequence of virulent SA14 strain of Japanese encephalitis and its attenuated derivative. Virology 1990; 177:541–552.

[139] Kofler RM, Aberle JH, Aberle SW, Allison SL, Heinz FX, Mandl CW. Mimicking live flavivirus immunization with a noninfectious RNA vaccine. Proc Natl Acad Sci USA 2004; 101:1951-1956.

[140] Chu JH, Chiang CC, Ng ML. Immunization of flavivirus West Nile recombinant envelope domain III protein induced specific immune response and protection against West Nile virus infection. J Immunol 2007; 178:2699-2705.

[141] Theiler M, Smith HH. The use of yellow fever virus modified by *in vitro* cultivation for human immunization. J Exp Med 2000; 65:787-800.

[142] Chang GJ, Davis BS, Hunt AR, Holmes DA, Kuno G. Flavivirus DNA vaccines: current status and potential. Ann NY Acad Sci 2001; 951:272-285.

[143] Chang GJ, Kuno G, Purdy DE, Davis BS. Recent advancement in flavivirus vaccine development. Expert Rev Vaccines 2004; 3:199-220.

[144] Davis BS, Chang GJ, Cropp B, *et al*. West Nile virus recombinant DNA vaccine protects mouse and horse from virus challenge and expresses *in vitro* a noninfectious recombinant antigen that can be used in enzyme-linked immunosorbent assays. J Virol 2001; 75:4040-4047.

[145] Putnak JR, Fuller J, VanderZanden L, Innis BL, Vaughn DW. Vaccination of rhesus macaques against dengue-2 virus with a plasmid DNA vaccine encoding the viral pre-membrane and envelope genes. Am J Trop Med Hyg 2003; 68:469-476.

[146] Raviprakash K, Porter KR, Kochel TJ, *et al*. Dengue virus type 1 DNA vaccine induces protective immune responses in rhesus monkeys. J Gen Virol 2000; 81:1659-1667.

[147] Mackenzie JM, Krohmykh AA, Jones MK, Westaway EG. Subcellular localization and some biochemical properties of the flavivirus Kunjin nonstructural proteins NS2A and NS4A. Virology 1998; 245:203-21.

[148] Sankarapandi S, Zweier JL, Mukherjee G, Quinn MT, Huso DL. Measurement and characterization of superoxide generation in microglial cells: evidence for an NADPH oxidase- dependent pathway. Arch Biochem Biophys 1998; 353:312-32.

[149] McDowell M, Gonzales SR, Kumarapperuma SC, Jeselink M, Arterburn JB, Hanley KA. A novel nucleoside analog, 1-beta-d-ribofuranosyl-3-ethynyl-[1,2,4]triazole (ETAR), exhibits efficacy against a broad range of flaviviruses *in vitro*. Antiviral Res 2010; 87:78-80.

[150] Wolf MC, Freiberg AN, Zhang T, *et al*. A broad spectrum antiviral activity targeting entry of enveloped viruses. Proc Natl Acad Sci USA 2010; 107:3157-3162.

[151] Sugrue RJ, Fu J, Howe J, Chan Y-C. Expression of the dengue virus structural proteins in *Pichia pastoris* leads to the generation of virus-like particles. J Gen Virol 1997; 78:1861-1866.

[152] Kaur R, Rauthan M, Vrati S. Immunogenicity in mice of a cationic microparticle-absorbed plasmid DNA encoding Japanese encephalitis virus envelope protein. Vaccine 2004; 22:2776-2782.

[153] Konishi E, Terazawa A, Fujii A. Evidence for antigen production in muscles by dengue and Japanese encephalitis DNA vaccines and a relation to their immunogenicity in mice. Vaccine 2003; 21:3713-3720.

[154] Raviprakash K, Ewing D, Simmons M, *et al*. Needle-free Biojector injection of a dengue virus type 1 DNA vaccine with human immunostimulatory sequences and the GM-CSF gene increases immunogenicity and protection from virus challenge in Aotus monkeys. Virology 2003; 315:345-352.

[155] Lin YL, Chen LK, Liao CL, Yeh CT, *et al*. DNA immunization with Japanese encephalitis virus nonstructural protein NS1 elicits protective immunity in mice. J Virol 1996; 72:191-200.

[156] Konishi E, Yamaoka M, Khin Sane W, Kurane I, Takada K, Mason PW. The anamnestic neutralizing antibody response is critical for protection of mice from challenge following vaccination with a plasmid encoding the Japanese encephalitis virus premembrane and envelope genes. J. Virol 1999; 73: 5527-5534.

[157]   Lin YL, Chen LK, Liao CL, *et al*. DNA immunization with Japanese encephalitis virus nonstructural protein NS1 elicits protective immunity in mice. J Virol 1998; 72:191–200.

[158]   Barrett AD. Japanese encephalitis and dengue vaccines. Biologicals 1997; 25:27-34.

[159]   Mason PW, Shustov AV, Frolov I. Production and characterization of vaccines based on flaviviruses defective in replication. Virology 2006; 351:432-443.

[160]   Nalca A, Fellows PF, Whitehouse CA. Vaccines and animal models for arboviral encephalitides. Antiviral Res 2003; 60:153-174.

[161]   Azevedo AS, Yamamura AMY, Freire MS, *et al*. DNA vaccines against Dengue virus type 2 based on truncate envelope protein or its domain III. PLoS ONE 2011; 6:e20528.

[162]   De Paula SO, Lima DM, de Oliveira Franca RF, *et al*. A DNA vaccine candidate expressing dengue-3 virus prM and E proteins elicits neutralizing antibodies and protects mice against lethal challenge. Arch Virol 2008; 153:2215–23.

[163]   Pugachev KV, Guirakhoo F, Trent DW, Monath TP. Traditional and novel approaches to flavivirus vaccines. Int J Parasitol 2003; 33:567–582.

[164]   Mandl CW, Aberle JH, Aberle SW, Holzmann H, Allison SL, Heinz FX. *In vitro*-synthesized infectious RNA as an attenuated live vaccine in a flavivirus model. Nat Med 1998; 4:1438-1440.

[165]   Gardner CL, Ryman KD. Yellow fever: a reemerging threat. Clin Lab Med 2010; 30:237–260.

[166]   Hombach J, Barrett AD, Cardosa MJ, *et al*. Review on flavivirus vaccine development: Proceedings of a meeting jointly organized by the World Health Organization and the Thai Ministry of Public Health, 26-27 April 2004, Bangkok, Thailand. Vaccine 2005; 23:2689-2695.

[167]   Pugachev KV, Guirakhoo F, Trent DW, Monath TP. Traditional and novel approaches to flavivirus vaccines. Int J Parasitol 2003; 33:567-582.

[168]   Monath TP. Yellow fever vaccine. In Vaccines. Fourth edition. Edited by Plotkin SAOWA. Philadelphia, W.B. Saunders; 2004:1095-1176.

[169]   Co MD, Terajima M, Cruz J, Ennis FA, Rothman AL. Human cytotoxic T lymphocyte responses to live attenuated 17D yellow fever vaccine: identification of HLA-B35-restricted CTL epitopes on nonstructural proteins NS1, NS2b, NS3, and the structural protein E. Virology 2002; 293:151-163.

[170]   Querec T, Bennouna S, Alkan S, *et al*. Yellow fever vaccine YF-17D activates multiple dendritic cell subsets *via* TLR2, 7, 8, and 9 to stimulate polyvalent immunity. J Exp Med 2006; 203:413-424.

[171]   Tao D, Barba-Spaeth G, Rai U, Nussenzweig V, Rice CM, Nussenzweig RS. Yellow fever 17D as a vaccine vector for microbial CTL epitopes: protection in a rodent malaria model. J Exp Med 2005; 201:201-209.

[172]   van der Most RG, Harrington LE, Giuggio V, Mahar PL, Ahmed R. Yellow fever virus 17D envelope and NS3 proteins are major targets of the antiviral T cell response in mice. Virology 2002, 296:117-124.

[173]   Bonaldo MC, Garratt RC, Caufour PS, Freire MS, Rodrigues MM, Nussenzweig RS, Galler R. Surface expression of an immuno dominant malaria protein B cell epitope by yellow fever virus. J Mol Biol 2002, 315:873-885.

[174]   Pugachev KV, Guirakhoo F, Monath TP. New developments in flavivirus vaccines with special attention to yellow fever. Curr Opin Infect Dis 2005, 18:387-394.

[175] Arroyo J, Miller C, Catalan J, *et al*. ChimericVax-West Nile virus live-attenuated vaccine: preclinical evaluation of safety, immunogenicity, and efficacy. J Virol 2004; 78:12497-12507.

[176] Monath TP, Guirakhoo F, Nichols R, *et al*. Chimeric live, attenuated vaccine against Japanese encephalitis (ChimeriVax- JE): Phase 2 clinical trials for safety and immunogenicity, effect of vaccine dose and schedule, and memory response to challenge with inactivated Japanese encephalitis antigen. J Infect Dis 2003; 188:1213-1230.

[177] Brandler S, Brown N, Ermak TH, *et al*. Replication of chimeric yellow fever virus-dengue serotype 1-4 virus vaccine strains in dendritic cells and hepatic cells. Am J Trop Med Hyg 2005; 72:74-81.

[178] Huang CY, Silengo SJ, Whiteman MC, Kinney RM. Chimeric dengue 2 PDK-53/West Nile NY99 viruses retain the phenotypic attenuation markers of the candidate PDK-53 vaccine virus and protect mice against lethal challenge with West Nile virus. J Virol 2005; 79:7300-7310.

[179] Wilder-Smith A, Deen JL. Dengue vaccines for travelers. Expert Rev Vaccines 2008; 7:569-578.

[180] Singh R, Rothman AL, Potts J, Guirakhoo F *et al*. Sequential immunization with heterologous chimeric flavivirus induces broad-spectrum cross-reactive CD8+ T-cell responses. J Infect Dis. 2010; 202:223–233.

[181] Wu S-F, Lee C-J, Liao C-L, Dwek R, Zitzmann N, Lin Y-L. Antiviral effects of an iminosugar derivative on flavivirus infections. J Virol 2002; 76:3596-3604.

[182] Courageot MP, Frenkiel MP, Dos Santos CD, Deubel V, Despres P. Alpha-glucosidase inhibitor reduce dengue virus production by affecting the initial steps of virion morphogenesis in the endoplasmic reticulum. J Virol 2000; 74:564-572.

[183] Zhou Y, Ray D, Zhao Y, Dong H, *et al*. Structure and function of flavivirus NS5 methyl-transferase. J Virol 2007; 81: 3891-3903.

[184] Pan YT, Ghidoni J, Elbein AD. The effects of castanospermine and swainosine on the activity and synthesis of intestinal sucrase. Arch Biochem Biophys 1983; 303:134-144.

[185] Courageot MP, Frenkiel MP, Dos Santos CD, Deubel V, Despres P. Alpha-glucosidase inhibitors reduce dengue virus production by altering the initial steps of virion morphogenesis in the endoplasmic reticulum. J Virol 2000; 74:564-572.

[186] Whitby K, Pierson TC, Geiss B, Lane K, *et al*. Castanospermine, a potent inhibitor of dengue virus infection *in vitro* and *in vivo*. J Virol 2005; 79: 8698-8706.

[187] Chu JJ, Yang PL. c-Src protein kinase inhibitors block assembly and maturation of dengue virus. Proc Natl Acad Sci USA 2007; 104:3520-3525.

[188] Anwar A, Hosoya T, Leong KM, *et al*. The Kinase Inhibitor SFV785 Dislocates Dengue Virus Envelope Protein from the Replication Complex and Blocks Virus Assembly. PLoS ONE 2011; 6:e23246.

[189] Shigeta S, Mori S, Kodama E, Kodama J, Takahashi K, Yamase T. Broad spectrum anti-RNA virus activities of titanium and vanadium substituted polyoxotungstates. Antiviral Res 2003; 58:265-271.

[190] Talarico LB, Pujol CA, Zibetti RG, *et al*. The antiviral activity of sulfated polysaccharides against dengue virus is dependent on virus serotype and host cells. Antiviral Res 2005;66: 103-110.

[191] Pujol CA, Estevez JM, Carlucci MJ, Ciancia M, Cerezo AS, Damonte EB. Novel DL-galactan hybrids from the red seaweed *Gymnogongrus torulosus* are potent inhibitors of herpes simplex virus and dengue virus. Antivir Chem Chemother 2002; 13: 83-89.

[192] Hrobowski YM, Garry RF, Michael SF. Peptide inhibitors of dengue virus and West Nile virus infectivity. Virology J 2005; 2:49.

[193] Schmidt AG, Yang PL, Harrison SC. Peptide Inhibitors of Flavivirus Entry Derived from the E Protein Stem. J Virol 2010; 84: 12549-12554.

[194] Wang Q, Patel SJ, Vangrevelinghe E, et al. A Small- Molecule Dengue Virus Entry Inhibitor. Antimicrob Agents Chemother 2009; 53:1823-1831.

[195] Borowski P, Lang M, Haag A, et al. Characterization of imidazo [4,5-d] pyridazine nucleosides as modulators of unwinding reaction mediated by West Nile virus nucleoside triphosphate/helicse: evidence for activity on the level of substrate and/or enzyme. Antimicrob Agents and Chemother 2002; 46:1231-1239.

[196] Borowski P, Deinert J, Schalinski S, et al. Halogenated benzimidazoles and benzotriazoles as inhibitors of the NTPase/helicase activities of hepatitis C and related viruses. Eur J Biochem 2003; 270:1645-1653.

[197] Zhang N, Chen HM, Kock V, et al. Ring-expanded ("fat") nucleoside analogues exhibit potent *in vitro* activity against flaviviridae NTPases/helicases, including those of the West Nile virus, hepatitis C virus, and Japanese encephalitis virus. J Med Chem 2003; 46:4149-4164.

[198] Zhang N, Chen HM, Kock V, et al. Potent inhibition of NTP/helicase of the West Nile virus by ring-expanded ("fat") nucleoside analogues. J Med Chem 2003; 46:4776-4789.

[199] Mastrangelo E, Pezzullo M, De Burghgraeve T, et al. Ivermectin is a potent inhibitor of flavivirus replication specifically targeting NS3 helicase activity: new prospects for an old drug. J Antimicrob Chemother 2012; In press.

[200] Ben-Nathan D, Lustig S, Tam G, Robinson S, Segal S, Rager-Zisman B. Prophylactic and therapeutic efficacy of human intravenous immunoglobulin in treating west nile virus infection in mice. J Infe Dis 2003; 188:5-12.

[201] Eagle M, Diamond MS. Antibody prophylaxis and therapy against west nile virus infection in wild type and immunodeficient mice. J Virol 2003; 77:12941-12949.

[202] Raviprakash K, Liu K, Matteucci M, Wagner R, Riffenburgh R, Carl M. Inhibition of dengue virus by novel, modified antisense oligonucleotides. J Virol 1995; 69:69-74.

[203] Adelman ZN, Sanchez-Varyas I, Travanty EA, et al. RNA silencing of dengue virus type 2 replication in transformed C6/36 mosquito cells transcribing an inverted-repeat RNA derived from the virus genome. J Virol 2002; 76:12925- 12937.

[204] McCown M, Diamond MS, Pekosz A. The utility of siRNA transcripts produced by RNA polymerase i in down regulating viral gene expression and replication of negative and positive strand RNA viruses. Virology 2003; 313:514-524.

[205] Bai F, Wang T, Pal U, Bao F, Gould L, Fikrig E. Use of RNA interference to prevent lethal murine West Nile virus infection. J Infect Dis 2005; 191:1148-1154.

[206] Geiss B, Pierson T, Diamond M. Actively replicating West Nile virus is resistant to cytoplasmic delivery of siRNA. Virol J 2005; 2:53.

[207] Uchil PD, Satchidanandam V. Architecture of the flaviviral replication complex: protease, nuclease, and detergents reveal encasement within double-layered membrane compartments. J Biol Chem 2003; 278:24388-24398.

[208] Deas TS, Binduga-Gajewska I, Tlyner M, et al. Inhibition of flavivirus infections by antisense oligomers specifically suppressing viral translation and RNA replication. J Virol 2005; 79:4599-4609.

[209] Leyssen P, Balzarini J, De Clercq E, Neyts J. The predominant mechanism by which ribavirin exerts its antiviral activity *in vitro* against flaviviruses and paramyxoviruses is mediated by inhibition of IMP dehydrogenase. J Virol 2005; 79:1943-1947.

[210] Leyssen P, Van Lommel A, Drosten C, Schmitz H, De Clercq E, Neyts J. A novel model for the study of the therapy of flavivirus infections using the Modoc virus. Virology 2001; 279: 27-37.

[211] Diamond MS, Zachariah M, Harris E. Mycophenolic acid inhibits dengue virus infection by preventing replication of viral RNA. Virology 2002; 304:211-221.

[212] Yin Z, Chen Y, Schul W, *et al*. An alternative nucleoside inhibitor of dengue virus. Proct Natl Acad Sci USA; 2009:20435-20439.

[213] Chen Y, Yin Z, Lakshminarayana SB, *et al*. Inhibition of dengue virus by ester prodrug of an adenosine analog. Antimicrob Agents Chemother 2010; 54:3255-3261.

[214] Dong H, Ray D, Ren S, *et al*. Distinct RNA elements confer specificity to flavivirus RNA cap methylation events. J Virol 2007; 81:4412-4421.

[215] Zhou Y, Ray D, Zhao Y, Dong H, *et al*. Structure and function of flavivirus NS5 methyl-transferase. J Virol 2007; 81: 3891-3903.

[216] Benarroch D, Egloff MP, Mulard L, Guerreiro C, Romette JL, Canard B. A structural basis for the inhibition of the NS5 dengue virus mRNA 2'-O-methyltransferase domain by ribavirin 5'-triphosphate. J Biol Chem 2004; 279:35638–35643.

[217] Puig-Basagoiti F, Qing M, Dong H, *et al*. Identification and characterization of inhibitors of West Nile virus. Antiviral Res 2009; 83:71–79.

[218] Dong H, Liu L, Zou G, *et al*. Structural and functional analyses of a conserved hydrophobic pocket of flavivirus methyltransferase. J Bio Chem 2010; 285:32586- 32595.

[219] Gu B, Ouzunov S, Wang L, *et al*. Discovery of small molecule inhibitors of west nile virus using a high throughput sub-genomic replicon screen. Antivir Res 2006; 70:39-50.

[220] Puig-Basagoiti F, Tilgner M, Forshey BM, *et al*. Triaryl pyrazoline compound inhibits flavivirus RNA replication. Antimicrob Agents Chemother 2006; 50:1320-1329.

[221] Noueiry AO, Olivo PD, Slomczynska U, *et al*. Identification of Novel Small-Molecule Inhibitors of West Nile Virus Infection. J Virol 2007; 81:11992–12004.

[222] Stahla-Beek HJ, April DG, Saeedi BJ, Hannah AM, Keenan SM, Geiss BJ. Identification of a Novel Antiviral Inhibitor of the Flavivirus Guanylyltransferase Enzyme. J Virol 2012; In press.

[223] Chang C, Ou Y, Raung S, Chen C. Antiviral effect of dehydroepiandrosterone on Japanese encephalitis virus infection. J Gen Virol 2005; 86:2513-2523.

[224] Yang N, Jeng KG, Ho W, Chou S, Hu M. DHEA inhibits cell growth and induces apoptosis in BV-2 cells and the effects are inversely associated with glucose concentration in the medium. J Steroid Biochem Mol Biol 2000; 75:159-166.

[225] Guo H, Pan X, Mao R, Zhang X, *et al*. Alkylated Porphyrins Have Broad Antiviral Activity against Hepadnaviruses, Flaviviruses, Filoviruses, and Arenaviruses. Antimicrob Agents Chemother 2011; 55:478–486.

[226] Burghgraeve TD, Kaptein SJF, Ayala- Nunez NV, *et al*. An analogue of the antibiotic Teicoplanin prevents flavivirus entry *in vitro*. PLos One 2012; 7:e37244.

[227] Yang J, Chen Y, Tu Y, Yen K, Yang Y. Combinatorial computational approaches to identify tetracycline derivatives as flavivirus inhibitors. PLos One 2007; 5:e428.

Send Order for Reprints to reprints@benthamscience.net

CHAPTER 5

Plant Derived Edible Vaccines and Therapeutics

Emrah Altindis[1,*]**, Sultan Gulce Iz**[2]**, Mehmet Ozgun Ozen**[2]**, Pinar Nartop**[2]**, Ismet Deliloglu Gurhan**[2] **and Aynur Gurel**[2]

[1]*Harvard Medical School, Microbiology and Immunobiology Department, 200 Longwood Avenue, Boston, 02115 MA, USA and* [2]*Ege University, Faculty of Engineering, Department of Bioengineering, 35100 Bornova, Izmir, Turkey*

Abstract: Defining the molecular basis of the infectious diseases by the highly accumulated data on genetics and molecular biology facilitated developing new prophylactic and therapeutic vaccination strategies against human and animal diseases. On the other hand, vaccine development process still has many technical and economical limitations. Therefore, low-income countries, which do not have powerful *healthcare infrastructure*, neither produce their own vaccines nor benefit enough from the current vaccination programs. In this aspect, plants are promising to be used as biofactories for producing edible vaccines and several therapeutics due to their easy manipulations and also low-cost manufacturing properties. In this chapter, the advantages, limitations of the edible vaccines, the studies related to plant-derived edible vaccines and therapeutics up to date are summarized.

Keywords: Plant derived edible vaccines, therapeutics, recombinant proteins.

INTRODUCTION

Defining the molecular basis of the infectious diseases by highly accumulated data on genetics and molecular biology facilitated developing new treatment and prophylactic vaccination strategies against human and animal diseases. On the other hand, vaccine development process still has many technical and economic limitations. Therefore, low income countries which do not have powerful *healthcare infrastructure* neither produce their own vaccines nor benefit enough from the current vaccination programs. Recently in 2008, World Health Organization (WHO) estimated that 1,5 million of deaths among children under 5 years were due to diseases that could have been prevented by routine vaccination.

Address correspondence to Emrah Altindis: Harvard Medical School, Microbiology and Immunobiology Department, Room 1039, 77 Avenue Louis Pasteur Boston, MA 02115, USA; Tel: + 1 6173193788; E-mail: Emrah_Altindis@hms.harvard.edu

This represents 17% of global total mortality in children under 5 years of age [1]. In this aspect, some international organizations are supporting developing countries as means of vaccine supply for the vaccine preventable diseases [2]. In 10 years time, immunizations supported by GAVI (Global Alliance for Vaccines and Immunization) prevented 5,000,000 deaths of infants and children worldwide [3].

In the 1990s, Charles Arntzen *et al.* put forward an idea to develop effective and safe subunit antijen formulations with the specific engineering techniques using plants for the diseases which are mostly seen in developing and un-developed countries with reduced transportation costs [4]. Transgenic plants are also convenient for producing antigens in large amounts (kg) for using in parenteral and oral applications [5]. Immunisation by directly eating the transgenic plant parts (fruit, tuber, seed) which produce vaccine antigens are thought to reduce the high production costs like purification, storing and transportation costs. In addition, WHO established an initiative for children vaccination (CVI; Children Vaccination Initiative) to support the projects related with the children vaccine development that (i) uses multiple antijens, (ii) are safe and heat stable, (iii) provides long-lasting protection and (iv) lastly easily applicable by oral route as transgenic plant vaccination projects [6].

EDIBLE VACCINES

Vaccine antigen expression in plant tissues are first demonstrated in tobacco by Streptococcus mutans surface protein antigen (SpaA) in 1990. In that study, after feeding the mice with transgenic tobacco, an antibody response was detected against SpaA [7]. After that, in 1998, the first human trial was done with raw potato which is expressing a part of *Escherichia coli* toxin that causes diarrhea. Ten of the 11 volunteers (91 percent) who ingested the transgenic potatoes had four fold rises in serum antibodies at some point after immunization, and six of the 11 (55 percent) developed four fold rises in intestinal antibodies. The potatoes were well tolerated and no one experienced serious adverse side effects which is the first demonstration of edible vaccines inducing immune responses in humans [8, 9]. After the first human trial edible vaccines are used as promising tools to develop vaccines against several infectious agents like Norwalk virus and Hepatitis B virus [43, 157, 158]. Strenghts, weakness, opportunities and threat analysis (SWOT) of edible vaccines is listed in Table **1**. The production of the vaccines in plant tissues eliminates some of the classical vaccine manufacturing limitations due to plants are easy to manufacture by only using sun light, water and minerals found in the soil [10]. If they are delivered orally, edible vaccines and biopharmaceuticals are advantageous that there is no need for expensive fermentation and purification systems and other expenses associated with cold storage, transportation and sterile delivery [159].

The advantages of manufacturing vaccines in plants could be summarized as follows:

i.   Plants are free of the animal viruses which is a big problem when producing vaccines with animal cells [11].

ii.   Edible vaccine application route is very easy compared to syringe based delivery [10, 12].

iii. Orally administration can induce broader immune responses including IgA [10].

iv.   There is no need to have cold transportation chains which is a problem for poor populations which do not have a powerful health care infrastructure [13].

v.   Edible vaccines have 1:100 - 1:1000 folds lower cost production systems compared to traditional vaccines; moreover, production costs in the plants are 50 fold lower than vaccine antigen manufacturing in *E. coli* [14].

vi.   Gene expression systems of the plants can be easily manipulated for higher production yields with tissue specific promoters and accumulation of the encoded proteins to specific compartments [10].

vii. Edible vaccine production and scale up processes are relatively easy compared to other production systems thus edible vaccines can also be used against biological weapons when excess amount of vaccine in a short manufacturing period is needed [10].

viii. Plant production systems can be used either for human or veterinary vaccines [10].

Edible vaccines have important disadvantages as well as their advantages. The vaccines produced in plants are subunit protein vaccines which are less immunogenic compared to live or inactivated vaccines. In order to solve low immunogenicity problem, periodic re-vaccination or high dose vaccine application is needed. Moreover to increase the immunogenicity of the antigen, specific adjuvants are used like cholera toxin or *E. coli* toxin [15, 16]. In oral applications, plant vaccines could be digested before an immune reaction is elicited, plant vaccines must be encapsulated to overcome this obstacle. Proteins are expressed differentially in several plant transformants using *Agrobacterium tumefaciens* which can insert the related protein gene into plant chromosomes randomly [17], this random insertion can also cause growth problems in plants [18]. One of the important challenges is the low expression ratios of the vaccine antigens in the plants. Generally the expression ratios of the vaccine antigens are ranged from 0.01% to 12% of the total soluble plant proteins [19]. The unknown consequences of genetically modified organism consumption as food are the other important limitation of edible vaccines [11]. In addition, there is not much information about the effects of plant development period and physiological differentiation on efficiency and quality of foreign proteins produced in trangenic plants depending on climatical conditions [5]. All these subjects must be in consideration while producing an edible vaccine.

The Mostly Used Plants for Edible Vaccine Production

Organs and tissues of different plant species are used for edible vaccine production. A plant for vaccine production must be selected according to some features such as transformation capability, adequate protein expression in an edible tissue and production of non-toxic compounds for target users. Leaves [29], fruits [30, 31], tubers [32, 33, 34] and seeds [35, 36] are commonly used as edible parts of plants for the expression of recombinant proteins.

Besides field cultivation, plant tissue and organ cultures including mainly cell [37] and hairy root cultures [38-40] are biotechnological systems for high expression of desired proteins. Seeds are very important for commercial production expressing recombinant proteins at sufficient amounts [41].

Table 1: SWOT analysis of edible vaccines

Strenghts	Weaknesses	Opportunities
• Have low cost, easy administration and highly scalable features [20, 21, 22, 23, 24, 25]. • Can be one of the alternatives of the traditional vaccines as they can overcome all the problems encountered with traditional vaccination against infectious diseases, autoimmune diseases and tumours [10, 21]. • Can be used as raw food or dry powder [10]. • Are targeted to elicit both mucosal and systemic immunity although in traditional vaccines mucosal immunity is not obtained [21]. • No need for "Cold chain" that results in low cost of storage, preparation, production and transportation [10, 21, 22, 23]. • Stable at room temperature unlike traditional vaccines [9, 21, 22] • Manufacturing cost is lower than traditional vaccines costs because no sterilization is needed [21, 22]. • Could be the source for new vaccines combining numerous antigens [second generation vaccines] [21].	• Degradation of protein components in the stomach and gut [20]. • The best plant species or tissue for commercial production is not clear [20, 23]. • Levels of pharmaceutical proteins produced in most transgenic plants are not adequate for total immunity and commercial feasibility [20, 22, 26]. • There are several successful reports of high-level expression of non-human proteins *via* the nuclear genome, but it is needed to increase expression levels of proteins to enable the commercial production of pharmacologically important proteins in plants [20]. • Effective dosage requirement is not clear [23]. • Consistency of dosage and stability differs from plant to plant, fruit to fruit, generation to generation according to protein content and patient's age [21, 23]. • The quantity of plant tissue constituting a vaccine dose must be at optimum size for consumption [20]. • They need to be stored in optimum conditions to prevent microbial spoilage [21]. • Short storage life and long production cycle make difficult vaccine production in some plants like tomatoes and bananas [26]. • It is expensive to extract the desired protein due to the presence of interfering compounds in plants [26]. • Commercial and economic successes are dependent on the protein amounts produced by plants [25].	• As an alternative to sterile needle injection, they can be consumed directly without isolation and purification [10, 20, 23, 26, 27]. • Chloroplast transformation is possible in which gene silencing has not been observed whereas it is a common phenomenon in nuclear transformation [20, 26]. • Some proteins can cause oral immunization when consumed as food, even though expression levels are too low for large-scale oral administration [20]. • There is no need for adjuvants as they are effective as delivery vehicles for immunization [21]. • Seeds of transgenic plants can be dried having more storage opportunities regarding their oil and aqueous extracts [21]. • Protein production in seeds means large amounts of protein can be stored after harvest in the dry seeds with protein integrity [26]. • Costly equipments and machines are not necessary to produce edible vaccines [21]. • Plants can be easily grown on nutrient-rich soils [21]. • The cost for growing plants is lower than cell cultures grown in fermenters [21].

Strenghts	Weaknesses	Opportunities
• Do not involve attenuated pathogens and there is no risk of proteins to reform into infectious organism [21, 24]. • Mass production is easier compared to an animal system [21, 23].		• There is no need for medical personnel [21, 23]. • Contamination risk is low [21]. • Free of animal pathogens and toxins [9, 23]. • Have a wider use in veterinary medicine as vaccines and food additives [10]. • Secretion into the extracellular area eases protein purification [26]. • Some proteins have intrinsic stability that eases commercially production [25]. • These products have to acquire regulatory approval before use, but this would not take longer time than for current technologies [26].

Table 2: Advantages and disadvantages of the plants which are mostly used for edible vaccine production

Plant	Advantage	Disadvantage
Tobacco	Easy and efficient to transform. Have high amounts of leaves and seeds as major source of biomass. Do not have a complex protein-lipid content which makes it easy to purify the protein.	Can not be consumed orally bec alkaloid content.
Potato	Model plant in edible vaccine production. Easy and efficient to transform. Tubers can be eaten and has specific promoters. Clonal propagation allows stable production of transgenic plant lines. Outcrossing risk is very low.	Can not be consumed uncooked.

Plant	Advantage	Disadvantage
	Has a well-defined process in food industry. Stored for long periods without refrigeration.	
Tomato	Grows fast. Can be consumed uncooked. Have fruit specific promoters. Easy and efficient to transform. Has a well-defined process in food industry. High content of vitamin A may boost immune response.	Protein content is low in the fruit. Acidic nature of the plant cannot b antigens.
Banana	High amounts can be found in Africa where economical vaccines are required. Can be consumed both by children and adults easily. Clonal propagation allows stable production of transgenic plant lines. Outcrossing risk is very low.	Protein content is low in the fruit. Needs big farming area to be produce
Clover	Easy and efficient to transform. Has high protein content in the leaves. Can be consumed uncooked.	Has a high outcrossing risk. Has a deep root system.
Corn	Proteins can be stable in the seeds for a long time due to the low water content. Has a well-defined process in food industry.	Can not be consumed uncooked by hu Animals can be vaccinated orally.
Lettuce	Grows fast and adaptable for cultivation in various conditions. Naturally free of harmful substances. Suitable for direct consumption and experimental studies. Simple for processing into formulas of potential oral vaccine. High expression levels of S-HBsAg antigen. Naturally free of harmful alkaloids.	Lower amount of harvested material than tobacco or potato. Spoils rapidly.
Strawberry	Can be consumed uncooked.	Spoils quickly. Requires special storage conditions
Rice	Due to low allergenic potential it's commonly used in baby food. High amounts of proteins/ antigens expressions	Grows slowly. Can not be consumed uncooked.

Potato and tobacco were used as model organisms in the beginning of the gene transfer studies to plants. Tobacco and alfalfa have leaves which are major source of biomass; banana, tomato, apple, guava and strawberry are the fruit crops; peanut, corn, soybean and chickpea are seed-based crops; cabbage, lettuce, potato and spinach are the vegetable plants which are used for the production of vaccine antigens [41, 42]. Furthermore, new studies are focused on the other plant species such as clover and arabidopsis which are consumed as uncooked. Some of the plants which are used for edible vaccine production and their properties are listed on Table **2**.

Expressions of Antigens in Transgenic Plants

Recently, several recombinant proteins such as antibodies, viral and bacterial antigens, human and animal therapeutic proteins have been expressed in transgenic plants obtained by genetic engineering advanced techniques [5]. Stable and transient gene expressions are two systems used for candidate vaccine antigen production in plant systems. Integration of the recombinant DNA into plant nucleus or chloroplast genome allows obtaining genetically stable plant lines which are producing recombinant proteins after sexual or vegetative propagation [41]. Integration into nucleus genome results in <u>Mendelian heredity</u> and into chloroplast genome results in <u>maternal heredity</u> which is an environmental friendly approach minimizes the risk of out-crossing In addition chloroplasts have high polypoid genomes allow them to produce edible vaccine antigens and biopharmaceuticals [162]. The stable gene expression requires well defined transformation protocols which are reviewed in Tiwari *et al.* (2009) [41]. Use of *Agrobacterium tumefaciens* for gene expression is one of the highly desired methods for edible plant derived vaccine production [5]. Biolistic and electroporation methods are also being used successfully in the production of edible vaccines [21]. Other convenient approaches are dependent on plant cell and hairy root cultures in order to produce antigens and generate stable cell lines *in vitro* [41]. Expressions of some different antigens in transgenic plants are listed on Table **3**.

Table 3: Expressions of some different antigens in transgenic plants

Antigens	Pathogens	Diseases	Plants	References
Hepatitis B surface antigen (HBsAg)	Hepatitis B virus (HBV)	Hepatitis	Tobacco Potato Tomato Banana	[4, 37, 46, 49, 50]

Table 3: contd...

Cholerae toxin B subunit (CTB)	*V. cholerae*	Cholerae	Tobacco Potato Tomato Rice	[51, 52, 53, 54, 55]
Heat labile toxin B subunit (LTB)	*E. coli*	Diarrhea	Maize Tobacco Tomato Soybean Carrot	[18, 56, 57, 58, 59, 60]
Norwalk virus capsid protein (NVCP)	Norwalk virus	Gastroenteritis	Potato Tobacco	[29]
Rotavirus capsid glycoprotein VP7	Rotavirus	Gastroenteritis	Potato	[45, 61]
Human papilloma virus (HPV) 11 major capsid protein	Human papilloma virus (HPV)	Cervical cancer	Potato Tobacco *A. thaliana*	[62, 63]
Rabies glycoprotein	Rabies virus	Rabies	Tomato Tobacco	[64, 65]
Measles virus glycoprotein	Measles virus	Measles	Tobacco Carrot	[66]
Human cytomegalovirus glycoprotein B	Human cytomegalovirus	Central nervous system disease	Tobacco	[67]
Streptococcus mutans surface protein antigen A	*Streptococcus mutans*	Bacteraemia	Tobacco	[7]
Bacillus anthracis protective antigen	*Bacillus anthracis*	Anthrax	Tobacco Spinach Potato	[68, 69, 70, 71]
Clostridium tetani vaccine antigen	*Clostridium tetani*	Tetanus	Tobacco	[72]
Vaccinia virus coat protein	Vaccinia virus	Smallpox	Tobacco	[73]
Human immunodeficiency virus (HIV) tat protein	HIV	AIDS	Tomato Spinach	[74, 75]
Foot and Mouth Disease virus (FMDV) VP1 protein	FMDV	FMD [Foot and Mouth Disease	Arabidopsis Alfa-lfa	[17, 76]
Swine transmissible gastroenteritis corona virus (TGEV) S glycoprotein	TGEV	TGE	Arabidopsis Potato Tobacco	[77, 78, 79]
Rabbit hemorrhagic syndrome virus (RHSV) structural protein 60 (VP60)	RHSV	RHS	Potato	[80]

Table 3: contd...

Avian reovirus (ARV) structural protein σ-C	ARV	Viral arthritis [tenosynovitis], runting/stunting, malabsorbtion syndrome and feed passage	Alfaalfa	[81]
Infectious bursal disease (IBDV) VP2	IBDV	Suppresses the immune system of young chickens	*Arabidopsis thaliana*	[82]
Canine parvovirus capsid protein 2L21	Canine parvovirus	Depression, vomiting and diarrhea	Tobacco	[83]
P. falciparum AMA1 and MSP1	Plasmodium falciparum	Malaria	Tobacco Lettuce	[163]

Mucosal Immunization and Edible Vaccines

Edible vaccines would induce mucosal immune response after consumption as food but there is a need for an adjuvant system for the transition of innate immune system to adaptive immune system. To overcome this problem adjuvants which can be either co-expressed or co-administared with the vaccine antigens are used [42]. The most used adjuvants are bacterial toxins [cholera toxin (CT) produced by several strains of *Vibrio cholera*e, heat labile enterotoxin (LT) produced by enterotoxigenic *E. coli*], oligodeoxynucleotides containing unmethylated CpG dinucleotides, interleukine 12, muramil dipeptid, auridin, alum salts, monophosphoryl lipid A (MPL) and MF 59 [13, 18, 32, 84]. Another problem in mucosal immunization route, vaccine antigens are faced with gastrointestinal degradation problem. Degradation can be prevented by encapsulation and some plant cells can protect vaccine antigens from the acidic environment of the stomach [85]. Rukavtsova *et al.* (2009) developed transformation vectors free of any selective antibiotic resistance markers to produce HBsAg in tomato cells to be used safely as vaccine candidates [86]. Oral tolerance is another problem for plant based edible vaccines or plant derived antigens since they have a risk to activate regulatory T cells and suppressing the antigen specific immune responses. However, the antigen concentration and plant tissue quantity can be optimized to induce no tolerance. Mice immunized with 100 ng HBsAg in 1 mg tobacco leaves orally did elicited a lineer regulatory T cell response but at low doses 0.5 ng HBsAg in 5 µg tobacco leaves did not induce. Therefore, it is important to optimize the concentration of the vaccine antigen and the quantity of the plant tissue not to induce an oral tolerance but efficient immune response [87].

Edible Vaccines for Human Use

Hepatitis B

Hepatitis B is one of the most common viral diseases of humans. The late manifestation of the disease is hepatocellular carcinoma and cirrhosis of the liver [4]. The current limitations of the hepatitis B vaccination of the world-wide are the cost of the vaccine and administration [42]. Plant-derived vaccines can at least partially replace present hepatitis B vaccines with a lower cost than the regular vaccine [164].

Hepatitis B surface antigen (HBsAg) is the first antigen expressed in *Nicotiana tabacum* [4]. In that study, a virus protein was expressed in plant system for the first time. The antigen quantity was 2-6 ng/mg total protein, subsequently the antigen quantity was improved to 66 ng/mg by different promoter systems [4, 42]. Due to the high alkaloid content of the tobacco, potato was used for HBsAg production to be used in oral applications further. The HBsAg protein was shown in the vesicles of the endoplasmic reticulum in the transgenic potato by transmission electron microscope (TEM) for the first time [43].

The antigen quality produced in the plant is also an important issue, raw consumption has no problems but potato can not be consumed raw by humans. So, a thermal degredation study was also performed by boiling the transgenic potato in order to compare the immune response elicited by uncooked and cooked potato. The antibody response was decreased in the groups vaccinated by cooked potatos. However, due to the encapsulation of the HBsAg by the potato reduced the risk of thermal degradation and stabilize the antigen [46, 88].

The ideal oral hepatitis B vaccine criterias are;

- the amount of the HBsAg (that can easily consumed by orally)

- the antigen concentration in the plant (should be uniform allowing for even dosing of subjects)

- the plant should be palatable (can easily be consumed);

- the HBsAg in the plant should be stable at ambient temperatures for prolonged storage time [88].

There are also studies focused on plant derived vaccine tablet formulation against hepatitis B in lettuce which are promising prototyapes for further candidates [161].

Enterotoxigenic Escherichia coli infection

Enterotoxic *E. coli* and *V. cholerae* colonization in the small intestine can cause acute diarrhea through producing enteretoxins. Protein structure and function of heatlabile enterotoxin of *E. coli* (ETEC) is nearly similar to cholera toxin [18]. ETEC causes 650 milion diarrhea cases with 800.000 death annually [56]. The toxin has two subunits, one is heat labile toxin A (LTA) and the other is heat labile toxin B (LTB) that is functioning on binding intestinal epithelial cells. LTB is non-toxic by itself which can be used as a potent mucosal adjuvant co-expressed with other antigens [41].

The first human trial of edible vaccines was done by LTB antigens produced by potato; after oral immunization studies; it was shown that the edible vaccine expressing LTB antigens elicited humoral and mucosal antibody responses [18]. LTB antigens were produced by using synthetically optimized gene cassette which was transformed by *A. tumefaciens*. Two different transformed clones were selected with 4μg LT-B/g potato tuber and 10 μg LT-B/g potato tuber, respectively. The differences were attributed to *A. tumefaciens* transformation. The antigenic gene cassette can be integrated to plant genome randomly which can inhibit plant growth [48].

Maize was also transformed by LTB antigenic cassette expressing *A. tumefaciens*. Mice were fed with 5 μg or 50 μg LTB containing maize. Anti-LTB antibodies were detected in serum and fecal samples of the mice even fed with 5 μg LTB containing maize [56]. Protein targeting strategies were also used to target synthezied protein to ER. The results of this study showed that protein targeting to ER did not increase the protein expression levels. Mice were fed with 5g tuber and injected with 0.1 mL purified LTB. Anti-LTB antibodies did not elicit with oral immunizations that can be due to the low LTB concentration of the tuber, immunization schedule or antigen degredation in the gastro intestinal system. However, anti-LTB antibodies were elicited by vaccinating mice with 0.1 mL purified LTB [32].

Another study focused on the maternal transfer of the antibodies from mother to fetuses. In that study, mice were fed with 37.8 µg/g LTB containing tomatoes and immune responses were analyzed. The IgA responses of the mother and newborn mice were correlated which is an indicator of maternal transfer of the antibodies [13].

Different targeting strategies are used to express LTB antigens in maize. The highest LTB accumulation is achieved in vacuole targeting signal as 12% of total soluble proteins. However nucleus and plastid targeting strategies did not increase the expression levels [19].

In clinical trials, human volunteers were fed with 50 g or 100 g raw potato which contains 375µg or 750µg LTB antigen during 3 weeks. Anti-LTB IgA level was increased four fold in 10 of 11 serum samples of volunteers [48].

Vibrio Cholerae Infection

V. cholera causes severe diarrhea in humans like enterotoxic *E. coli*. Cholera toxin (CT) is a protein complex secreted by *V. cholerae* and responsible for the massive, watery diarrhea characteristic of cholera infection. CT is also composed of two subunits. Cholera toxin B subunit (CTB) is used in plant expression studies. The results of this study showed that CTB has high immunogenicity and high affinity to intestinal epithelial cells as carrier protein and adjuvant like LTB [71]. Rice was used for expression of CTB and mice fed with this vaccine showed neutralizing antibodies and protection was conferred after oral challenge [54].

Measles

Measles is a highly contagious viral disease which may cause pneumonia, encephalitis even death. The conventional measles vaccine is a live attenuated vaccine which is used since 1960s. In 2010, there were 139 300 measles deaths globally [89], in addition to this, measles rates are increasing in elderly populations at developed countries that can be due to immunosenecence [90]. Carrot was also used successfully for measles hemaglutinin production which also elicited neutralizing antibody response [91].

Bacillus Anthracis

Developing a vaccine against *Bacillus anthracis* gain more importance after its possible usage as biological weapon was realized. Conventional vaccine against *B. anthracis* has serious limitations for massive vaccinations as it requires repeated vaccinations and has severe adverse effects. Thus, to elicit mucosal immune responses, oral and nasal vaccines are needed [70]. Tobacco and spinach are the plants which were used to produce *B.anthracis* protective antigen [68, 69, 70]. Additionally, Kim *et al.* used cholera toxin LTB as a carrier protein to produce *Bacillus anthracis* protective antigen in potatos [71].

Rabies

WHO reports 50000 deaths annually from rabies which is a severe threat for humans and animals. In USA, the cause of the disease is generally wild animals including bat, raccoon, fox, on the other hand, in Asia, Africa and South America dogs are the vectors for transmision to humans. Rabies glycoprotein and nucleoprotein are used for rabies vaccine production. The gene transfer to plant viruses was done by antigenic gene fusing to viral envelope protein of the plant virus [24].

Alfalfa mosaic virus was used for rabies vaccine production. Briefly, the B cell epitope of rabies glycoprotein (G5-24) and T cell epitope of rabies nucleoprotein (31D) were introduced into alfa-lfa mosaic virus genome and tobacco were transformed and recombinant protein were purified two weeks later. Mice were immunized by the purified protein intraperitonelly, and immune responses were elicited by this strategy [24].

Nicotiana benthamiana (wild type tobacco) and *Spinacia oleracea* (spinach) were also used for expression of rabies virus proteins. The mice were immunized intraperitonally and orally, promising responses were obtained and human trials were initiated. In human trials, volunteers were fed three times with 150g spinach leaves/dose and rabies specific IgG and IgA responses were elicited [24].

Rotavirus Infection

Human rotavirus is the causative agent of severe diarrhea in children. In developing countries, 6 % of children deaths below 5 years old and 25% of the diarrhea related deaths are caused by rotavirus [45]. Plant derived rotavirus vaccines were produced by using potato. Rotavirus VP7 glycoprotein was fused to endoplasmic tragetting SEKDEL sequence and the vector was transformed to potato by *A. tumafaciens*. Mice were fed with potato tubers and as a result serum IgG and mucosal IgA responses were elicited against rotavirus VP7 glycoprotein [45]. The full-length rotavirus VP7 gene expressing potatoes were propagated for over fifty generations and they showed a stable transcription of the VP7 gene [61].

Malaria

Malaria is a global health problem caused by the *Plasmodium* genus. The disease is endemic in 90% sub-Saharan African countries but still a problem for Asia, Latin America, Middle East, and Europe [165]. *Plasmodium falciparum* and vivax are the ones that infect humans thus the research is focused on their proteins. The complex life cycle of the parasite makes the selection of the vaccine candidate antigen very challenging. There are several studies targeting the preerythrocytic stage of the parasite [166], asexual blood [167] and sexual stages [168] of the parasite. In addition, vaccination mosquitos to prevent the disease transmission is another approach. Considering the poverty in malaria endemic areas, transgenic plants are promising for malaria vaccine development. The most Asexual plasmodium antigens produced in Tobacco are *P. falciparum* AMA1 and MSP1. The expression levels are varying regarding to transformation method, chloroplast transformation results in higher yields compared to nuclear transformation [163, 169].

Edible Vaccines for Veterinary Use

Using the edible vaccines has also many advantages for livestock industry. Animals will be immunized against many pathogens easily while feeding, without any expertise, in a short time and cheaper than the conventional vaccines that need cold chain and injection [10, 12, 13]. These studies can only be evaluated with *in vivo* studies because the nature of the immune response needs an animal model or volunteer to see the results and success. Here are some of the recent studies focused on veterinary field.

Avian Reovirus Infection

Huang *et al.* (2006) reported their research for avian reovirus [ARV] structural protein σ-C, which is the first candidate for an edible vaccine against ARV infection. They used two promoters *i.e* CaMV 35S and rice actin promoter and transformed them *via Agrobacterium* to alfalfa. The highest expression levels of σ-C protein was 0,008 % for the strains with CaMV 35 S and 0,007 % for the strains with actin promotor, of the total soluble protein in cellular extracts [81].

Lue *et al.* (2011) extended their research with avian reovirus (ARV) structural protein σ-C because of the importance of avian infection in poultry industry. They constructed four nuclear expression vectors with or without codon modification of S1 gene, and used CaMV 35S promoter and rice actin 1 promoter. Cytosol and chloroplasts were the targets for expressed σ-C. The highest expression was detected with rice actin promotor constructs 0,021 %. According to the author, in post-translational level the expression of σ-C may be limited [92].

Infectious Bursal Disease (IBDV)

Infectious bursal disease is one of the important pathogens of chickens. Wu *et al.* (2004a) developed transgenic *Arabidopsis thaliana* plants to express the major host protective protein (VP2). Electroporation was used to send the constructs, which have VP2 to *A. tumefaciens*. Total soluble protein of the selected plants varied from 0,5 to 4,8 %. Chickens orally immunized with recombinant VP2 protein generated antibodies against IBDV [82].

Rabies

The effectiveness of cholera subunit toxin (CTB) as a mucosal adjuvant has been reported by many of the researchers. Roy *et al.* (2010) has fused rabies surface glycoprotein (G protein) with synthetic CTB and expressed in tobacco plants. Quantity of the 80,3 kDa fused protein in tobacco leaves were 0,4 %. And a 403 kDa pentameric protein has been reported also that can actively bind to GM1 receptor. Anti-cholera toxin antibodies and anti-rabies antibodies were able to detect pentameric fusion protein. As the author reported the next step for this protein is to check its' immunoprtective effectiveness in challenge studies [93].

Yusibov *et al.* (2002) constructed a chimeric peptide containing rabies virus glycoprotein (G), nucleoprotein (N) and fused with alfalfa mosaic virus (AIMV) coat protein (CP) with *Nicotiana tabacum* cv. Samsun NN and used *Nicotiana benthamiana* and spinach *(Spinacia oleracea)* with tobacco mosaic virus (TMV) which do not have CP. Mice were immunized parenterally with recombinant rabies protein and in the challenge studies recombinant protein's protection has survived mice. After success of this study, mice were also immunized orally with recombinant proteins in transgenic spinach. And in human trials, raw transgenic spinach leaves were given to human volunteers in two groups. In first group, 3 of 5 volunteers who have been vaccinated against rabies with a commercial vaccine got infection after they had transgenic spinach leaves. Not vaccinated volunteers were given recombinant spinach leaves in the second group, and five of nine them had detectable level of anti-rabies or anti- AIMV antibodies. Second group vaccinated with commercial rabies after one dose of transgenic spinach and neutralizing antibodies were detected in three of nine volunteers serum samples. Also there weren't any antibody detected in serum samples of control group (five volunteers) [24].

According to the Yusibov *et al.* (2002) oral immunizations are more suitable than parenteral usage. *E. coli* heat labile protein, Norwalk virus capsid protein and hepatitis B surface antijen has been searched in human volunteers. Because of the low peptide used in immunizations, virus antibody responses are low. The authors reported that the most important points in their research are high yield of antigen expression and absence of unwanted immune reactions in volunteers who has consumed transgenic spinach [24].

Foot and Mouth Disease [FMDV]

Santos *et al.* (2002) expressed highly immunogenic epitope [VP1] of foot and mouth disease virus [FMDV] in alfalfa plants. Intraperitoneally immunized mice's sera were screened for anti-VP1 antibodies. Antibodies against synthetic peptide, to native VP protein and purified FMDV particles were detected in these sera. Also mice survived after challenge tests. It was reported to be the first example of a peptide-based vaccine produced in transgenic plants, which induces a complete protection in experimental hosts [94].

Wu *et al.* (2003) expressed FMDV serotype O in tobacco plants using recombinant tobacco mosaic virus [TMV]. Tobacco plant derived FMDV particles were used to immunize guinea pigs. Mice and swine were also used in animal challenge studies. Three of eight animals have survived from orally fed guinea pigs with 3 mg of tobacco plants [45].

Zhang *et al.* (2011) reported a novel model for an oral FMDV vaccine. They combined VP1 protein of two serotypes of FMDV, O and Asia 1 serotype and transferred it to maize with genes of Cholera toxin B subunit (CTB) and *Escherichia coli* heat-labile enterotoxin B subunit (LTB) as adjuvants. Maize plantlets were analyzed for the expression of VP1 genes. This is the first study on combining of two serotypes of FMDV expressed together as a potential feed or feed additive [95].

Very recent study of Wang *et al.* (2012) used transgenic rice, which contains capsid precursor polypeptide (P1) gene of FMDV with dual cauliflower mosaic virus (CaMV 35S) promoter. Especially in the leaves of the rice, expression level of P1 was 0.6 - 1.3 µg/mg of total soluble protein. Mice were immunized intraperitoneally and orally with transgenic rice derived P1 protein, *E. coli* derived P1 protein and a commercial inactivated FMDV vaccine [96].

PLANT DERIVED THERAPEUTICS

Production of Antibodies and Antibody Fragments

Monoclonal antibodies, which are used in passive immunizations, form a big portion of the therapeutics produced in biotechnology. In passive immunization, it is estimated that used antibody targets a spesific epitope and takes part as a therapeutic agent. On the other hand, in the active immunization it is expected from immunized subject to produce specific antibodies against the injected antigen [97].

Transgenic plants are used to produce monoclonal antibodies (plant made antibodies, PmAbs) against tooth decay, rheumatid arthritis, cholera, diarrhea, measles, a few kinds of cancers, Norwalk virus, HIV, rhinovirus, influenza, hepatitis B virus and herpes simplex virus. The most developed product is against *S. mutans* related to tooth decay. Plants serve as an alive, scaled-up (large-scale) production systems for antibodies [98, 26, 171]. Antibodies expressed in plants were firstly reported at the end of 1980s [109]. Even tobacco leaves are the most popular expressing systems for recombinant antibodies, different research groups are interested in other plants, seeds for expressions and also large-scale production [99, 100]. Pharmaceutic antibodies derived from transgenic plants are reviewed in Table **4**.

Also a new class of antibody was discovered by Hamers-Casmetrman in 1993 in camels and llamas, which is without light chain, with variable region named V_HH or nanobody [178]. Unless they require glycosylation they can be produced *via* both prokaryotic and eukaryotic hosts with molecular weight of 14-15 kD [179]. Their pharmacological and biophysical properties let these antibodies be used in drug development, diagnose and treatment of diseases and biosensor development [180, 181, 171]. After these improvements, Teh and Kavanagh showed the production possibility of nanobodies (anti-hen egg white lyzozome antibody) in *Nicotiana* leaves [182].

In a short time, antibody production by using transgenic plants reached to clinical trials and market from being a theory. Recombinant antibodies are the most popular products of biotechnology in pharmaceutical industry since they are used in;

- Diagnosis and therapy,

- Detection and elimination of environmental contaminants,

- Control of pathogens,

- Industrial purification processes [5].

Table 4: Pharmaceutic antibodies derived from transgenic plants

Antigen	Host Plant	Antibody type	References
Carcinomaembryonic antigen	Tobacco, pea, Tomato, rice, wheat	scFv, diabody	[101]
Colon cancer antigen	Tobacco	IgG	[102]
Hepatitis B virus	Tobacco	IgG	[103] [100]
Human IgG	Alfalfa	IgG	[104]
HIV-1	Tobacco [cv. BY-2] BY-2	IgG [2G12]	[105, 106,107, 17
Herpes simplex virus	Soybean, rice	IgG	[108]
Low molecular weight phosphonate ester	Tobacco	IgG	[109]
Nematode	Tobacco	IgG	[110]
Non-Hodgkin's lymphoma idiotypes	Tobacco	Virus vector scFv	[97]
Rabies virus	Tobacco	IgG	[111]
Respiratory syncytial virus	Tobacco	IgG	[112]
Streptococcus surface antigen SAI	Tobacco	sIgA/G	[113]
Samonella erterica	Tobacco	scFv	[175]
Skin cancer (TeraCIM)	Tobacco	humanized antibody	[173]
Carcinoma (colorectal and breast)	Tobacco	IgG2a	[174]
Prevent greft rejection (LO-BM2)	Tobacco	IgG	[176]
Tumor-associated antigen tenascin-C (TNC	Tobacco	mAb H10	[177]

Therapeutic Protein, Peptide and Enzyme Production

Plant derived antibodies and edible vaccines, important pharmaceuticals like peptides, proteins and enzymes were produced from transgenic plants with biological or economical advantages over other production systems (extraction from human/animal body fluids, tissues, recombinant microorganisms and animal cell cultures) [115, 5].

High level and low cost of protein production in transgenic plants and animals caused the emergence of a new industry. Plants and animals are used as 'bioreactors' in these production systems. There are some problems to be solved in each type of production systems. The necessity of long rearing periods and prion contamination of purified proteins in transgenic animal systems, are leading problems. Recent studies also show, that plant production systems in large scale, tends to have lower costs for both industrial and pharmaceutical recombinant proteins [116].

Transgenic proteins can be expressed in whole plant, specific plant parts (seed, leaves, stem, *etc.*) or can be accumulated in an organelle in spesific cells. Transgenic protein accumulations have been shown in seeds of maize, soy, tobacco and barley [12].

Recombinant human somatotrophin (hST) was expressed in transgenic tobacco chloroplasts as much as 7% of total soluble protein, which was processed correctly [117]. Another research group also reported the expression level of recombinant protein to total soluble protein, as between 0.02 % to 11.0 % [98]. Potential therapeutics expressed in plant systems are listed in Table **5**.

Table 5: Potential therapeutics expressed in plant systems

Therapeutics	Host Plants	Reference
Angiotensin-converting enzyme	Tobacco, tomato	[118]
Avidin	Maize	[119]
α-trichosanthin	Tobacco	[120]
β-glucuronidase	Corn	[121] [122]
Erythropoietin	Tobacco	[123]
Glucocererbosidase	Tobacco	[124]
Glutamic acid decarboxylase	Tobacco	[183]
Granulocyte-macrophage colony stimulating factor	Tobacco Rice	[125] [126]
Growth hormone	Tobacco, BY-2 Tobacco, BY-2 *O. sativa* L. cv. Donjin	[117] [127] [128]
Hirudin	Brassica seeds	[129]
Human alkaline phosphatase	Tobacco [cvNT-1]	[130]
Human aprotinin	Maize	[118]
Human collagen I	Tobacco	[131]

Table 5: contd...

Human α-antitrypsin	*O. sativa* cv Taipei 309 Rice	[132, 133, 134, 50]
Human enkephalins	*Arabidopsis*	[135]
Human epidermal growth factor	Tobacco	[136]
Human hemoglobin α, β	Tobacco	[136]
Human interferon-α	Rice,	[137]
Human interferon-α-2b	Tobacco [BY-2]	[138]
Human interlukin-12 [IL-12]	Rice	[139]
Human α-iduronidase	Tobacco [BY-2]	[140]
Human lysozyme	*O. sativa* cv Taipei 309	[50]
Human somatotrophin [hST]	Tobacco	[98, 117]
Human serum albumin	Rice	[50]
Lactoferrin	*Acanthopanax senticosus*, Tobacco [BY-2]	[141] [142]
Lactoferrin and β-casein	Potato	[143]
Protein-C	Tobacco	[124]
Phytase	Canola	[144]
Ribosome-inactivating protein	Tobacco	[145]
Thermostable endo-1,4 β-D-glucanase	Arabidopsis	[146]

Plant Derived Therapeutics on Market

For three decades, studies for plant derived proteins are being done. But only a small percentage of these studies are conducted to phase trials. Here are some of the examples of the products which are approved by authorities and which are on the market, listed in Table **6** [114, 147].

IMPORTANCE OF PLANT CELL SUSPENSION CULTURES FOR PRODUCING PLANT-DERIVED VACCINE ANTIGENS AND THERAPEUTICS

In plant cell suspension culture systems, there is no possibbility of contamination with agrochemicals and fertilizers faced in whole plants, they do not contain human pathogens or produce endotoxins. As a result of increasing demand, plant cell suspension cultures are also used for the production of recombinant proteins. Plant cell suspension cultures are formed from mainly calli cells to produce high biomass and secondary metabolites in liquid culture media. *Nicotiana tabacum, Oryza sativa* and *Glycine max* cell suspension cultures are mostly used for expression of therapeutic proteins and they can be cultivated in bioreactors for large-scale production [148, 149].

Table 6: Plant derived therapeutics on market/approved [114, 147]

Aim	Plants	Product	Status
Newcastle disease	Tobacco [Suspension culture]	HN antigen of Newcastle disease	USDA approved
Dental caries	Tobacco	CaroRX	EU approved
Vaccine Purification	Tobacco	Antibody against Hepatitis B	On market
Gaucher's disease	Carrot [Suspension culture]	Elelyso [Taliglucerase alfa]	USDA approved
Human growth factor	Barley	ISOkine™ , DERMOkine™	On market
Vitamin B12 deficiency	Arabidopsis	Human intrinsic factor	On market
Anti-infection, anti-inflammatory	Rice	Human lactoferrin	On market
		Human lysozyme	

Bioreactors provide an environment to cells and this environment is able to optimize the productivity, cell property and desired cell growth. Physiological and biological properties of plant cells in suspension culture are distinct from the bacterial and mammalian cells. Cell growth, nutrient and oxygen mass transfer, heat transfer, production kinetics and fluid hydrodynamics are need to be considered to design an effective bioreactor and bioprocessing for plant cell cultures. The major design considerations are cell growth and oxygen supply, shear stress, cell aggregation and culture rheology, foaming and wall growth [150, 151].

Advanced cell culture operations have been developed for plant suspension cell cultures. These operation modes are; batch, fed-batch, two-stage, perfusion, semicontinuous and continuous culture [152, 153]. Choosing an appropriate process mode is very important because of its influences on product yield and cellular productivity. There are several parameters that must be considered while choosing an operation mode and these are characteristics of the host plant, properties of recombinant protein expression systems, inherent stability [150], properties of expressed proteins, promoter properties that is used for production of recombinant protein and protein localization [148].

Purification of these proteins, antibodies, pharmaceuticals from plants and plant tissues were often underestimated. Downstream costs are nearly % 80 of total costs, in a large-scale recombinant protein production from transgic plants [154]. Large-scale production of transgenic proteins by plants in large fields requires easy and cheaper purification methods. During the cultivation of suspension plant cells in bioreactor for recombinant protein production, secreted proteins' proteolytic degredation or instability is an important problem [155]. This leads the researchers for new methods to alter the functions of recombinant proteins and new strategies for purification [150].

Wilken and Nikolov (2012) reviewed the methods used so far about recovery and purification of recombinant plant derived proteins. As reviewed downstream processes of plant derived recombinant proteins have to be done respectively with primary recovery and then purification [156]. Wilken and Nikolov (2012) declared key points of choosing the plant and purification systems. From the purification cost view, following considerations should be taken in account;

- Choose the target organelle for easier extraction,

- Choose a reasonable plant product and purification method, for ex.: oilseeds and oleosin fusion technology.

- Choose Generally Regarded as Safe (GRAS) plants for oral vaccines and stabile methods for industrial enzyme productions.

- Develop secreted product producing systems.

- Focus on soluble forms of the products expressed in chloroplasts.

- Develop more specific methods for intracellular products.

- Do not use affinity tags for biopharmaceuticals.

- Use models or computer based programs for determining the significant points of the whole system.

- Try to simplify purification steps; number or method [156].

CONCLUSIONS

According to WHO global estimations in 2008, 23.5 million infants did not receive all three doses of DPT (diphteria, pertussis, tetanus vaccine), one of the cheapest, safest and simplest vaccines available. The majority of those infants were in Africa and Southern Asia where most countries have economical obstacles, very low income, ongoing wars and week health infrastructure. Moreover, 13,000 new borns, children and mothers die everyday from vaccine-preventable diseases in Africa [170]. We need a sustained and concerted effort to overcome problems of access to vaccines in all over the world. Affordable vaccines and bioproducts that could be easily produced in these low income countries may contribute solving this problem.

Plant genetic engineering technology offers opportunities and new ways for producing low-cost plant-based vaccines, antibodies and biopharmaceuticals. As a second significant point, the administration of these products are easier than most of the traditional implementations. Additionally, high production levels of proteins in some plant (or plant cell) systems and low cost manufacturing of the product are also very remarkable adventages of plant-based systems over others. As summarized in this chapter, there are many useful utilizations of this systems. Different edible vaccine studies were performed against important human and veterinary pathogens, including both bacteria and viruses, and promising results were obtained. Moreover antibodies, recombinant proteins and secondary metobolites are produced by plants and research efforts are going on. As described in the last part of the chapter, affordable and easy to manufacture bioprocesses could be based on plant cell systems too.

On the other hand, this biotechnological innovation also have some limitations in technical and physiological aspects as described above. First of all, plant-based biopharmaceuticals are also genetically modified organisms and they carry same risks similar to transgenic plants. In order to avoid possible risks, adequate cautions must be taken into account to protect public health and environment .The products are need to be approved by an independent authority (like FDA) and basic regulatory guidelines must be implemented before their regular usage.

Improvements about plant expression systems, such as chloroplast transformation, nuclear transgenes, *etc.* may help to overcome these concerns. New approaches are also needed to increase mucosal immune response, avoid oral tolerance and degredation of plant antigens in gastrointestinal tract.

These new systems let the researchers to have different options from the conventional methods and center upon newly research areas like bioactive peptides and protein production, antibody production for passive immunization therapy and edible vaccines. Using GM-plants to produce potentially important therapeutics may turn in near future to useful therapies for both human and animal health. Overall, this biotechnological innovation has some limitations in technical and physiological aspects, studies on edible vaccines are still increasing remarkably and plant-based biopharmaceuticals will probably be very strong alternatives to traditional vaccine manufacturing technologies in the near future.

ACKNOWLEDGEMENTS

None Declare.

CONFLICT OF INTEREST

The authors confirm that this chapter contents have no conflict of interest.

REFERENCES

[1]   http://www.who.int/immunization_monitoring/diseases/en/, March 2013.
[2]   Francis DP. Successes and failures: Worldwide vaccine development and application. Biologicals 2010; 38(5):523-8.
[3]   http://www.gavialliance.com, March 2013.

[4]   Mason HS, Lam DM, Arntzen CJ. Expression of hepatitis B surface antigen in transgenic plants. Proceedings National Academcy of Sciences USA. 1992; 89(24):11745-9.

[5]   TeliNP and TimkoMP. Recent developments in the use of transgenic plants for the production of human therapeutics and biopharmaceuticals. Plant Cell, Tissue and Organ Culture 2004; 79: 125–145.

[6]   Mor TS, Gomez-Lim MA, Palmer KE. Perspective: edible vaccines-a concept coming of age. Trends Microbiology1998; 6(11):449-53.

[7]   Curtiss RI, Cardineau CA (1990) Oral immunization by transgenic plants. World Patent Application, WO 90/02484.

[8]   Tacket CO, Mason HS, Losonsky G, Clements JD, Levine MM and Arntzen CJ. Immunogenicity in humans of a recombinant bacterial antigen delivered in a transgenic potato. *Nature Medicine* 1998;4(5):607-09.

[9]   Arntzen CJ. Pharmaceutical foodstuffs-oral immunization with transgenic plants. *Nature Medicine* (vaccine supplement) 1998;4(5):502-03.

[10]   Sala F, Manuela Rigano M, Barbante A, Basso B, Walmsley AM, Castiglione S. Vaccine antigen production in transgenic plants: strategies, gene constructs and perspectives. Vaccine 2003; 30;21(7-8):803-8.

[11]   Arntzen CJ (1996) Plant vaccines: Edible but how credible? CVI Forum No.12: 10-13.

[12]   Larrick JW, Thomas DW. Producing proteins in transgenic plants and animals. Current Opinion Biotechnology 2001;12(4):411-8.

[13]   Walmsley AM, Alvarez ML, Jin Y, Kirk DD, Lee SM, Pinkhasov J, *et al*. Expression of the B subunit of *Escherichia coli* heat-labile enterotoxin as a fusion protein in transgenic tomato. Plant Cell Rep. 2003; 21(10):1020-6.

[14]   Giddings G. Transgenic plants as protein factories. Current Opinion Biotechnology 2001; 12(5):450-4

[15]   Stewart-Tull The use of adjuvants in experimental vaccines. In: Methods in Molecular Medicine: Vaccine Protocols. (1996) Tacket CO.

[16]   Francis, MJ (1993) Carriers for peptides: Theories and Technology. In: Gregoriadis G, Allison AC, Poste G (eds), Proc. NATO Advanced Study Institute on New Generation Vaccines; The Role of Basic Immunology, June 24-July 5, 1992 at Cape Sounion Beach, Greece. Plenum Pres, New York.

[17]   Dus Santos MJ, Wigdorovitz A, Trono K, Rios RD, Franzone PM, Gill F, Carillo C, Escribano JM, Borca M A novel methodology to develop a foot and mouth disease virus peptide-based vaccine in transgenic plants. Vaccine 2002; 20: 1141-1147.

[18]   Mason HS, Tacket CO, Richter LJ, Arntzen CJ Subunit vaccines produced and delivered in transgenic plants as "edible vaccines". Res Immunol 1998; 149:71-74.

[19]   Streatfield SJ, Lane JR, Brooks CA, Barker DK, Poage ML, Mayor JM, Lampher BJ, Drees CF, Jilka JM, Hood EE, Howard JA. Corn as a production system for human and animal vaccines. Vaccine 2003, 21: 812-815.

[20]   Daniell H, Streatfield SJ, Wycoff K. Medical molecular farming: production of antibodies, biopharmaceuticals and edible vaccines in plants. Trends Plant Sci 2001; 6(5):219-26.

[21]   Shah C, Trivedi MN, Vachhani UD, Joshi VJ. Edible vaccine: A better way for immunization, International Journal of Current Pharmaceutical Research 2011; 3 (1): 53-56. ISSN- 0975-7066.

[22]   Goldstein DA, Thomas JA. Biopharmaceuticals derived from genetically modified plants, Q.J. Med 2004; 97:705-716.

[23]   Mishra N, Gupta PN, Khatri K, Goyal AK, Vyas SP. Edible vaccines: A new approach to oral immunization, Indian Journal of Biotechnology 2008; 7: 283-294.

[24]   Yusibov VM, Mamedov TG. Plants as an alternative system for expression of vaccine antigens, Proceedings of ANAS (Biological Sciences) 2010; 65 (5-6): 195-200.

[25]   Fischer R, Schillberg S, Hellwig S, Twyman RM, Drossard J. GMP issues for recombinant plant-derived pharmaceutical proteins, Biotechnology Advances 2012; 30: 434-439.

[26]   Thomas BR, Deynze AV, Bradford KJ. Production of Therapeutic Proteins in Plants, Agricultural Biotechnology in California Series 2002; 1-12.

[27]   Han M, Su T, Zu YG, An ZG. Research advances on transgenic plant vaccines, Acta Genetica Sinica 2006; 33(4): 285-293.

[28]   Roy N, Agarwal S. Therapeutic protein production-an overview, Business Briefing: Future Drug Discovery 2003; 79-82.

[29]   Mason HS, Ball JM, Shi JJ, Jiang X, Estes MK, Arntzen CJ. Expression of Norwalk virus capsid protein in transgenic tobacco and potato and its oral immunogenicity in mice. Proceedings National Academcy of Sciences USA 1996; 93: 5335-5340.

[30]   Sandhu JS, Krasnyanski SF, Domier LL, Korban SS, Osadjan MD, Buetow DE. Oral immunization of mice with transgenic tomato fruit expressing respiratory syncytial virus-F protein induces a systemic immune response. Transgenic Research 2000; 9: 127-135.

[31]   Waghulkar VM. Fruit derived edible vaccines: Natural way for the vaccination. International Journal of Pharmaceutical Technical Research 2010; 2(3): 2124-2127.

[32]   Lauterslager TGM, Florack DEA, van der Wal TJ, Molthoff JW, Langeveld JPM, Bosch D, Boersma WJA, Hilgers LA. Oral immunisation of naive and primed animals with transgenic potato tubers expressing LT-B. Vaccine 2001; 19: 2749-2755.

[33]   Biemelt S, Sonnewald U, Galmbacher P, Willmitzer L, Müller M. Production of human papillomavirus type 16 virus-like particles in transgenic plants. Journal of Virology 2003; 77(17):9211-9220.

[34]   Tripurani SK, Reddy NS, Rao KRSS. Green revolution vaccines, edible vaccines. African Journal of Biotechnology 2003; 2(12):679-683.

[35]   Rossi L, Baldi A, Dell'orto V, Fogher C. Antigenic recombinant proteins expressed in tobacco seeds as a model for edible vaccines against swine oedema. Veterinary Research Communications 2003; 27: 659-661.

[36]   Takagi H, Hiroi T, Yang L, Tada Y, Yuki Y, Takamura K, Ishimitsu R, Kawauchi H, Kiyono H, Takaiwa F. A rice-based edible vaccine expressing multiple T cell epitopes induces oral tolerance for inhibition of Th2-mediated IgE responses. Proceedings National Academcy of Sciences USA 2005; 102, 17525-17530.

[37]   Kumar GBS, Ganapathi TR, Revathi CJ, Srinivas L, Bapat VA. Expression of hepatitis B surface antigen in transgenic banana plants. Planta 2005; 222: 484-493.

[38]   Ko S, Liu JR, Yamakawa T, Matsumoto Y, Expression of the protective antigen (SpaA) in transgenic hairy roots of tobacco. Plant Molecular Biology Reporter 2006; 24:251.

[39]   Woffenden BJ, Nopo LH, Cramer CL, Dolan MC, Medina-Bolivar F. Expression of a ricin B: F1:V fusion protein in tobacco hairy roots: steps toward a novel pneumonic plaque vaccine. Electronic Journal of Integrative Biosciences 2008; 3(1): 10-19.

[40]   De Guzman G, Walmsley AM, Webster DE, Hamill JD. Hairy roots cultures from different Solanaceous species have varying capacities to produce *E. Coli* B-subunit heat-labile toxin antigen. Biotechnol Lett 2011; 33:2495-2502.

[41] Tiwari S, Verma PC, Singh PK, Tuli R. Plants as bioreactors for production of vaccine antigens. Biotechnology Advances 2009; 27: 449-467.

[42] Kumar GB, Ganapathi TR, Bapat VA. Production of hepatitis B surface antigen in recombinant plant systems: an update. Biotechnol Prog. 2007; 23(3):532-9.

[43] Thanavala Y, Yang YF, Lyons P, Mason HS, Arntzen C. Immunogenicity of transgenic plant-derived hepatitis B surface antigen. Proc Natl Acad Sci U S A. 1995; 92(8):3358-61.

[44] Khandelwal A, Sita GL, Shaila MS. Expression of hemagglutinin protein of rinderpest virus in transgenic tobacco and immunogenicity of plant-derived protein in a mouse model. Virology 2003; 308(2):207-15.

[45] Wu YZ, Li JT, Mou ZR, Fei L, Ni B, Geng M, et al. Oral immunization with rotavirus VP7 expressed in transgenic potatoes induced high titers of mucosal neutralizing IgA. Virology 2003; 313(2):337-42.

[46] Kong Q, Richter L, Yang YF, Arntzen CJ, Mason HS, Thanavala Y. Oral immunization with hepatitis B surface antigen expressed in transgenic plants. Proceedings National Academcy of Sciences USA 2001; 98(20):11539-44.

[47] Larkin JC, Oppenheimer DG, Lloyd AM, Paparozzi ET, Marks MD. Roles of the Glabrous1 And Transparent Testa Glabra Genes in Arabidopsis Trichome Development. Plant Cell. 1994; 6(8):1065-76.

[48] Mason HS, Warzecha H, Mor T, Arntzen CJ. Edible plant vaccines: applications for prophylactic and therapeutic molecular medicine. Trends Molecular Medicine. 2002; 8(7):324-9.

[49] Richter LJ, Thanavala Y, Arntzen CJ, Mason HS. Production of hepatitis B surface antigen in transgenic plants for oral immunization. Nature Biotechnology 2000; 18(11):1167-71.

[50] Huang Z, Elkin G, Maloney BJ, Beuhner N, Arntzen CJ, Thanavala Y. Virus-like particle expression and assembly in plants: hepatitis B and Norwalk viruses. Vaccine. 2005; 23(15):1851-8.

[51] Hein MB, Yeo TC, Wang F, Sturtevant A. Expression of cholera toxin subunits in plants. Ann N Y Acad Sci. 1996; 792:50-6

[52] Jani D, Meena LS, Rizwan-ul-Haq QM, Singh Y, Sharma AK, Tyagi AK. Expression of cholera toxin B subunit in transgenic tomato plants. Transgenic Research 2002;11(5):447-54.

[53] Sharma MK, Singh NK, Jani D, Sisodia R, Thungapathra M, Gautam JK, et al. Expression of toxin co-regulated pilus subunit A (TCPA) of Vibrio cholerae and its immunogenic epitopes fused to cholera toxin B subunit in transgenic tomato (Solanum lycopersicum). Plant Cell Rep. 2008; 27(2):307-18.

[54] Nochi T, Takagi H, Yuki Y, Yang L, Masumura T, Mejima M, et al. Rice-based mucosal vaccine as a global strategy for cold-chain- and needle-free vaccination. Proceedings National Academcy of Sciences USA. 2007; 104(26):10986-91.

[55] Oszvald M, Kang TJ, Tomoskozi S, Jenes B, Kim TG, Cha YS, et al. Expression of cholera toxin B subunit in transgenic rice endosperm. Molecular Biotechnology. 2008; 40(3):261-8.

[56] Streatfield SJ, Jilka JM, Hood EE, Turner DD, Bailey MR, Mayor JM, et al. Plant-based vaccines: unique advantages. Vaccine. 2001; 19(17-19):2742-8.

[57] Chikwamba R, Cunnick J, Hathaway D, McMurray J, Mason H, Wang K. A functional antigen in a practical crop: LT-B producing maize protects mice against *Escherichia coli* heat labile enterotoxin (LT) and cholera toxin (CT). Transgenic Research 2002; 11(5):479-93.

[58]   Walmsley AM, Arntzen CJ. Plant cell factories and mucosal vaccines. Current Opinion Biotechnology 2003;14(2):145-50.

[59]   Moravec T, Schmidt MA, Herman EM, Woodford-Thomas T. Production of *Escherichia coli* heat labile toxin (LT) B subunit in soybean seed and analysis of its immunogenicity as an oral vaccine. Vaccine 2007; 25(9):1647-57.

[60]   Rosales-Mendoza S, Soria-Guerra RE, Lopez-Revilla R, Moreno-Fierros L, Alpuche-Solis AG. Ingestion of transgenic carrots expressing the *Escherichia coli* heat-labile enterotoxin B subunit protects mice against cholera toxin challenge. Plant Cell Rep 2008; 27(1):79-84.

[61]   Li JT, Fei L, Mou ZR, Wei J, Tang Y, He HY. Immunogenicity of a plant-derived edible rotavirus subunit vaccine transformed over fifty generations. Virology 2006; 356(1-2):171-8.

[62]   Warzecha H, Mason HS, Lane C, Tryggvesson A, Rybicki E, Williamson AL, *et al.* Oral immunogenicity of human papillomavirus-like particles expressed in potato. Journal of Virol 2003; 77(16):8702-11.

[63]   Kohl TO, Hitzeroth, II, Christensen ND, Rybicki EP. Expression of HPV-11 L1 protein in transgenic Arabidopsis thaliana and Nicotiana tabacum. BMC Biotechnology 2007;7:56.

[64]   McGarvey PB, Hammond J, Dienelt MM, Hooper DC, Fu ZF, Dietzschold B, *et al.* Expression of the rabies virus glycoprotein in transgenic tomatoes. Biotechnology (NY). 1995; 13(13):1484-7.

[65]   Ashraf S, Singh PK, Yadav DK, Shahnawaz M, Mishra S, Sawant SV, *et al.* High level expression of surface glycoprotein of rabies virus in tobacco leaves and its immunoprotective activity in mice. Journal of Biotechnology 2005; 119(1): 1-14.

[66]   Marquet-Blouin E, Bouche FB, Steinmetz A, Muller CP. Neutralizing immunogenicity of transgenic carrot (Daucus carota L.)-derived measles virus hemagglutinin. Plant Molecular Biology 2003; 51(4): 459-69.

[67]   Tackaberry ES, Dudani AK, Prior F, Tocchi M, Sardana R, Altosaar I, *et al.* Development of biopharmaceuticals in plant expression systems: cloning, expression and immunological reactivity of human cytomegalovirus glycoprotein B (UL55) in seeds of transgenic tobacco. Vaccine 1999; 17(23-24): 3020-9.

[68]   Aziz MA, Singh S, Anand Kumar P, Bhatnagar R. Expression of protective antigen in transgenic plants: a step towards edible vaccine against anthrax. Biochem Biophys Res Commun 2002; 299(3): 345-51.

[69]   Singh S, Kumar PA, Bhatnagar R, Aziz MA. Expression of protective antigen in transgenic plants: a step towards edible vaccine against anthrax. Biochemical and Biophysical Research Communications 2002; 299: 345-351.

[70]   Sussman HE. Spinach makes a safer anthrax vaccine. Drug Discov Today 2003; 8(10): 428-30.

[71]   Kim TG, Galloway DR, Langridge WH. Synthesis and assembly of anthrax lethal factor-cholera toxin B-subunit fusion protein in transgenic potato. Mol Biotechnol 2004; 28(3): 175-83.

[72]   Tregoning JS, Nixon P, Kuroda H, Svab Z, Clare S, Bowe F, *et al.* Expression of tetanus toxin Fragment C in tobacco chloroplasts. Nucleic Acids Res 2003; 31(4): 1174-9.

[73]   Golovkin M, Spitsin S, Andrianov V, Smirnov Y, Xiao Y, Pogrebnyak N, *et al.* Smallpox subunit vaccine produced in Planta confers protection in mice. Proc Natl Acad Sci U S A. 2007; 104(16): 6864-9.

[74] Ramirez YJ, Tasciotti E, Gutierrez-Ortega A, Donayre Torres AJ, Olivera Flores MT, Giacca M, *et al.* Fruit-specific expression of the human immunodeficiency virus type 1 tat gene in tomato plants and its immunogenic potential in mice. Clin Vaccine Immunol. 2007; 14(6): 685-92.

[75] Karasev AV, Foulke S, Wellens C, Rich A, Shon KJ, Zwierzynski I, *et al.* Plant based HIV-1 vaccine candidate: Tat protein produced in spinach. Vaccine. 2005; 23(15): 1875-80.

[76] Carrillo C, Wigdorovitz A, Oliveros JC, Zamorano PI, Sadir AM, Gomez N, *et al.* Protective immune response to foot-and-mouth disease virus with VP1 expressed in transgenic plants. J Virol. 1998 Feb;72(2): 1688-90.

[77] Gomez N, Carrillo C, Salinas J, Parra F, Borca MV, Escribano JM. Expression of immunogenic glycoprotein S polypeptides from transmissible gastroenteritis coronavirus in transgenic plants. Virology 1998; 249(2): 352-8.

[78] Gomez N, Wigdorovitz A, Castanon S, Gil F, Ordas R, Borca MV, *et al.* Oral immunogenicity of the plant derived spike protein from swine-transmissible gastroenteritis coronavirus. Arch Virol 2000; 145(8): 1725-32.

[79] Tuboly T, Yu W, Bailey A, Degrandis S, Du S, Erickson L, *et al.* Immunogenicity of porcine transmissible gastroenteritis virus spike protein expressed in plants. Vaccine 2000; 18(19): 2023-8.

[80] Castanon S, Marin MS, Martin-Alonso JM, Boga JA, Casais R, Humara JM, *et al.* Immunization with potato plants expressing VP60 protein protects against rabbit hemorrhagic disease virus. J Virol 1999; 73(5): 4452-5.

[81] Huang LK, Liao SC, Chang CC, Liu HJ. Expression of avian reovirus sigmaC protein in transgenic plants. J Virol Methods 2006; 134(1-2): 217-22.

[82] Wu H, Singh NK, Locy RD, Scissum-Gunn K, Giambrone JJ. Immunization of chickens with VP2 protein of infectious bursal disease virus expressed in Arabidopsis thaliana. Avian Dis 2004a; 48(3): 663-8.

[83] Molina A, Hervas-Stubbs S, Daniell H, Mingo-Castel AM, Veramendi J. High-yield expression of a viral peptide animal vaccine in transgenic tobacco chloroplasts. Plant Biotechnol J 2004; 2(2):141-53.

[84] Engers H, Kieny MP, Malhotra P, Pink JR (2003) Meeting report. Third meeting on Novel Adjuvants Currently in or Close to Clinical Testing World Health Organisation. Annecy, France, 7-9 January, Vaccine 2002, 21: 3503-3524.

[85] Lauterslager TG, Hilgers LA. Efficacy of oral administration and oral intake of edible vaccines. Immunol Lett 2002; 84(3): 185-90.

[86] Rukavtsova EB, Gaiazova AR, Chebotareva EN, Bur'ianova Ia I. Production of marker-free plants expressing the gene of the hepatitis B virus surface antigen. Genetika 2009; 45(8): 1055-60.

[87] Kostrzak A, Cervantes Gonzalez M, Guetard D, Nagaraju DB, Wain-Hobson S, Tepfer D, *et al.* Oral administration of low doses of plant-based HBsAg induced antigen-specific IgAs and IgGs in mice, without increasing levels of regulatory T cells. Vaccine 2009; 27(35): 4798-807.

[88] Streatfield SJ. Delivery of plant-derived vaccines. Expert Opin Drug Deliv. Jul, 2005; 2(4): 719-28.

[89] http://www.who.int/mediacentre/factsheets/fs286/en/

[90] Hendriksz T, Malouf P, Foy JE. Vaccines for measles, mumps, rubella, varicella, and herpes zoster: immunization guidelines for adults. J Am Osteopath Assoc, 2011; 111(10 Suppl 6):S10-2.

[91]   Muller CPF, Fack, B, Damien, F, Bouche B, Immunogenic measles antigens expressed in plants: role as an edible vaccine for adults. Vaccine 2003; 21: 816-819.

[92]   Lu SW, Wang KC, Liu HJ, Chang CD, Huang HJ, Chang CC. Expression of avian reovirus minor capsid protein in plants. J Vir Methods 2011; 173: 287–293.

[93]   Roy S, Tyagi A, Tiwari S, Singh A, Sawant SV, Singh PK, Tuli R. Rabies glycoprotein fused with B subunit of cholera toxin expressed in tobacco plants folds into biologically active pentameric protein. Pro Exp and Pur 2010; 70: 184 – 190.

[94]   Santos MJD, Wigdorowitz A, Trono K, Rios RD, Franzone PM, Gil F, Moreno J, Carillo C, Escribano JM, Borca MV. A novel methodology to develop a foot and mouth disease virus (FMDV) peptide-based vaccine in transgenic plants. Vacc 2002; 20: 1141–1147.

[95]   Zhang S, Zhang G, Rong T, Pan L, Zhou P, Zhang YG. Transformation of Two VP1 Genes of O- and Asia 1-Type Foot-and-Mouth Disease Virus into Maize. Agricultural Science in China 2011; 10(5): 661-667.

[96]   Wang Y, Shen Q, Jiang Y, Song Y, Fang L, Xiao S, Chen H. Immunogenicity of foot-and-mouth disease virus structural polyprotein P1 expressed in transgenic rice. J Vir Methods 2012; 181: 12– 17.

[97]   McCormick AA, Kumagai MH, Hanley K, Turpen TH, Hakim I, Grill LK, Tuse D, Levy S and Levy R. Rapid production of specific vaccines for lymphoma by expression of tumor-derived single-chain Fv epitopes in tobacco plants. Proc Natl Acad Sci USA 1999; 96: 703–708.

[98]   Daniell, H. Molecular strategies for gene containment in transgenic crops. Nature Biotechnol 2002; 20(6): 581-586.

[99]   Fischer R, Drossard J, Commandeur U, Schillberg S and Emans, N. Towards molecular farming in the future: moving from diagnostic protein and antibody production in microbes to plants. Biotechnol Appl Biochem 1999; 30: 101–108.

[100]   Valdes R, Gomez L, Padilla S, Brito J, Reyes B, Alvarez T, Mendoza O, Herrera O, Ferro W, Pujol M, Leal V, Linares M, Hevia Y, Garcia C, Mila L, Garcia O, Sanchez R, Acosta A, Geada D, Paez R, Luis Vega J, Borroto C. Large-scale purification of an antibody directed against hepatitis B surface antigen from transgenic tobacco plants. Biochem Biophys Res Commun 2003; 308: 94-100.

[101]   Stoger E, Sack M, Fischer R and Christou P. Plantibodies: applications, advantages and bottlenecks. Curr Opin Biotechnol 2000; 13: 161- 166.

[102]   Verch T, Yusibov V, Koprowski H, Expression and assembly of a full-length monoclonal antibody in plants using a plant virus vector. J Immunol Methods 1998; 220: 69–75.

[103]   Yano A, Maeda F, Takekoshi M. Transgenic tobacco cells producing the human monoclonal antibody to hepatitis B virus surface antigen. J Med Virol 2004; 73: 208–215.

[104]   Khoudi, H., Laberge, S., Ferullo, J.M., Bazin, R., Darveau, A., Castonguay, Y., Allard, G., Lemieux, R., Vezina, L.P. Production of a diagnostic monoclonal antibody in perennial alfalfa plants. Biotechnol Bioeng 1999; 64: 135–143.

[105]   Rademacher T, Sack M, Arcalis E, Stadlmann J, Balzer S, Altmann F, Quendler H, Stiegler G, Kunert R, Fischer R. Recombinant antibody 2G12 produced in maize endosperm efficiently neutralizes HIV-1 and contains predominantly single-GlcNAc N-glycans. Plant Biotechnol J 2008; 6(2): 189–201.

[106]   Ramessar K, Rademacher T, Sack M, Stadlmann J, Platis D, Stiegler G, Labrou N, Altmann F, Ma J, Stoger E. Cost-effective production of a vaginal protein microbicide to prevent HIV transmission. Proc Natl Acad Sci USA 2008: 105(10); 3727–3732.

[107] Holland T, Sack M, Rademacher T, Schmale K, Altmann F, Stadlmann J. Optimal nitrogen supply as a key to increased and sustained production of a monoclonal full-size antibody in BY-2 suspension culture. Biotechnol Bioeng 2010; 107: 278–89.

[108] Zeitlin L, Olmsted SS, Moench TR, Co MS, Martinell BJ, Paradkar VM, Russell DR, Queen C, Cone RA and Whaley KJ. A humanized monoclonal antibody produced in transgenic plants for immunoprotection of the vagina against genital herpes. Nature Biotechnol 1998; 16: 1361–1364.

[109] Hiatt A, Cafferkey R, Bowdish K. Production of antibodies in transgenic plants. Nat 1989; 342: 76–8.

[110] Baum TJ, Hiatt A, Parrott WA, Pratt LH, Hussey RS, Expression in tobacco of a functional monoclonal antibody specific to stylet secretions of the root-knot nematode. Mol Plant-Microbe Int. 1996; 9: 382–387.

[111] Girard LS, Fabis MJ, Bastin M, Courtois D, Petiard V, Koprowski H. Expression of a human anti-rabies virus monoclonal antibody in tobacco cell culture. Biochem Biophys Res Commun 2006; 345: 602–7.

[112] Belanger H, Fleysh N, Cox S, Bartman G, Deka D, Trudel M, Koprowski H, Yusibov V. Human respiratory syncytial virus vaccine antigen produced in plants. FASEB J 2000; 14: 2323–2328.

[113] Ma JK, Hikmat B, Wycoff K, Vine M, Chargelegue D, Yu L, Hein M, Lehner T. Characterization of a recombinant plant monoclonal secretary antibody and preventive immunotherapy in humans. Nature Med 1998; 4: 601–606.

[114] Obembe OO, Popoola JO, Leelavathi S, Reddy SV. Advances in plant molecular farming. Biotechnol Adv 2011;29:210–222.

[115] Desai UA, Sur G, Daunert S, Babbitt, Li Q. Expression and affinity purification of recombinant proteins from plants. Protein Expression and Purification 2002; 25: 195–202.

[116] Wycoff KL. Secretory IgA antibodies from plants. Curr Pharmaceu Design 2004; 10: 0.

[117] Staub JM, Garcia B, Graves J, Hajdukiewicz PT, Hunter P, Nehra N, Paradkar V, Schlittler M, Caroll JA, Spatola L, Ward D, Ye GN, Russell DA. High-yield production of human therapeutic protein in tobacco chloroplast. Nature Biotechnol 2000;18:333–338.

[118] Giddings G, Allison G, Brooks D, Carter A. Transgenic plants as factories for biopharmaceuticals. Nat Biotechnol 2000;18:1151–1155.

[119] Kramer KJ, Morgan TD, Throne JE, Dowell FE, Bailey M, Howard JA. Transgenic avidin maize is resistant to storage insect pests. Nat Biotech 2000;18:670–674.

[120] Kumagai MH, Turpen TH, Weinzenl N, Della-Cioppa G, Turpen AM, Donson JD, Hilf ME, Grantham GL, Dawson WO, Chow TP, Piatak M, Grill LK. Rapid, highlevel expression of biologically active alpha-trichosanthin in transfected plants by an RNA viral vector. Proc Natl Acad Sci USA 1993;90:427–430.

[121] Evangelista RL, Kusnadi AR, Howard JA, Nikolov ZL. Process and economic evaluation of the extraction and purification of recombinant glucuronidase from transgenic corn. Biotechnol Prog 1998;14:607–614.

[122] Shaaltiel Y, Hashmueli S, Bartfeld D, Baum G, Ratz T, Mizrachi E. System and method for production of antibodies in plant cell culture. United States Patent Publication No. 20090082548, 2009.

[123] Matsumoto S, Ikura K, Ueda M, Sasaki R. Characterization of a human glycoprotein (erythropoietin) produced in cultured tobacco cells. Plant Mol Biol 1995;27:1163–1172.

[124] Cramer CL, Weissenborn DL, Oishi KK, Grabau EA, Bennett S, Ponce E, Grabowski GA, Radin DN. Bioproduction of human enzymes in transgenic tobacco. Ann NY Acad Sci 1996;792:62–71.

[125] Lee JS, Choi SJ, Kang HS, Oh WG, Cho KH, Kwon TH, Kim DH, Jang YS, Yang MS. Establishment of a transgenic tobacco cell-suspension culture system for producing murine granulocyte-macrophage colony stimulating factor. Mol Cell 1997;7:783–787.

[126] Shin YJ, Hong SY, Kwon TH, Jang YS, Yang MS. High level of expression of recombinant human granulocyte-macrophage colony stimulating factor in transgenic rice cell suspension culture. Biotechnol Bioeng 2003;82:778–783.

[127] Xu JF, Okada S, Tan L, Goodrum KJ, Kopchick JJ, Kieliszewski MJ. Human growth hormone expressed in tobacco cells as an arabinogalactan-protein fusion glycoprotein has a prolonged serum life. Transgenic Res 2010;19(5):849–867.

[128] Kim TG, Baek MY, Lee EK, Kwon TH, Yang MS. Expression of human growth hormone in transgenic rice cell suspension culture. Plant Cell Rep 2008;27:885–891.

[129] Parmenter DL, van Rooijen GJH, Moloney MM. Oleosins as Carriers of the Anti-Coagulant Protein, Hirudin in Brassica Seeds. Abstracts 4th International Congress of Plant Molecular Biology, Amsterdam, The Netherlands, 1994.

[130] Becerra-Arteaga A, Mason HS, Shuler ML. Production, secretion, and stability of human secreted alkaline phosphatase in tobacco NT1 cell suspension cultures. Biotechnol Prog 2006;22:1643–1649.

[131] Ruggiero F, Exposito JY, Bournat P, Gruber V, Perret S, Comte J, Olagnier B, Garrone R, Theisen M. Triple helix assembly and processing of human collagen produced in transgenic tobacco plants. FEBS Lett 2000;469:132–136.

[132] Terashima M, Murai Y, Kawamura M, Nakanishi S, Stoltz T, Chen L. Production of functional human alpha (1)-antitrypsin by plant cell culture. Appl Microbiol Biotechnol 1999;52:516–523.

[133] Trexler MM, McDonald KA, Jackman AP. Bioreactor production of human alpha (1)-antitrypsin using metabolically regulated plant cell cultures. Biotechnol Prog 2002;18:501–508.

[134] Trexler MM, McDonald KA, Jackman AP. A cyclical semicontinuous process for production of human alpha (1)-antitrypsin using metabolically induced plant cell suspension cultures. Biotechnol Prog 2005;21:321–328.

[135] Kusnadi AR, Nikolov ZL, Howard JA. Production of recombinant proteins in transgenic plants: Practical considerations. Biotechnol Bioeng 1997;56(5):473-484.

[136] Cramer CL, Weissenborn DL, Oishi KK, Grabau EA, Bennett S, Ponce E, Grabowski GA, Radin DN. Bioproduction of human enzymes in transgenic tobacco. Ann NY Acad Sci 1996;792:62–71.

[137] Zhu Z, Hughes KW, Huang L, Sun B, Liu C, Li Y, Hou Y, Li X. Expression of human-interferon cDNA in transgenic rice plants. Plant Cell Tiss Org Cult 1994;36:197–204.

[138] Xu JF, Tan L, Goodrum KJ, Kieliszewski MJ. High-yields and extended serum half-life of human interferon alpha 2b expressed in tobacco cells as Arabinogalactan-protein fusions. Biotechnol Bioeng 2007;97:997-1008.

[139] Shin YJ, Lee NJ, Kim J, An XH, Yang MS, Kwon TH. High-level production of bioactive heterodimeric protein human interleukin-12 in rice. Enzyme Microb Technol 2010;46:347–351.

[140] Fu LH, Miao YS, Lo SW, Seto TC, Sun SSM, Xu ZF, Clemens S, Clarke LA, Kermode AR, Jiang LW. Production and characterization of soluble human lysosomal enzyme alpha-iduronidase with high activity from culture media of transgenic tobacco BY-2 cells. Plant Sci 2009;177:668–675.

[141] Jo SH, Kwon SY, Park DS, Yang KS, Kim JW, Lee KT. High-yield production of functional human lactoferrin in transgenic cell cultures of Siberian ginseng (*Acanthopanax senticosus*). Biotechnol Bioprocess Eng 2006;11:442–448.

[142] Choi SM, Lee OS, Kwon SY, Kwak SS, Yu DY, Lee HS. High expression of a human lactoferrin in transgenic tobacco cell cultures. Biotechnol Lett 2003;25:213–218.

[143] Chong DKX and Langridge WHR. Expression of full length bioactive antimicrobial human lactoferrin in potato plants. Transgenic Res 2000;9:71–78.

[144] Zhang ZB, Kornegay ET, Radcliffe JS, Wilson JH, Veit HP. Comparison of phytase from genetically engineered Aspergillus and canola in weaning pig diets. J Anim Sci 2000;78:2868–2878.

[145] Francisco JA, Gawlak SL, Miller M, Bathe J, Russell D, Chace D, Mixan B, Zhao L, Fell HP, Siegall CB. Expression and characterization of bryodin 1 and a bryodin 1-based single-chain immunotoxin from tobacco cell culture. Bioconjugate Chem 1997;8:708–713.

[146] Ziegler M, Thomas S, Danna K. Accumulation of a thermostable endo-1,4-b-D-glucanase in the apoplast of Arabidopsis thaliana leaves. Mol Breed 2000;6:37–46.

[147] Fox JL. First plant-made biologic approved. Nat Biotechnol 2012;30(6);472.

[148] Xu J, Ge X, Dolan MC. Towards high-yield production of pharmaceutical proteins with plant cell suspension cultures. Biotechnol Adv 2011;29(3):278-299

[149] Hellwig S, Drossard J, Twyman RM, Fischer R. Plant cell cultures for the production of recombinant proteins. Nat Biotechnol 2004;22(11):1415-1422.

[150] Huang TK and McDonald KA. Bioreactor engineering for recombinant protein production in plant cell suspension cultures. Biochem Eng J 2009;45:168–184.

[151] Eibl R, Werner S, Eibl D. Bag bioreactor based on wave-induced motion: Characteristics and applications. Adv Biochem Engin/Biotechnol 2009:115;55–87.

[152] Fong W, Zhang Y and Yung, P. Optimization of monoclonal antibody production: Combined effects of potassium acetate and perfusion in a stirred tank bioreactor. Cytotechnol 1997; 24: 47-54.

[153] Lee SY, Kim DI, Perfusion Cultivation of transgenic Nicotiana tabacum suspensions in bioreactor for recombinant protein production. J Microbiol Biotechnol. 2006;16(5):673–677.

[154] Schillberg S, Twyman RM, Fischer R. Opportunities for recombinant antigen and antibody expression in transgenic plants–technology assessment. Vacc 2005:23;1764–1769.

[155] Doran PM. Foreign protein degradation and instability in plants and plant tissue cultures. Trends Biotechnol. 2006:24(9);426–432.

[156] Wilken LR and Nikolov ZL. Recovery and purification of plant-made recombinant proteins. Biotechnol Adv 2012;30:419–433.

[157] Tacket CO, Mason HS, Losonsky G, Estes MK, Levine MM, Arntzen CJ. Human immune responses to a novel Norwalk virus vaccine delivered in transgenic potatoes. Journal of Infectious Diseases 2000;182:302-5.

[158] Thanavala Y, Mahoney M, Pal S, Scott A, Richter L, Natarajan N. Immunogenicity in humans of an edible vaccine for hepatitis B. Proceedings Natinal Academy of Sciences USA 2005;102:3378-82.

[159] Daniell H, Singh ND, Mason H, Streatfield SJ. Plant-made vaccine antigens and biopharmaceuticals. Trends in Plant Sciences 2009;12:669-79.

[160] Lal B, Ramachandran P, Goyal VG, Sharma R. Edible vaccines: Current status and future, Indian Journal Of Medical Microbiology 2007; 25(2): 93-102.

[161] Pniewski A, Kapusta T, Bociag J, Wojciechowicz P, Kostrzak J, Gdula A, Fedorowicz-Stronska M, Wojcik O, Otta P, Samardakiewicz H, Wolko S, Plucienniczak B. Low-dose oral immunization with lyophilized tissue of herbicide-resistant lettuce expressing hepatitis

B surface antigen for prototype plant-derived vaccine tablet formulation, Journal of Applied Genetics 2011; 52:125-136.

[162] Daniell H, Khan MS, Allison L. Milestones in chloroplast genetic engineering: an environmentally friendly era in biotechnology. Trends in Plant Sciences 2002; 2:84-91.

[163] A. Davoodi-Semiromi, M. Schreiber, S. Nalapalli *et al.*,"Chloroplast-derived vaccine antigens confer dual immunity against cholera and malaria by oral or injectable delivery," Plant Biotechnology Journal, vol. 8, no. 2, pp. 223–242, 2010.

[164] Pniewski T. The twenty-year story of a plant-based vaccine against hepatitis B: Stagnation or promising prospects? Internatinonal Journal of Molecular Sciences 2013; 14:1978-1998.

[165] Clemente M and Coriglino MG. Overview of plant-made vaccine antigens against malaria. Hindawi Publishing Corporation Journal of Biomedicine and Biotechnology 2012 ; 2012, Article ID 206918, 8 pages.

[166] Greenwood BM and Targett GAT. Malaria vaccines and the new malaria agenda. Clinical Microbiology and Infection 2011;17 (11), 1600-1607.

[167] Anders RF, Adda CG, Foley M, and Norton RS. Recombinant protein vaccines against the asexual blood-stages of Plasmodium falciparum. Human Vaccines 2010; 6 (1)39-53.

[168] Arevalo-Herrera M, Solarte Y, Marin C. Malaria transmission blocking immunity and sexual stage vaccines for interrupting malaria transmission in Latin America. Memorias Instituto Oswaldo Cruz 2011; 106 (1): 202-211.

[169] Ghosh S, Malhotra P, Lalitha PV, Guha-Mukherjee S, Chauhan VS. Expression of Plasmodium falciparum C terminal region of merozoite surface protein (PfMSP119), a potential malaria vaccine candidate, in tobacco. Plant Science 2002; 162:335-343.

[170] Altindis E and Izulla P. An overview of African health: this time for Africa? Building on 2010 FIFA World Cup success. Expert Review of Anti-infective Therapy 2010; 8 (12):1325-1328.

[171] Sarrion-Perdigones A, Juarez P, Granell A and Orzaez D. Production of antibodies in plants. Antibody expression and production. Cell Engineering Vol. 7: Antibody Expression and Production 2011;143-164

[172] Sack M, Patez A, Kunert R, Bomble M, Hesse F, Stiegler G, Fischer R, Katinger H, Stoeger E and Rademacher T. Functional analysis of the broadly neutralizing human anti-HIV-1 antibody 2F5 produced in transgenic BY-2 suspension cultures. FASEB Journal 2007;21(8):1655-1664.

[173] Rodríguez M, Ramírez NI, Ayala M, Freyre F, Pérez L, Triguero A, Mateo C, Selman-Housein G, Gavilondo JV and Pujol M.. Transient expression in tobacco leaves of an aglycosylated recombinant antibody against the epidermal growth factor receptor. Biotechnology and Bioengineering 2005; 89(2):188-194.

[174] Brodzik R, Glogowska M, Bandurska K, Okulicz M, Deka D, Ko K, van der Linden J, Leusen JH, Pogrebnyak N, Golovkin M, Steplewski Z, and Koprowski H. Plant-derived anti-Lewis Y mAb exhibits biological activities for efficient immunotherapy against human cancer cells. Proceedings of the National Academy of Sciences of the United States of America 2006;103(23):8804-8809.

[175] Makvandi-Nejad S, McLean MD, Hirama T, Almquist KC, Mackenzie CR, and Hall JC. Transgenic tobacco plants expressing a dimeric single-chain variable fragment (scfv) antibody against Salmonella enterica serotype Paratyphi B. Transgenic Research 2005;14(5):785-792.

[176] De Muynck B, Navarre C, Nizet Y, Stadlmann J and Boutry M. Different subcellular localization and glycosylation for a functional antibody expressed in Nicotiana tabacum plants and suspension cells. Transgenic Research 2009;18(3):467-482.

[177] Villani ME, Morgun B, Brunetti P, Marusic C, Lombardi R, Pisoni I, Bacci C, Desiderio A, Benvenuto E and Donini M. Plant pharming of a full-sized, tumour-targeting antibody using different expression strategies. Plant Biotechnology 2009;7(1);59-72.

[178] Hamers-Casterman C, Atarhouch T, Muyldermans S, Robinson G, Hamers C, Songa EB, Bendahman N and Hamers R. Naturally occurring antibodies devoid of light chains. Nature 1993;363:446–448.

[179] Muyldermans, S. Single domain camel antibodies: current status. J Biotechnology 2001;74:277–302.

[180] Van Bockstaele F, Holz JB and Revets H. The development of nanobodies for therapeutic applications. Current Opinion in Investigetion of Drugs 2009;10:1212–1224.

[181] Muyldermans S, Baral TN, Retamozzo VC, De Baetselier P, De Genst E, Kinne J, Leonhardt H, Magez S, Nguyen VK, Revets H, Rothbauer U, Stijlemans B, Tillib S, Wernery U, Wyns L, Hassanzadeh-Ghassabeh G and Saerens, D. Camelid immunoglobulins and nanobody technology. Vet Immunol Immunopathol 2009;128:178–183.

[182] Teh YH and Kavanagh TA. High-level expression of Camelid nanobodies in Nicotiana benthamiana. Transgenic Res 2010;19:575–586.

[183] Avesani A, Vitale A, Pedrazzini E, de Virgilio M, Pompa A, Barbante A, Gecchele E, Dominici P, Morandini F, Brozzetti A, Falorni A and Mario Pezzotti M. Recombinant human GAD65 accumulates to high levels in transgenic tobacco plants when expressed as an enzymatically inactive mutant. Plant Biotechnolog Journal 2010;8(8):862–872.

Send Orders for Reprints to reprints@benthamscience.net
Frontiers in Clinical Drug Research: Anti-Infectives, Vol. 1, 2014, 237-262   237

CHAPTER 6

Recent Advances in the Discovery and Development of New Drugs Against Gram-Negative Pathogens

Ashok Rattan[1], V. Samuel Raj[2] and Kulvinder S. Saini[3,4,*]

[1]*Department of Laboratory Medicine, Medanta-The Medicity, Sector-38, Gurgaon, Haryana-122 001, India;* [2]*Department of Microbiology, Daiichi-Sankyo India Pharma Pvt. Ltd. (DSIN), Sector-18, Udyog Vihar Industrial Area, Gurgaon 122 015, Haryana, India;* [3]*Research & Development, Eternal University, Baru Sahib-173 101, Himachal Pradesh, India and* [4]*Department of Biological Sciences, Faculty of Science, King Abdulaziz University, Jeddah 21589, Saudi Arabia*

Abstract: In the 20th century, particularly in the 1950s and 1960s, the discovery and development of antimicrobial agents along with stringent vaccination schedule and improved hygiene led to tremendous improvements in life expectancy globally. However, in the last 20 years, it is distressing to observe that bacterial resistance has emerged as a major threat and some authorities are warning about a return to the pre-antibiotic era, where even trivial infections could prove life threatening. In developing countries, neglected diseases kill more than half a million people annually, and inflict severe economic, psychological and physical damage in much larger populations. In the last decade, advances in our understanding of molecular physiology and genomics of these pathogens along with significant advances in Medicinal Chemistry (*e.g.,* computer–aided drug design) and Biotechnology have ushered an exciting era of new drug discovery research. For a long time, the major pharmaceutical companies focused their attention towards discovering new drugs primarily against Gram-positive pathogens, and recently there appears to be renewed research and development efforts focused on multidrug resistant Gram-negative pathogens. The ESKAPE pathogens (*Enterococcus faecium, Staphylococcus aureus, Klebsiella pneumoniae, Acinetobacter baumannii, Pseudomonas aeruginosa*, and *Enterobacter species*) are responsible for multidrug resistant nosocomial infections and resistance of these clinical isolates to antimicrobial agents presents serious therapeutic dilemmas for physicians. Infectious diseases can be difficult to diagnose, the causative agent may not be clinically apparent and untreated infections can have dire consequences. Latest strategies to prevent the emergence and spread of antimicrobial resistance and to "re-engineer" the effective life of available drugs are urgently required. Medical practitioners and researchers should optimize clinical outcomes while minimizing unintended consequences of antimicrobial use, including toxicity, selection of pathogenic organisms and emergence of resistance.

*Address correspondence to Kulvinder S. Saini: Department of Biological Sciences, Faculty of Science, King Abdulaziz University, Jeddah 21589, Saudi Arabia; Tel:-966-555880173; Fax: 966-2-6400 736; E-mail: sainikulvinder@gmail.com

Atta-ur-Rahman (Ed)

Given the association between antimicrobial use and the selection of resistance pathogens, the frequency of inappropriate use of antimicrobials is often used as a surrogate marker for the avoidable impact of antibiotic resistance. This review will outline recent progress made globally in the discovery and development of new drugs against Gram-negative pathogens.

Keywords: Gram negative pathogens, New drug discovery, bacterial pathogens, *Enterococcus faecium*, *Staphylococcus aureus*, *Klebsiella pneumoniae*, *Acinetobacter baumannii*, *Pseudomonas aeruginosa*, *Enterobacter species*.

INTRODUCTION

Discovery and development of antimicrobials is one of the greatest scientific achievements of the 20th century. The success of antibacterial drugs is reflected by their continued use and the consequential decrease in morbidity and mortality from bacterial infections over the past 50 years. The increase in multidrug-resistant bacteria has led to the prediction that we might be reentering the pre-antibiotic era. In reality, the situation would be far worse because today's bacterial strains are not only resistant to commonly available antibiotics, but more importantly may also have acquired "lethal mutations", thereby having virulent genes in their genomes. As a result, even commonly occurring bacteria have been transformed into invasive and toxin-producing pathogens. This precarious setting has been exacerbated by the cessation, or at least downsizing, of antibacterial drug discovery efforts by large pharmaceutical companies in the last decade [1]. Taken together, these scientific observations further highlight the fact that in the last 50 years, only one new class of antibiotic, the narrow-spectrum daptomycin, has actually progressed through various steps of discovery and development, and was able to reach clinical practice. Soon after each broad spectrum antibacterial agent entered into clinical practice, resistance was reported in at least against one bacterial pathogen (Table 1). Because the first antibiotics, excluding the synthetic sulfa drugs, were all identified or derived from natural products, resistance determinants had already accumulated in the environments from which these agents were isolated. From subsequent studies, it became apparent that only a short period of exposure time was required before selection pressures allowed these environmental resistance determinants to become incorporated into the pathogenic bacteria that were being treated with these new antibiotics. Selection of resistant strains occurred so quickly for some bacteria, antibiotic combinations that clinical utility of the antibiotic was severely diminished within a 5-year time span after its launch. The first documented example of this was the increase in penicillin resistance from 8% to almost 60% in *Staphylococcus aureus* from 1945 to 1949. In many cases, chromosomal mutations led to class resistances to other closely related agents, allowing the emergence of clonal strains when subjected to antibiotic pressures. Examples of chromosomal mutations leading to class resistances include mutations in type II topoisomerase (Gyrase and Topoisomerase IV) selected by quinolones and later fluroroquinolones, and selection of derepressed AmpC β-lactamases by any of a number of second and third generation cephalosporins. Perhaps even graver consequences are the plasmid-mediated resistances, whereby a resistant strain selected by one antibiotic could result in resistance to other antibiotic classes due to the presence of multiple resistance determinants in the same operon. Most important observation was that these plasmids can be transferred among species, resulting in multi-factorial resistance in organisms such as *Escherichia coli, Klebsiella pneumoniae and Pseudomonas aeruginosa* [2].

Table 1: Reported resistance to new agents following approval for clinical use

Agent for which Resistance was Observed	FDA Approval	Resistance Reported	Mechanism of Resistance
Penicillin G	1943	1940	Penicillinase production
Streptomycin	1947	1947	Mutation in ribosomal protein S12
Tetracycline	1952	1952	Efflux
Penicillin and tetracycline (Neisseria gonorrhea and enterobacteriaceae)	1943 & 1952	1976–80	Plasmid-encoded broad spectrum b-lactamases and tetracycline efflux pump
Methicillin (and eventually all b-lactams in *S. aureus*)	1960	1961	MecA (penicillin-binding protein 2a)
Nalidixic acid	1964	1966	Topoisomerase mutations
Gentamicin	1967	1969	Aminoglycoside modifying enzyme
Cefotaxime	1981	1981 1983	AmpC β-lactamase Extended-spectrum b-lactamases (ESBLs)
Linezolid	2000	1999	23S RNA mutation

ANTIBACTERIAL DRUG DISCOVERY AND DEVELOPMENT

Antibacterial drug discovery research, accompanied by clinical development, has historically been conducted by large pharmaceutical companies. Although the earliest antibiotics were first discovered in academic laboratories such as those of Alexander Fleming (penicillin) and Selman Waksman (streptomycin), pharmaceutical companies were responsible for successful optimization, compound scale-up, formulation and clinical development activities that allowed anti-infective drug research to gain prominence as a viable area for corporate investment [2]. The earlier antibiotics were sulfonamides, penicillin and streptomycin and later followed by tetracyclines, isoniazid, macrolides, glycopeptides, cepahalosporins, nalidixic acid and rifampicin. Linezolid has become one of the first new class (oxazolidinones) of antibacterial agent to be marketed after rifampicin. Since 1999, it appears that the pharma industries have once again pulled back from anti-infective research in a concerted manner, with 10 of the 15 largest companies significantly scaling down their discovery research [3].

In determining research priorities, initially pharmaceutical companies were driven by the market forces prevalent at the time of project launch, such as the unmet medical need, the potential patient population need to be treated by the new drug, and how the new drug is differentiated in the competition scenario, the price for the new drug would fetch in different markets, and the investment required to bring the product to the market including the cost to promote the new drug in the market. In 2004, Tufts Centre survey indicated that the cost to bring the new product to the market from the discovery and development was between US $ 800 million to $ 1 billion. Therefore, the pharmaceutical companies are more aware of the market potential for each new drug and prioritize their resources accordingly [3].

LESS FREQUENT DOSING

In addition to a medical need to treat a resistant infection, there can be less obvious medical advantages offered by the new agents. Frequency of dosing is one area in which new drugs have made an important difference. An early example was the replacement of ampicillin by amoxicillin because the dosing regimen changed from four times a day to three times a day. Ceftriaxone with once-a-day dosing became a more highly used drug than cefotaxime dosed three times a day, in spite of the fact that cefotaxime was introduced to the market first and had an interchangeable spectrum of activity with ceftriaxone. Also, levofloxacin with once a day dosing replaced twice-a-day ciprofloxacin as the leading hospital fluoroquinolone. These changes in dosing regimens obviously resulted in higher patient compliance, especially in the outpatient population where once-a-day dosing has become the standard of care. In the hospital setting, reduced dosing frequencies could lead to lower medical costs due to the resources involved in delivering parenteral agents. There also are advantages offered by providing both oral and parenteral agents so that hospitalized patients can be transferred from intravenous (*i.v.*) therapy to oral therapy on the same drug, and subsequently released from the hospital earlier. These attributes are shared by many of the fluoroquinolones and by linezolid where both oral and *i.v.* forms are available, making switch therapy possible [2].

IMPROVED SAFETY PROFILE

Improvement in safety can also provide an opportunity for development of a second- or third-in class agent. Although some adverse events are inherently associated with specific antibiotic classes, the degree to which these manifest can vary greatly. Nephrotoxicity with cephalosporins can be a major problem, with drugs like cephaloridine removed from the market because of their renal toxicity. However, the plethora of 'cephawhatchamacallums' in the 1980s can be attributed to the fact that the newer cephalosporins exhibited good safety profiles with minimal nephrotoxicity when normal doses were given. Fluoroquinolones generally exhibit binding to the hERG ion channel as a class, but the IC50 values for binding in the patch clamp assay can vary by as much as two orders of magnitude, from 18 lM for sparfloxacin to 1420 lM for ofloxacin. Thus, there may be room to improve safety profiles and drug–drug interactions for currently existing antibacterial agents such as linezolid with its reversible monoamine oxidase inhibition and telithromycin with its reversible cytochrome P450 interactions [2].

IMPACT OF GENOMICS ON ANTIBACTERIAL DRUG DISCOVERY

The antibacterial drug discovery faces unique challenges. First and significant challenge is the scientific challenge. Even after many years of the emergence of bacterial genomics, there are no promising antibacterial agents in the market or clinical development by using this genomics technology. Recent data suggest that the target based drug discovery has come to some light and starting to pay off. Peptide deformylase (pdf) inhibitors as well as novel fatty acid synthesis inhibitors are entering clinical development [4].

By combining traditional small molecule drug discovery efforts with the modern biotechnology tools, such as the knowledge of bacterial genomes (and their mutant counterparts), may prove to be more rewarding.

PRESENT ANTIBACTERIAL DRUGS

Oxazolidinones

Oxazolidinones are considered to be the first new class of antibacterial drugs introduced in the past three decades. Linezolid was approved by the FDA in 2000 [5]. Oxazolidinones have activity against Gram-positive bacteria, such as MRSA, vancomycin-resistant enterococci (VRE), *Streptococcus pneumoniae* and *Mycobacterium tuberculosis* [6, 7]. The linezolid inhibits bacterial protein synthesis at the initiation/elongation step. The drug–ribosome cross-linking studies reveal that oxazolidinones bind to the peptidyl transferase center (PTC) of the 50S ribosomal subunit [8, 9]. Thus, oxazolidinones compete with the PTC-binding drugs chloramphenicol and lincomycin. However, it is unclear whether oxazolidinones are the initiation or elongation inhibitors [10]. It is interesting to note that oxazolidinones bind only to the mitochondrial 70S ribosomes and not to the cytoplasmic 80S ribosomes, explaining the myelosuppression observed in patients treated with linezolid is little and also reversible [11, 12]. In 2009, studies from our laboratory reported that the linezolid binding site (or active site) of *E. coli* 50S ribosomes was found to lie within the important residues G2053, A2054,C2055, G2056, G2057, A2058, A2059, G2061, A2062, C2452, A2451, A2503, U2504, G2505, U2506, G2576, A2577, U2585, C2611, C2612, *etc.* [13]. In an attempt to discover newer oxazolidinones with improved potency, better aqueous solubility and reduced toxicity, novel modifications of linezolid have been reported [14-16]. The potency of another oxazolidinone, torezolid against linezolid-resistant microbes was explained by additional hydrogen-bond interactions with 23S rRNA (residues A2451 and U2484) and lesser requirement for residues associated with linezolid resistance [14]. Torezolid (TR-700), a new oral oxazolidinone, is four- to eight-fold active than linezolid in linezolid-susceptible and linezolid-resistant strains of staphylococci and enterococci, and have shown up to four-fold higher activity against anaerobes [14]. Radezolid (RX-1741) is a new oxazolidinone, in a Phase II clinical trial for uncomplicated skin and skin structure infections (uSSSI) [15]. In this study, RWJ-416457, pyrrolopyrozolyl-substituted oxazolidinone, exhibited two-to-four fold higher *in vitro* potency than linezolid against susceptible and multi-drug-resistant staphylococci, enterococci, and streptococci, *Haemophilus influenzae* and *Moraxella catarrhalis*. In addition, RWJ-416457 showed activity against atypical intracellular respiratory tract pathogens Chlamydia pneumoniae, *Legionella pneumophila*, and *Mycoplasma pneumoniae* [15]. RWJ-416457 had two-to-four fold lower MICs than linezolid against linezolid-resistant *Staphylococcus aureus* and enterococci [16]. Ranbezolid (RBx 7644), an investigational oxazolidinone, showed activity against Gram positive pathogens and Gram negative anaerobes [17, 18]. Linezolid resistance is selected *in vitro* as well as *in vivo* studies by prolonged drug treatment or increasing the drug dose from low to higher concentration. Most of the clinical resistant mutants have the G2576T mutation in domain V of 23S rRNA. The resistance to oxazolidinones resulted from mutations in ribosomal RNA (rRNA), all of which map near the peptidyl transfer centre. In addition, mutations in ABC transporter genes causing over-expression, as well as RNA methyltransferase mutations, have led to linezolid resistance in *S. pneumoniae* [19].

Ketolides

Drug resistance in community-acquired respiratory tract infections (CARTIs) has driven the discovery and development of ketolides. These compounds are derived from 14-member red ring macrolides and have a carbonyl group at the C3-position, which is crucial in conferring sensitivity to macrolide-resistant strains [20]. Telithromycin was approved by the FDA in 2004 for community-acquired pneumonia, chronic bronchitis and acute sinusitis [5]. Crystal structure of macrolide bound to 50S ribosomal subunit revealed that this antibacterial drug occupies and blocks the peptide exit tunnel without affecting the peptidyl transferase activity [21]. The macrolides form hydrogen bonds with the 23S rRNA at domains II and V and inhibit bacterial protein synthesis. Ketolides are known to act against macrolide-resistant mechanisms by improving ribosome binding affinity, especially binding to domain II, and evading macrolide efflux mechanisms [22, 23]. Telithromycin and cethromycin are two ketolides derived from erythromycin. Cethromycin has a 67-fold higher binding affinity to 50S ribosomes from S. pneumonia than erythromycin and also accumulates at a higher rate within both susceptible and resistant cells due to the differences in efflux mechanisms [22]. Cethromycin has an MIC_{50} of $0.008 - 0.016$ µg/ml for different isolates of *S. pneumoniae* and was more potent than telithromycin against macrolide-susceptible strains of *S. pneumoniae*, Streptococcus pyogenes, *S. aureus, Staphylococcus epidermidis, Enterococci, Helicobacter pylori*, and *Mycobacterium avium* complex, and also against *M. pneumoniae, Chlamydia trachomatis, H. influenzae, M. catarrhalis, C. pneumoniae* and *Toxoplasma gondii* [24, 25]. Cethromycin is potent against macrolide-resistant streptococci and enterococci. In a Phase III double-blind, randomized, multicenter clinical trial in 584 patients with cethromycin, as compared to clarithromycin, for mild to moderate community acquired pneumonia, a non-inferior clinical cure rate was observed, with no safety concerns [26]. Ketolides do not induce MLSB (Macrolide Lincosamide Streptogramin B) resistance phenotype. The first ketolide resistant mutation was identified by *in vitro* selection assay of mutated rRNA operon, and the U2609C mutation in 23S rRNA was found to be resistant to cethromycin [27]. Dimethylation of 23S rRNA at position A2058 confers ketolide resistance in *S. pyogenes* [28]. In a macrolide resistance surveillance study in Europe, 13% of erythromycin-resistant *S. pneumoniae* isolates were telithromycin-resistant by agar diffusion method, but this incidence of telithromycin resistance has not been found in other studies [29]. Laboratory-generated mutants of *S. aureus* that are telithromycin-resistant were found to have mutations in rplV, which led to amino acid duplications in L22 ribosomal protein and also a few gene conversion events between rplV and rplB (encoding ribosomal protein L2) were observed [30].

Glycopeptides

Glycopeptide antibacterial drugs are natural products, first introduced in the 1950s. Due to the emergence of vancomycin resistance, newer derivatives of glycopeptides with potency against the resistant pathogens are now being developed. The *in vitro* activities of glycopeptides are given in Table **2**. Oritavancin is a phenyl glycopeptide derivative and inhibits peptidoglycan biosynthesis by inhibiting trans-glycosylation and transpeptidation [31]. This antibacterial drug blocks utilization of D-Ala-D-Ala containing PG precursors and the selective action of glycopeptide antibacterial drugs is due to strong intramolecular interaction with D-amino acid-containing PG residues [32]. Telavancin is another semi-synthetic glycopeptides which showed potency against MRSA, streptococci and also against VISA, hVISA, and VRE [33, 34]. Telavancin has additional mode of action of depolarization and permeabilization of the bacterial membrane, causing rapid bactericidal activity. Two Phase III clinical trials were conducted with telavancin in patients with a focus on MRSA cSSTIs and were found to be equal or better than vancomycin. The telavancin clinical cure rate was significantly higher for patients infected with *S. aureus*.

These above mentioned new glycopeptides offer significant advantages over vancomycin. In addition to the improved *in vitro* potency, their pharmacokinetics allowed less frequent dosing and possibly improved distribution. Oritavancin, a semi-synthetic glycopeptide and dalbavancin, a second-generation lipoglycopeptide, are two new glycopeptides in clinical development. These new glycopeptides show good potency towards *S. pneumoniae*, and staphylococci, though Dalbavancin is highly potent against *S. aureus* (MIC$_{50}$ 0.03 – 0.12 µg/ml) and coagulase-negative staphylococci (MIC$_{50}$ 0.03 µg/ml) [35, 36]. Oritavancin is also potent against MRSA, but not as potent against vancomycin-intermediate *S. aureus* (VISA) [35, 36]. Two randomized, double-blind, multicenter, Phase III trials have been completed for oritavancin in cSSTIs. The clinical cure rate in the clinically evaluable patients was 77.7% for the oritavancin group and 75.8% for the vancomycin group. Oritavancin, however, exerts concentration-dependent cell-killing activity against vancomycin-intermediate isolates of *S. aureus* (VISA), including heterogeneous VISA (hVISA) [37], and against vancomycin-resistant staphylococci and *Enterococcus faecium* (VRE), which correlates to disruption of membrane integrity [38]. Oritavancin, however, due to its hydrophobic side chain, and binding to lipid II on the membrane, binds equally well to both substrates and is therefore useful against vancomycin-resistant bacteria. Oritavancin-resistant clinical isolates have not been reported. Dalbavancin is developed for treatment of skin and soft-tissue infections (SSTIs) and catheter-related blood stream infections. Dalbavancin is not potent against vanA vancomycin resistant enterococci (VRE), but highly potent (MIC_{50} of 0.06 µg/ml, compared with MIC_{50} of 1 µg/ml for vancomycin) towards vancomycin-susceptible enterococci [39]. Dalbavancin is potent against strains of VRE expressing vanB and vanC gene products, but inactive against VRE expressing vanA [39]. Dalbavancin resistance did not develop in staphylococci by direct selection and serial passage at sub-therapeutic concentrations. In a Phase II clinical trial for catheter-related bloodstream infection caused by staphylococci, a success rate of 87% in 75 adult patients was observed with dalbavancin compared to 50% with vancomycin [40].

In a Phase III randomized, double-blind study, dalbavancin was better than or equal to linezolid in a total of 854 patients with complicated SSTIs (cSSTIs) [41]. Glycylcyclines: Glycylcyclines are the recent members in the tetracycline class of antibacterial drugs, and one member (tigecycline) was approved by the FDA in 2005 for cSSTIs and also for abdominal infection [5]. The novelty about glycylcyclines is their ability to subvert the common tetracycline resistance mechanisms acquired by genetically mobile element encoding the tet genes. Tigecycline, the new version of glycylcycline, exhibits potent antibacterial activity and has also shown potency against tetracycline-resistant bacteria. The glycylcyclines bind with a 5- to 100-fold higher affinity to 30S and 70S ribosomes, respectively, than tetracyclines. Protein synthesis inhibition by tigecycline is 20-fold more efficient than tetracycline [42]. Glycylcyclines have a broad spectrum of antimicrobial activity ranging from aerobic to anaerobic bacteria, Gram-positive and Gram-negative. Tigecycline is active against MRSA, vancomycin resistant enterococci; drug-resistant *S. pneumonia*, and respiratory Gram negative pathogens, such as *H. influenzae*, *M. pneumoniae*, *M. catarrhalis* and *Enterobacteriaceae*. Tigecycline is active against pathogens like *Neisseria gonorrhoeae*, Mycobacteria, and anaerobes such as *Clostridium* spp. and *Bacteroides* spp [43]. *Pseudomonas aeruginosa* is, however, resistant to tigecycline (MIC > 4 µg/ml). Tigecycline is equally potent against both tetracycline-susceptible and resistant clinical isolates. Tigecycline was also effective against some carbapenemase producing *Acinetobacter baumannii* and

Enterobacteriaceae. The safety and efficacy of tigecycline was evaluated in hospitalized patients with serious infections caused by resistant Gram-negative pathogens such as *A. baumannii, Klebsiella pneumoniae* and *E. coli*, which were unresponsive to previous antimicrobial therapy [44]. In these studies, tigecycline was found to be effective in hospitalized patients with serious infections. However, side effects such as nausea/vomiting were common in tigecycline treated patients. PTK0796 is a new aminomethylcycline in development that has shown oral activity. In a Phase II trial (randomized, double-blinded, multicenter) for cSSSI, PTK0796 had a clinical success rate of 98% compared with 93.2% for linezolid in the clinically selected population [45].

Table 2: Glycopeptide *in vitro* activity*

Organism	MIC$_{50}$ (µg/ml)				
	Vanco	**Teicoplanin**	**Oritavancin**	**Telavancin**	**Dalbavancin**
S. aureus	1 – 2	1 – 4	1	0.5	0.06
VISA	8	-	1 – 8	2	2
S. epidermidis	1 – 2	4 – 8	2	0.5	0.25
Enterococci	1	0.5	-	0.5	0.1
Van A		32 - > 128	1 – 4	4 – 8	>128
Van B		2	0.1	-	1
S. pneumoniae		0.06	0.008	0.01	0.03

*Unpublished data-Rattan, A. Also see references [33, 34].

Carbapenems

Carbapenems are β-lactam antibacterial drugs related to penicillin (penam) and cephalosporin (cephem). They differ from the penams by the presence of a carbon at position-1 instead of a sulfur, and unsaturation in the 5-membered ring. Carbapenems bind to penicillin-binding proteins (PBPs) and are required for elongation and cross-linking the peptide glycan of the cell wall, in both Gram-positive and Gram-negative bacteria. Carbapenems are capable of passing through porins in the outer wall of Gram-negative bacteria, have a high affinity to PBPs, and are stable against β-lactamases. Carbapenems are susceptible to hydrolysis by serine carbapenemases and metallo-β-lactamases (MBLs). KPC carbapenemases (Class A) are most prevalent and are plasmid mediated in *K. pneumoniae*, Enterobacter cloacae and other Enterobacteriaceae. Doripenem is the new member of the carbapenems, and received FDA approval in 2007 for complicated urinary tract infections and intra-abdominal infections. Doripenem preferentially binds to PBP-2 of *E. coli* and PBP-2 & PBP-3 of *P. aeruginosa*, similar to meropenem [46]. Doripenem is unique in that it has a spectrum against Gram positive cocci similar to that of imipenem and activity against Gram-negative bacilli, similar to meropenem [46]. The presence of a side chain at position 2 leads to greater activity against Gram-negative multidrug-resistant bacilli such as *P. aeruginosa* and Acinetobacter. Doripenem is active against staphylococci, enterococci and streptococci. Doripenem is also active against the Enterobacteriaceae such as *E. coli* (including producers of extended-spectrum β-lactamases [ESBLs]), *Klebsiella* spp. (including producers of ESBLs), Enterobacter spp., *Salmonella* spp. and *Shigella* spp. [46]. Multicenter, randomized, open-label Phase III trials for ventilator-associated pneumonia (VAP) and hospital-acquired pneumonia (HAP) involving 1000 patients have shown that intravenous doripenem was as effective as imipenem (68 *vs.* 64% clinical cure rate) and piperacillin/tazobactam (81 *vs.* 80% clinical cure rate), respectively [47]. In addition to clinical studies conducted for diseases for which doripenem got FDA approval, additional trials have been conducted for VAP and HAP.

Meropenem is a potent carbapenem and is more potent than doripenem in respiratory tract pathogens [46]. Razupenem (PZ-601) is a novel carbapenem with activity against multi-drug resistant Gram-positive and Gram-negative (ESBL producers) bacteria and is currently in Phase II clinical trial for cSSSI [48]. Sulopenem is an orally active penem in clinical development and is potent against multi-drug-resistant pathogens including penicillin-resistant S. pneumonia and ESBL-producing Enterobacteriaceae (MIC_{50} 0.015 –0.125 µg/ml). Novel pro-drugs of sulopenem exhibited *in vivo* efficacy against ESBL producing *K. pneumoniae*, *E. coli* and *H. influenzae*. Ceftobiprole and ceftaroline are new cephalosporins and these two have significantly improved mechanism of action compared to the older cephalosporins. Ceftobiprole is a novel cephalosporin with activity against MRSA and penicillin-resistant streptococci. It was modified to bind strongly to PBP2a of methicillin-resistant staphylococci [49]. Ceftobiprole, due to its strong binding to *S. pneumoniae* PBP-2x has a MIC of 0.5 µg/ml against penicillin-resistant *S. pneumoniae* [50]. Ceftobiprole is stable against some enzymes (non-ESBL class A) due to its C7 side chains, but is hydrolyzed by ESBLs and carbapenemases. In two Phase III clinical trials for cSSSI, ceftobiprole had 82% activity against all baseline pathogens, demonstrating its broad spectrum profile, and was as effective as competitor vancomycin–ceftazidime or vancomycin alone [51]. Ceftaroline is another novel cephalosporin in Phase III development with broad-spectrum activity against MRSA and multi-drug-resistant *S. pneumoniae*. Ceftaroline inhibits PBP-2a, explaining its potency against MRSA [52]. The *in vitro* activity of ceftaroline against Staphylococcus spp is given in Table **3**. A double-blind, randomized, Phase III clinical trial has been conducted in 700 patients with cSSTI for the safety and efficacy of intravenous ceftaroline against vancomycin and aztreonam. Clinical cure rates were similar for both groups, and ceftaroline was non-inferior to vancomycin–aztreonam combination [53]. The *in vitro* activity of ceftaroline against the Gram-negative pathogens is given in Table **4**. Ceftaroline is synergistic with the β-lactamase inhibitor tazobactam against multi-drug-resistant Gram-negative pathogens such as ESBL-producing *E. coli* and *K. pneumoniae* [54].

Other class of small molecules: Novel DHFR (dihydrofolate reductase) inhibitors (*e.g.*, iclaprim, a diamino pyrimidine) that inhibit DNA/RNA synthesis are antibacterial agents designed with the knowledge of trimethoprim resistance in mind and have improved affinity to *S. aureus* DHFR [55]. Iclaprim has a broad spectrum of activity, including trimethoprim- and methicillin-resistant as well as vancomycin-intermediate *S. aureus* (MIC$_{90}$ is 0.5µg/ml), penicillin-resistant *S. pneumoniae* and Gram-negative bacteria such as *Enterobacter, Salmonella, L. pneumophila, H. influenzae, C. pneumonia, etc.* [56]. In a large surveillance study of about 4500 *S. aureus* isolates (MSSA and MRSA), iclaprim was 16-fold more potent than trimethoprim and had activity similar to TMP/SMZ (trimethoprim–sulfamethoxazole) combination [57]. Two Phase III trials for cSSSI with iclaprim and comparator linezolid were conducted in 2700 patients and clearly demonstrated that iclaprim has higher efficacy as compared to linezolid [58]. NXL103 (XRP 2868, from Novaxel) is a mixture of modified forms of quinupristin–dalfopristin streptogramins, making it water-soluble and permitting oral administration. NXL103 is also more potent as compared to nine other antibacterial drugs tested against Gram-positive clinical isolates including vancomycin, daptomycin, linezolid, clarithromycin, telithromycin, clindamycin, ampicillin, quinupristin–dalfopristin and pristinamycin [59]. It is more effective (two- to five fold lower MIC_{50} values) in inhibiting erythromycin-resistant *S. pneumoniae*, MRSA and β-lactamase-positive *H. influenzae*, when compared to the combination of quinupristin–dalfopristin [60]. Mutations in ribosomal proteins L4 and L22 in *S. aureus* and streptococci reduce the potency of NXL103 [60]. NXL104 is a small molecule that inhibits serine β-lactamases and is potent in combination with extended-spectrum cephalosporins and aztreonam against Gram-negative infections [61]. It is in clinical trials in combination with ceftazidime in patients with complicated urinary tract infections. NXL104 is the first β-lactamase inhibitor to be studied in clinical trials since tazobactam in the 1980s. New β-lactamase inhibitors such as imidazole-substituted 6-methylidene-penem molecules have high *in vitro* activity against Class A and C β-lactamases, and can be used with β-lactams as combination therapy. BAL30376 is an antimicrobial combination of monobactam BAL19764, Class C β-lactamase inhibitor BAL29880, and clavulanic acid (an oxapenem, a class A, β-lactamase inhibitor). It has shown *in vitro* activity against carbapenem-resistant strains of *P. aeruginosa*, among other multi-drug-resistant Gram-negative bacteria, and against *in vivo* murine sepsis models [62].

BAL30072 is a new siderophore monobactam that bypasses porin mutations and inhibits PBPs and it has broad-spectrum Gram-negative activity including multi-drug resistant Acinetobacter and Pseudomonas [63]. JNJ-Q2 is the newest member of the fluoroquinolone family of type II topoisomerase inhibitors. It is eight fold more potent than moxifloxacin against Gram-positive pathogens, including MRSA and levofloxacin-resistant *S. pneumoniae*. It also has shown Gram-negative activity against *H. influenzae*, *E. cloacae* and *K. pneumoniae* [64]. Finafloxacin is a novel 8-cyano fluoroquinolone exhibiting optimal activity at acidic pH and it is potent against urinary tract pathogens, and can be used for treating *H. pylori* infection as well [65]. To discover new drugs, bacterial cell division inhibition remains an unexploited antibacterial target. PC190723 was identified by structure-based molecular docking studies, and it binds to FtsZ, inhibiting septum production during cell division and has activity against *S. aureus* in mice [66]. Pyrrolamides are novel DNA gyrase inhibitors discovered by structure-guided drug design and are bactericidal against MRSA, *S. pneumoniae* and respiratory tract pathogens, making pyrrolamides a potential drug class for the treatment of nosocomial pneumonia [67]. Several peptides and peptide mimetics are in commercial development, and resolving their issues of poor pharmacokinetics and toxicity are very important. Some other attractive antibacterial compounds include quorum-sensing blockers (such as LED209), lipid II binding compounds, bacterial efflux pump inhibitors, and bacterial 2-component signal transduction inhibitors [68]. Additional promising antibacterial drugs in the pipeline of natural products produced by certain microorganisms to survive in their natural environment (secondary metabolites) have been the "goldmine" of antibacterial drug discovery. Apart from soil bacteria, deep sea micro-organisms are also being profiled to look for new chemotypes and new biologics. Due to rampant drug resistance observed for virtually all antimicrobial drugs, most of the antimicrobial R & D efforts have led to the development of new drugs with novel mechanisms and increased potency towards resistant microbes. Some of these new compounds have potent Gram-negative activity.

Table 3: *In vitro* activity of Ceftaroline against *Staphylococcus***

Organism	No. of Isolates	MIC$_{50}$ (µg/ml)	MIC$_{90}$ (µg/ml)	MIC Range (µg/ml)
S. aureus: MSSA	2199	0.12- 0.25	0.25 – 0.5	<0.008 – 1
MRSA	2082	0.5 – 1	1.0 - 2	0.12 – 4
CA MRSA	244	0.5	0.5 – 1	0.25 – 1
VISA & hVISA	123	1	2	0.25 – 4
VRSA	9	-	-	0.12 – 1
CONS MSSE	251	0.06 – 0.12	0.12 – 0.25	<0.16 – 0.5
MRSE	379	0.25 – 0.5	0.5 - 1	<0.016 – 2

** Unpublished data-Rattan, A. Also see References [52, 53].

ESKAPE Pathogens

Though there are several drugs in the market or in pipeline, still there remains an unmet medical need in the field of anti-infectives [69]. Infections caused by

Table 4: *In vitro* activity of Ceftaroline against Gram-negative pathogens***

Organism	No. of Isolates	MIC$_{50}$ (µg/ml)	MIC$_{90}$ (µg/ml)	MIC Range (µg/ml)
Non fermentors				
Pseudomonas aeruginosa	20	16	>32	4 - >32
Acinetobacter spp	20	16	>32	2 to >32
K.pneumoniae wild type	21	0.06	0.5	0.03 – 4
ESBL +	15	>32	>32	> 32

*** Unpublished data-Rattan, A. Also see References [54-57].

antibiotic-resistant bacteria continue to challenge physicians. There is growing resistance among Gram-positive and Gram-negative pathogens that cause infection in the hospital and in the community [70]. Recently there were reports on "ESKAPE" pathogens *Enterococcus faecium, Staphylococcus aureus, Klebsiella pneumoniae, Acinetobacter baumanii, Pseudomonas aeruginosa,* and *Enterobacter species*) to emphasize that they currently cause the majority of hospital infections around the globe, and have effectively devised mechanisms to "escape" the effects of antibacterial drugs. Data from the Center for Disease Control and Prevention (CDC), USA, have shown rapidly increasing rates of infection due to MRSA, vancomycin-resistant *E. faecium* (VRE), and fluoroquinolone-resistant *P. aeruginosa* [71]. More people now die of MRSA infection in US hospitals than of HIV/AIDS and tuberculosis combined. Furthermore, pan-antibiotic-resistant infections now occur frequently. Recently, several highly resistant Gram-negative pathogens namely Acinetobacter species, multidrug-resistant (MDR) *P. aeruginosa,* and carbapenem-resistant Klebsiella species and *E. coli* have emerged as significant pathogens around the world. At the moment, the therapeutic options for these pathogens appears to be extremely limited, which resulted in clinicians taking out older, previously discarded drugs, such as colistin, that were associated with significant toxicity and for which there is a lack of robust data to guide selection of dosage regimen or duration of therapy [71-73]. The growing number of elderly patients and patients undergoing surgery, transplantation and chemotherapy and further increases in populations in neonatal intensive care units produce greater number of immune-compromised individuals at risk of these infections. Over the past several years, the Infectious Diseases Society of America (IDSA) has worked with US Congress, USFDA, the National Institutes of Health, the CDC, and other groups to highlight this problem of lower investments into new drug discovery & development. Despite ongoing efforts and some successes, only one new antibacterial doripenem has been approved. Certainly current situation is alarming, as the number of new antibacterial drugs approved for marketing in the world continues to decrease with a simultaneous closing of anti-infective discovery programs by the large pharmaceutical companies.

ANTIMICROBIAL DRUG DEVELOPMENT NEEDS

E. faecium (VRE)

E. faecium is identified as the third most frequent cause of nosocomial bloodstream infection (BSI) in the United States, and also enterococcal BSIs remain a significant problem. Vancomycin resistance continues to increase with a rate of ~60% among *E. faecium* isolates [74]. Despite growing incidence, there is a scarcity of data that allows us to address the efficacy of newer agents, such as linezolid, daptomycin, and tigecycline, in the therapy of these infections, and also tolerability remains problematic [75].

S. aureus (MRSA)

Despite the addition of several new agents to treat MRSA infection, clinicians are routinely faced with treatment challenges involving patients with invasive disease. Although criteria for treating skin and skin-structure infection due to community associated MRSA are evolving, the need is now even greater for oral agents for step-down therapy for the patients requiring initial parenteral therapy [76]. Novel classes of antibiotics are needed for MRSA, because currently marketed drugs exhibit treatment-limiting toxicities and have emerging resistance [77, 78]. Unfortunately, many studies using biotech drugs, including vaccines and antibodies, have failed to demonstrate efficacy against *S. aureus* and other ESKAPE pathogens signifying the challenges involved.

K. pneumonia Carbapenemase-Hydrolyzing β-Lactamases

In the last few years, infections due to ESBL-producing *E. coli* and *Klebsiella* species have continued to increase in frequency and severity. The number of enzymes and the number of organisms that exhibit cross-resistance to other classes of antimicrobials is growing, which makes selection of therapy even more challenging. Despite this growing & serious problem, the molecules in late stages of development, as well as the recently approved doripenem, represent only incremental advances over existing carbapenems [69]. Carbapenem resistant Enterobacteriaceae are increasingly recognized as the cause of sporadic and outbreak infections in the United States and Europe. Plasmid mediated carbapenemases were initially described in *K. pneumoniae* and were later recognized in *E. coli* and other *Enterobacteriaceae* [79]. These organisms cause severe infections among residents of long-term-care facilities and in old-age nursing homes and are not easily detected in the clinical microbiology laboratory. Tigecycline and the polymyxins, including colistin, have been used in individual cases with variable success [80]. Aggressive infection-control practices are required in controlling these outbreaks, and there are currently no new drugs in advanced clinical development for these resistant pathogens.

A. baumannii

The incidence of infection due to MDR *Acinetobacter* species continues to increase globally. Recent studies of patients in the intensive care unit with blood stream infection and burn infections due to carbapenem-resistant Acinetobacter species demonstrate an increased mortality, as well as increased morbidity and longer length of stay in the intensive care units [81]. Tigecycline shows *in vitro* activity against Gram-positive and Gram-negative organisms, including MRSA, and Acinetobacter isolates. Although successful treatment of *A. baumannii* infection has been reported, reports of breakthrough infections have led to some caution with regard to the use of this newer agent to treat infection caused by this pathogen [82, 83]. Tigecycline received FDA approval in 2005 for the treatment of cSSSI and complicated intra-abdominal infections. Unfortunately, there is no back-up candidate in late-stage development for the treatment of MDR Acinetobacter infection. This pathogen is emblem of the mismatch between unmet medical needs and the current antimicrobial research and development pipeline [81].

P. aeruginosa

The rates of infection due to resistant *P. aeruginosa* continue to increase globally and becoming resistance to both the quinolones and carbapenems.

Aminoglycoside resistance is emerging as a significant problem [71, 84]. There are recent reports on the resistance to the polymyxins. Patients at risk include those in intensive care units, particularly those who are ventilator dependent, and individuals with cystic fibrosis [85, 86]. To date, no drugs in clinical development address the issue of carbapenem resistance or MDR or alternatively offer a less toxic alternative to the polymyxins.

Enterobacter species

Enterobacter species cause an increasing number of health care–associated infections and are increasingly resistant to multiple anti-bacterial regimes [87]. Infection due to *Enterobacter species*, especially blood stream infections, is associated with significant morbidity and mortality. As with other members of the Enterobacteriaceae, resistance occurs *via* ESBLs and carbapenemases including *K. pneumoniae* and inducible chromosomal cephalosporinases [88]. Other than colistin and perhaps tigecycline, few antibacterials are active against these resistant organisms, and there appears to be no drug in late stage development for these pathogens [89].

PERSPECTIVES & PROSPECTIVE

In the last 20 years, new drug discovery R&D in the area of anti-infective has remained a challenge for the scientist as well as for the clinician. Relatively fewer drugs, whether small molecule or biologics, for the treatment of infections due to problematic pathogens, such as MRSA, have been taken up for clinical development by the pharmaceutical or biotech companies. Importantly, no new drugs have reached clinical stages of development for the control of infections due to MDR gram-negative bacilli, such as *A. baumannii* and *P. aeruginosa*. MRSA has become even more aggressive by moving from hospital's setting to community based infections. Fewer molecules in the discovery pipeline just highlight an incremental advancement over currently available therapies. How can we beat these "clever pathogens" by designing targeted therapies? We need to go back to nature, deep sea organisms and soil bacteria, to look for new chemotypes or new biologics against specific molecular targets for these deadly ever evolving pathogens. One approach could be to do a comparative gene expression analysis of bacteria exposed to new molecules *vs.* currently available antibiotics. DNA sequencing technologies, particularly Next Generation Sequencing, have come a long way to become affordable and less time-consuming leading to the availability of specific genomes for a number of pathogenic bacteria in the public databases. However, from therapeutic point of view, there remain an important challenge and a major bottleneck perhaps, as how to penetrate the bacterial cell wall [90]. Simultaneously, new and novel nano-delivery systems need to be developed for some of the existing successful drug molecules, including those which failed in phase-3 trials (where proof-of-concept has been established in human phase-2 trials), exhibiting toxicity issues because of high doses. In addition, "pro-drug" approaches require critical evaluation, as these seem to hold a lot of promise for the development of anti-bacterials [90]. IDSA supports strengthening modern approaches to antimicrobial resistance, to protect effectiveness of the drugs currently available. We must "re-visit" and critically examine current hospital infection-control practices, to limit the spread of resistance and the "overuse" of certain antibiotics. Last but not the least, the United States, Europe and other developed economies must make the development of sustainable antibacterial drug R & D infra-structure a priority. Academic institutes, World Health Organisation (WHO), Bill and Melinda Gates Foundation (other NGOs), government and private companies need to work together to find & fund sustainable Public-Private-Partnership models addressing these issues. Only this will ensure a steady stream of new anti-bacterials to meet the needs of both the current populations and future generations. We can't afford to lose this battle, particularly against the antibiotic-resistant pathogens.

ACKNOWLEDGEMENTS

We would like to extend sincere thanks and gratitude to our colleagues for stimulating discussions. KSS also wish to extend sincere thanks to Dr. M.S. Atwal, Vice Chancellor, Eternal University, Baru Sahib, India for his support and encouragement.

CONFLICT OF INTEREST

The authors confirm that this chapter contents have no conflict of interest.

REFERENCES

[1]   Fernandes P. Antibacterial discovery and development–the failure of success? Nature Biotechnology 2006; 24(12): 1497-1503.

[2]   Bush K. Antibacterial drug discovery in the 21st century. Clin Microbiol Infect 2004; 10 (Suppl 4): 10 -17.

[3]   Projoan SJ and Shlaes DM: Is it all downhill from here? Clin Microbiol Infect 2004; 10 (Suppl 4): 18 -22.

[4]   Payne DJ, Gwynn MN, Holmes DJ, Pomliano DL. Drugs for bad bugs: confronting the challenges of antibacterial discovery. Nature Reviews 2007; 6: 29-40.

[5]   Outterson K, Samora JB, Keller-Cuda K. Will longer antimicrobial patents improve global public health? Lancet Infect Dis 2007; 7(8):559-66.

[6]   Brickner SJ, Hutchinson DK, Barbachyn MR, *et al.* Synthesis and antibacterial activity of U-100592 and U-100766, two oxazolidinone antibacterial agents for the potential treatment of multidrug-resistant gram-positive bacterial infections. J Med Chem 1996; 39(3):673-9

[7]   Jorgensen JH, McElmeel ML, Trippy CW. *In vitro* activities of the oxazolidinone antibiotics U-100592 and U-100766 against Staphylococcus aureus and coagulase-negative Staphylococcus species. Antimicrob Agents Chemother 1997; 41(2):465-7

[8]   Colca JR, McDonald WG, Waldon DJ, *et al.* Cross-linking in the living cell locates the site of action of oxazolidinone antibiotics. J Biol Chem 2003; 278(24):21972-9

[9]   Lin AH, Murray RW, Vidmar TJ, Marotti KR. The oxazolidinone eperezolid binds to the 50S ribosomal subunit and competes with binding of chloramphenicol and lincomycin. Antimicrob Agents Chemother 1997; 41(10):2127-31

[10]   Wilson DN, Nierhaus KH. The oxazolidinone class of drugs find their orientation on the ribosome. Mol Cell 2007; 26(4):460-2

[11]   French G. Safety and tolerability of linezolid. J Antimicrob Chemother 2003; 51(Suppl. 2): 45-53

[12]   Lee E, Burger S, Shah J, *et al.* Linezolid-associated toxic optic neuropathy: a report of 2 cases. Clin Infect Dis 2003; 37(10):1389-91

[13]   Kalia V, Miglani R, Purnapatre KP, *et al.* Mode of action of Ranbezolid against staphylococci and structural modeling studies of its interaction with ribosomes. Antimicrob Agents Chemother. 2009; 53(4):1427-33.

[14]   Schaadt R, Sweeney D, Shinabarger D, Zurenko G. *In vitro* activity of TR-700, the active ingredient of the antibacterial prodrug TR-701, a novel oxazolidinone antibacterial agent. Antimicrob Agents Chemother 2009; 53(8):3236-39.

[15]   Foleno BD, Abbanat D, Goldschmidt RM, *et al. In vitro* antibacterial activity of the pyrrolopyrazolyl-substituted oxazolidinone RWJ-416457. Antimicrob Agents Chemother 2007; 51(1):361-65.

[16]   Livermore DM, Warner M, Mushtaq S, *et al. In vitro* activity of the oxazolidinone RWJ-416457 against linezolid-resistant and -susceptible staphylococci and enterococci. Antimicrob Agents Chemother 2007; 51(3):1112-14

[17]   Hoellman DB, Lin G, Ednie LM, Rattan A, Jacobs MR, Appelbaum PC. Antipneumococcal and antistaphylococcal activities of ranbezolid (RBX 7644), a new oxazolidinone, compared to those of other agents. Antimicrob Agents Chemother. 2003; 47(3):1148-50.

[18] Ednie LM, Rattan A, Jacobs MR, Appelbaum PC. Antianaerobe activity of RBX 7644 (ranbezolid), a new oxazolidinone, compared with those of eight other agents. Antimicrob Agents Chemother. 2003; 47(3):1143-47.

[19] Feng J, Lupien A, Gingras H, *et al.* Genome sequencing of linezolid-resistant Streptococcus pneumoniae mutants reveals novel mechanisms of resistance. Genome Res 2009; 19(7):1214-23.

[20] Nilius AM, Ma Z. Ketolides: The future of the macrolides? Curr Opin Pharmacol 2002; 2(5):493-500.

[21] Schlunzen F, Harms JM, Franceschi F, *et al.* Structural basis for the antibiotic activity of ketolides and azalides. Structure 2003; 11(3):329-38.

[22] Capobianco JO, Cao Z, Shortridge VD, *et al.* Studies of the novel ketolide ABT-773: transport, binding to ribosomes, and inhibition of protein synthesis in Streptococcus pneumoniae. Antimicrob Agents Chemother 2000; 44(6):1562-67.

[23] Champney WS, Pelt J. The ketolide antibiotic ABT-773 is a specific inhibitor of translation and 50S ribosomal subunit formation in Streptococcus pneumonia cells. Curr Microbiol 2002; 45(3):155-60.

[24] Davies TA, Ednie LM, Hoellman DM, *et al.* Antipneumococcal activity of ABT-773 compared to those of 10 otheragents. Antimicrob Agents Chemother 2000; 44(7):1894-9

[25] Shortridge VD, Zhong P, Cao Z, *et al.* Comparison of *in vitro* activities of ABT-773 and telithromycin against macrolide-susceptible and –resistant streptococci and staphylococci. Antimicrob Agents Chemother 2002; 46(3):783-6

[26] Milanesio NA, English ML, Fredericks CE, *et al.* A comparative study of the safety and efficacy of cethromycin (CER) to clarithromycin (CLR) for the treatment of Community Acquired Pneumonia (CAP) in adults (CL05-001). 48th Annual Interscience Conference on Antimicrobial Agents and Chemotherapy (ICAAC) and Infectious Disease Society of America (IDSA). 46th Annual Meeting; 25 – 28 Oct 2008; Washington DC

[27] Garza-Ramos G, Xiong L, Zhong P, Mankin A. Binding site of macrolide antibiotics on the ribosome: new resistance mutation identifies a specific interaction of ketolides with rRNA. J Bacteriol 2001;183(23):6898-907.

[28] Douthwaite S, Jalava J, Jakobsen L. Ketolide resistance in Streptococcus pyogenes correlates with the degree of rRNA dimethylation by Erm. Mol Microbiol 2005; 58(2):613-22

[29] Rantala M, Haanpera-Heikkinen M, Lindgren M, *et al.* Streptococcus pneumoniae isolates resistant to telithromycin. Antimicrob Agents Chemother 2006; 50(5):1855-8

[30] Gentry DR, Holmes DJ. Selection for high-level telithromycin resistance in Staphylococcus aureus yields mutants resulting from an rplB-to-rplV gene conversion-like event. Antimicrob Agents Chemother 2008; 52(3):1156-8

[31] Allen NE, Hobbs JN Jr, Nicas TI. Inhibition of peptidoglycan biosynthesis in vancomycin-susceptible and –resistant bacteria by a semisynthetic glycopeptides antibiotic. Antimicrob Agents Chemother 1996; 40(10):2356-62

[32] Allen NE, LeTourneau DL, Hobbs JN Jr. Molecular interactions of a semisynthetic glycopeptide antibiotic with D-alanyl-D-alanine and D-alanyl-D-lactate residues. Antimicrob Agents Chemother 1997; 41(1):66-71

[33] King A, Phillips I, Kaniga K. Comparative *in vitro* activity of telavancin (TD-6424), a rapidly bactericidal, concentration-dependent anti-infective with multiple mechanisms of action against Gram-positive bacteria. J Antimicrob Chemother 2004; 53(5):797-803

[34] Krause KM, Renelli M, Difuntorum S, *et al. In vitro* activity of telavancin against resistant gram-positive bacteria. Antimicrob Agents Chemother 2008; 52(7):2647-52

[35] Candiani G, Abbondi M, Borgonovi M, *et al.* In-vitro and in-vivo antibacterial activity of BI 397, a new semi-synthetic glycopeptide antibiotic. J Antimicrob Chemother 1999; 44(2):179-92.

[36] Lin G, Credito K, Ednie LM, Appelbaum PC. Anti-staphylococcal activity of dalbavancin, an experimental glycopeptide. Antimicrob Agents Chemother 2005; 49(2):770-72.

[37] Arhin FF, Sarmiento I, Parr TR Jr, Moeck G. Comparative *in vitro* activity of oritavancin against Staphylococcus aureus strains that are resistant, intermediate or heteroresistant to vancomycin. J Antimicrob Chemother 2009; 64(4):868-70.

[38] McKay GA, Beaulieu S, Arhin FF, *et al.* Time-kill kinetics of oritavancin and comparator agents against Staphylococcus aureus, Enterococcus faecalis and Enterococcus faecium. J Antimicrob Chemother 2009; 63(6):1191-99.

[39] Streit JM, Sader HS, Fritsche TR, Jones RN. Dalbavancin activity against selected populations of antimicrobial-resistant Gram-positivepathogens. Diagn Microbiol Infect Dis 2005; 53(4):307-10.

[40] Raad I, Darouiche R, Vazquez J, *et al.* Efficacy and safety of weekly dalbavancin therapy for catheter-related bloodstream infection caused by Gram-positive pathogens. Clin Infect Dis 2005;40(3):374-80.

[41] Jauregui LE, Babazadeh S, Seltzer E, *et al.* Randomized, double-blind comparison of once-weekly dalbavancin *versus* twice-daily linezolid therapy for the treatment of complicated skin and skin structure infections. Clin Infect Dis 2005; 41(10):1407-15.

[42] Olson MW, Ruzin A, Feyfant E, *et al.* Functional, biophysical, and structural bases for antibacterial activity of tigecycline. Antimicrob Agents Chemother 2006; 50(6):2156-66.

[43] Petersen PJ, Jacobus NV, Weiss WJ, *et al. In vitro* and *in vivo* antibacterial activities of a novel glycylcycline, the 9-t-butylglycylamido derivative of minocycline (GAR-936). Antimicrob Agents Chemother 1999; 43(4):738-44.

[44] Vasilev K, Reshedko G, Orasan R, *et al.* A Phase 3, open-label, non-comparative study of tigecycline in the treatment of patients with selected serious infections due to resistant Gram-negative organismsincluding *Enterobacter species*, Acinetobacter baumannii and Klebsiella pneumoniae. J Antimicrob Chemother 2008; 62(Suppl 1):i29-40.

[45] Arbeit RD, Roberts JA, Forsythe AR, *et al.* Safety and efficacy of PTK 0796: results of the phase 2 study in complicated skin and skin structure infections following iv and oral step-down therapy (L-1515b). 48th Annual Interscience Conference on Antimicrobial Agents and Chemotherapy (ICAAC) and the Infectious Disease Society of America (IDSA). 46th Annual Meeting; 25 – 28 Oct 2008; Washington, DC.

[46] Fritsche TR, Stilwell MG, Jones RN. Antimicrobial activity of doripenem (S-4661): a global surveillance report (2003). Clin Microbiol Infect 2005; 11(12):974-84.

[47] Chastre J, Wunderink R, Prokocimer P, *et al.* Efficacy and safety of intravenous infusion of doripenem *versus* imipenem in ventilator-associated pneumonia: a multicenter, randomized study. Crit Care Med 2008; 36(4):1089-96.

[48] Livermore DM, Mushtaq S, Warner M. Activity of the anti-MRSA carbapenem razupenem (PTZ601) against Enterobacteriaceae with defined resistance mechanisms. J Antimicrob Chemother 2009; 64(2):330-35.

[49] Hebeisen P, Heinze-Krauss I, Angehrn P, *et al. In vitro* and *in vivo* properties of Ro 63-9141, a novel broad-spectrum cephalosporin with activity against methicillin-resistant staphylococci. Antimicrob Agents Chemother 2001; 45(3):825-36.

[50] Kosowska K, Hoellman DB, Lin G, *et al.* Antipneumococcal activity of ceftobiprole, a novel broad-spectrum cephalosporin. Antimicrob Agents Chemother 2005; 49(5):1932-42.

[51] Amsler KM, Davies TA, Shang W, *et al. In vitro* activity of ceftobiprole against pathogens from two phase 3 clinical trials of complicated skin and skin structure infections. Antimicrob Agents Chemother 2008; 52(9):3418-23.

[52] Villegas-Estrada A, Lee M, Hesek D, *et al.* Co-opting the cell wall in fighting methicillin-resistant Staphylococcus aureus: potent inhibition of PBP 2a by two anti-MRSA beta-lactam antibiotics. J Am Chem Soc 2008; 130(29):9212-13.

[53] Corey GR, Wilcox M, Talbot GH, *et al.* CANVAS-1: randomized, double-blinded, phase 3 study (p903-06) of the efficacy and safety of ceftaroline *vs.* vancomycin plus aztreonam in Complicated Skin and Skin Structure Infections (cSSSI) (L-1515a). 48th Annual Interscience Conference on Antimicrobial Agents and Chemotherapy (ICAAC) and the Infectious Disease Society of America (IDSA). 46th Annual Meeting; 25 – 28 Oct 2008; Washington DC.

[54] Vidaillac C, Leonard SN, Sader HS, *et al. In vitro* activity of ceftaroline alone and in combination against clinical isolates of resistant gram-negative pathogens, including beta-lactamase-producing Enterobacteriaceae and Pseudomonas aeruginosa. Antimicrob Agents Chemother 2009; 53(6):2360-66.

[55] Oefner C, Bandera M, Haldimann A, *et al.* Increased hydrophobic interactions of iclaprim with Staphylococcus aureus dihydrofolate reductase are responsible for the increase in affinity and antibacterial activity. J Antimicrob Chemother 2009; 63(4):687-98.

[56] Kohlhoff SA, Roblin PM, Reznik T, *et al. In vitro* activity of a novel diaminopyrimidine compound, iclaprim, against Chlamydia trachomatis and C. pneumoniae. Antimicrob Agents Chemother 2004; 48(5):1885-66.

[57] Sader HS, Fritsche TR, Islam K, *et al.* Antimicrobial activity of iclaprim tested against recent *S. aureus* clinical isolates: results from the International Study of Iclaprim susceptibility. 47th Annual Interscience Conference on Antimicrobial Agents and Chemotherapy (ICAAC); 17 – 20 Sept 2007; Chicago, IL (abstract E-902).

[58] Hadvary P, Stevens D, Solonets M, *et al.* Clinical efficacy of iclaprim in Complicated Skin and Skin Structure Infection (cSSSI): results of combined ASSIST Phase III studies (L-1512). 48th Annual Interscience Conference on Antimicrobial Agents and Chemotherapy (ICAAC) and the Infectious Disease Society of America (IDSA). 46th Annual Meeting; 25 – 28 Oct 2008; Washington DC.

[59] Goldstein EJ, Citron DM, Merriam CV, *et al.* Comparative *in vitro* activities of XRP 2868, pristinamycin, quinupristin-dalfopristin, vancomycin, daptomycin, linezolid, clarithromycin, telithromycin, clindamycin, and ampicillin against anaerobic gram-positive species, actinomycetes, and lactobacilli. Antimicrob Agents Chemother 2005; 49(1):408-13.

[60] Dupuis M, Leclercq R. Activity of a new oral streptogramin, XRP2868, against gram-positive cocci harboring various mechanisms of resistance to streptogramins. Antimicrob Agents Chemother 2006; 50(1):237-42.

[61] Endimiani A, Choudhary Y, Bonomo RA. *In vitro* activity of NXL104 in combination with beta-lactams against Klebsiella pneumoniae isolates producing KPC carbapenemases. Antimicrob Agents Chemother 2009; 53(8):3599-601

[62] Hermesh O, Page MGP, Carmeli Y, Navon-Venezia S. Efficacy of BAL30376, a New Monobactam/ beta-Lactamase Inhibitor Combination, against Pseudomonas aeruginosa

(PA) (F1-1166). 48th Annual Interscience Conference on Antimicrobial Agents and Chemotherapy (ICAAC) and Infectious Disease Society of America (IDSA). 46th Annual Meeting; 25 – 28 Oct 2008; Washington DC.

[63]   Hofer B, Miller C, Desarbre E, Page MGP. *In vitro* activity of the siderophore monobactam BAL30072 against multi-resistant non-fermenting gram-negative Pathogens (F1-1175). 48th Annual Interscience Conference on Antimicrobial Agents and Chemotherapy (ICAAC) and Infectious Disease Society of America (IDSA). 46th Annual Meeting; 25 – 28 Oct 2008; Washington DC.

[64]   Foleno BD, Morrow BJ, Wira E, *et al.* Broad spectrum *in vitro* activity of JNJ-Q2, a new fluoroquinolone (F1-2033). 48th Annual Interscience Conference on Antimicrobial Agents and Chemotherapy (ICAAC) and the Infectious Disease Society of America (IDSA). 46th Annual Meeting; 25 – 28 Oct 2008; Washington DC.

[65]   Kresken M, Korber-Irrgang B, Labischinski H, Stubbings W. Effect of pH on the *in vitro* activity of finafloxacin against gram-negative and gram-positive bacteria (F1-2037). 48th Annual Interscience Conference on Antimicrobial Agents and Chemotherapy (ICAAC) and the Infectious Disease Society of America (IDSA). 46th Annual Meeting; 25 – 28 Oct 2008; Washington DC.

[66]   Haydon DJ, Stokes NR, Ure R, *et al.* An inhibitor of FtsZ with potent and selective anti-staphylococcal activity. Science 2008; 321(5896):1673-75.

[67]   Green O, Ni H, Singh A, *et al.* Novel DNA gyrase inhibitors: structure-guided discovery and optimization of pyrrolamides (F1-2025). 48th Annual Interscience Conference on Antimicrobial Agents and Chemotherapy (ICAAC) and the Infectious Disease Society of America (IDSA). 46th Annual Meeting; 25 – 28 Oct 2008; Washington DC.

[68]   Su Z, Honek JF. Emerging bacterial enzyme targets. Curr Opin Investig Drugs 2007; 8(2):140-49.

[69]   Devasahayam G, Scheld WM, Hoffman PS. Newer antibacterial drugs for a new century. Expert Opin. Investig. Drugs 2010; 19(2):215-234.

[70]   Boucher HW, Talbot GH, Bradley JS, Edwards JE Jr, Gilbert D, Rice LB, Scheld M, Spellberg B, Bartlett J. Bad Bugs, No Drugs: No ESKAPE! An Update from the Infectious Diseases Society of America. Clinl Infect Dis 2009; 48:1–12.

[71]   National Nosocomial Infections Surveillance System Report, data summary from January 1992 through June 2004, issued October 2004. Am J Infect Control 2004; 32:470–85.

[72]   Bradford PA, Bratu S, Urban C, *et al.* Emergence of carbapenem resistant Klebsiella species possessing the class A carbapenem-hydrolyzing KPC-2 and inhibitor-resistant TEM-30 β-lactamases in New York City. Clin Infect Dis 2004; 39:55–60.

[73]   Urban C, Bradford PA, Tuckman M, *et al.* Carbapenem-resistant Escherichia coli harboring Klebsiella pneumoniae carbapenemase β-lactamases associated with long-term care facilities. Clin Infect Dis 2008; 46:e127–30.

[74]   Wisplinghoff H, Bischoff T, Tallent SM, Seifert H, Wenzel RP, Edmond MB. Nosocomial bloodstream infections in US hospitals: analysis of 24,179 cases from a prospective nationwide surveillance study. Clin Infect Dis 2004; 39:309–17.

[75]   Erlandson KM, Sun J, Iwen PC, Rupp ME. Impact of the more potent antibiotics quinupristin-dalfopristin and linezolid on outcome measure of patients with vancomycin-resistant Enterococcus bacteremia. Clin Infect Dis 2008; 46:30–6.

[76]   Liu C, Graber CJ, Karr M, *et al.* A population-based study of the incidence and molecular epidemiology of methicillin-resistant Staphylococcus aureus disease in San Francisco, 2004–2005. Clin Infect Dis 2008; 46:1637–46.

[77] Boucher HW, Sakoulas G. Perspectives on daptomycin resistance, with emphasis on resistance in Staphylococcus aureus. Clin Infect Dis 2007; 45:601–8.

[78] Tsiodras S, Gold HS, Sakoulas G, *et al.* Linezolid resistance in a clinical isolate of Staphylococcus aureus. Lancet 2001; 358:207–8.

[79] Bratu S, Tolaney P, Karumudi U, *et al.* Carbapenemase-producing Klebsiella pneumoniae in Brooklyn, NY: molecular epidemiology and *in vitro* activity of polymyxin B and other agents. J Antimicrob Chemother 2005; 56:128–32.

[80] Urban C, Bradford PA, Tuckman M, *et al.* Carbapenem-resistant Escherichia coli harboring Klebsiella pneumoniae carbapenemase β-lactamases associated with long-term care facilities. Clin Infect Dis 2008; 46:e127–30.

[81] Maragakis LL, Perl TM. Acinetobacter baumannii: epidemiology, antimicrobial resistance, and treatment options. Clin Infect Dis 2008; 46:1254–63.

[82] Perez F, Hujer AM, Hujer KM, Decker BK, Rather PN, Bonomo RA. Global challenge of multidrug-resistant Acinetobacter baumannii. Antimicrob Agents Chemother 2007; 51:3471–84.

[83] Noskin GA. Tigecycline: a new glycylcycline for treatment of serious infections. Clin Infect Dis 2005; 41(Suppl 5):S303–14.

[84] Neuhauser MM, Weinstein RA, Rydman R, Danziger LH, Karam G, Quinn JP. Antibiotic resistance among gram-negative bacilli in US intensive care units: implications for fluoroquinolone use. JAMA 2003; 289:885–88.

[85] Giske CG, Monnet DL, Cars O, Carmeli Y. Clinical and economic impact of common multidrug-resistant gram-negative bacilli. Antimicrob Agents Chemother 2008; 52:813–21.

[86] Jacoby GA, Munoz-Price LS. The new β-lactamases. N Engl J Med 2005; 352:380–91.

[87] Pfaller MA, Sader HS, Fritsche TR, Jones RN. Antimicrobial activity of cefepime tested against ceftazidime-resistant gram-negative clinical strains from North American hospitals: report from the SENTRY Antimicrobial Surveillance Program (1998–2004). Diagn Microbiol Infect Dis 2006; 56:63–8.

[88] Bratu S, Landman D, Alam M, Tolentino E, Quale J. Detection of KPC carbapenem-hydrolyzing enzymes in Enterobacter spp. from Brooklyn, New York. Antimicrob Agents Chemother 2005; 49:776–8.

[89] Pintado V, San Miguel LG, Grill F, *et al.* Intravenous colistin sulphomethate sodium for therapy of infections due to multidrug-resistant gram-negative bacteria. J Infect 2008; 56:185–90.

[90] Lewis K. Recover the lost art of drug discovery. Nature 2012; 485: 439-440.

Send Orders for Reprints to reprints@benthamscience.net

Frontiers in Clinical Drug Research: Anti-Infectives, Vol. 1, 2014, 263-307 263

CHAPTER 7

Current Status of Antimicrobial Resistance in Enteric Bacterial Pathogens

Yasra Sarwar[1], Aamir Ali[1], Asma Haque[2] and Abdul Haque[1,*]

[1]*Human Enteric Pathogens Group, Health Biotechnology Division, National Institute for Biotechnology and Genetic Engineering, Faisalabad, Pakistan and* [2]*Department of Bioinformatics and Biotechnology, GC University, Faisalabad, Pakistan*

Abstract: Bacterial enteric pathogens are by far the most dominant scourge of mankind. There are more than 200 million cases and 3 million deaths caused by these bacteria every year. Before the antimicrobial era, there were pandemics of enteric diseases which sometimes swept away whole populations. Advent of antimicrobial era provided a tool in the hand of mankind to fight this menace. In the beginning the results were promising and there was optimism of a decisive victory against disease causing bacteria. But the reality dawned within a couple of decades when antimicrobial resistance started to emerge and every new antimicrobial was generally knocked out in a couple of years. It became apparent that these bacteria held a distinct advantage because of very fast evolution rate due to relatively simple and small genome and short generation time. Currently, we are always playing a catch up game because the enemy is always ahead. The emergence of multiple drug resistance (MDR) has aggravated the situation and there is a distinct possibility that some of these menacing bugs may get out of control and situation of pre-antimicrobial era may return. Recently, a new term extreme-drug resistance (XDR) has been coined. This refers to bacteria resistant to all available drugs. This aptly summarizes the situation we are facing today. This catastrophe can only be avoided by putting more efforts in developing new concepts and products. This chapter is an effort to encompass the properties of these pathogens, the antimicrobials currently in use and the mechanisms of drug resistance evolved by these formidable bacteria.

Keywords: Antimicrobial drug resistance, human enteric pathogens, molecular mechanisms.

INTRODUCTION TO BACTERIAL ENTERIC PATHOGENS

Most of the bacterial enteric pathogens belong to family *Enterobacteriaceae*. In

***Address correspondence to Abdul Haque:** Human Enteric Pathogens Group, Health Biotechnology Division, National Institute for Biotechnology and Genetic Engineering (NIBGE), P.O. Box 577, Jhang Road, Faisalabad, Pakistan; Tel: (92-41) 2651475-79; Fax: (92-41) 2651472; E-mail: ahaq_nibge@yahoo.com*

addition, *V. cholerae*, and some microaerophilic/anaerobic bacteria especially *Campylobacter jejuni* are important enteric pathogens. *Although Enterobacteriaceae* includes nearly 50 genera, the significant members are *Salmonellae, Shigellae*, and pathogenic *Escherichia coli.* Members of the *Enterobacteriaceae* family are rod-shaped, and are typically 1-5 μm in length, Gram-negative, and facultative anaerobes. Many members of this family are a normal part of the gut flora found in intestines of humans and other animals, while others are found in water or soil, or are parasites on a variety of different animals and plants.

Members of *Enterobacteriaceae* family not only cause enteric diseases but are also important causes of urinary tract infections (UTIs), respiratory tract infections, bloodstream infections, hospital and healthcare associated pneumonias, and various intra-abdominal infections. Lower respiratory tract and bloodstream infections are the most lethal and UTIs are the most common [1]. The emergence and spread of resistance in *Enterobacteriaceae* are complicating the treatment of serious nosocomial infections and threatening to create strains resistant to all currently available agents [2].

Salmonellae --- Cause of Typhoidal Diseases

The majority of disease-associated *Salmonella* are serovars of *S. enterica* subspecies *enterica* that accounts for 99% of all human and animal infections [3]. *S. enterica* serovars *Typhi, Paratyphi A, Paratyphi B* and *Paratyphi C* are collectively referred to as typhoidal *Salmonella* serovars [4].

Typhoid fever is a potentially fatal bacterial infection caused primarily by *Salmonella enterica* serovar Typhi (hereafter referred to as *S. Typhi*). The estimated incidence is approximately 33 million cases each year. In the developed countries, the incidence is much lower, and most cases are usually from travelers returning from endemic areas. Humans are the natural host and reservoir for *S. Typhi* which can survive for days in groundwater or seawater and for months in contaminated eggs and frozen oysters. The infectious dose varies between 10^3-10^6 organisms when taken orally. Transmission of infection occurs by ingestion of food or water contaminated with feces.

In Asian countries, the incidence of typhoid fever among children (5-15 years) appears to be highest in South Asia (400-500 cases per 100,000 persons per year), intermediate in Southeast Asia (100-200 cases per 100,000 persons per year) and lowest in Northeast Asia (<100 cases per 100,000 persons per year). Similar high fever incidence rates have been reported from Bangladesh [5]. These data, together with the data from Nepal [6], seem to suggest that the sub-continental nations, including India, Pakistan, Bangladesh and Nepal, are at a very high risk for typhoid fever. These data are consistent with the previous observations [4].

Despite the role of the Vi antigen as a distinguishing feature of serovar *Typhi*, *Vi* negative isolates are not uncommon. Vi negative *S. Typhi* have been reported from various locations. In the 1970's Vi negative isolates were encountered in Jamaica [7], Indonesia [8], New Zealand, and Malaysia [9]. In 2000, there was a report from India showing prevalence of Vi negative strains [10]. We have also isolated and reported these strain from Faisalabad, Pakistan [11].

S. Paratyphi A is more prevalent in war torn and developing countries. A report from New Delhi, India, demonstrated a significant increase in *S. Paratyphi* A isolation from 1.7% in 2001 to 18% in 2005 to 2006. There was an increase of 3.8% in patients requiring hospitalization. One large sample size report from Nepal also indicated an increased incidence of *S. Paratyphi* A (from 23% during 1993 to 1998 to 34% during 1999 to 2003) [12]. Some reports from China also demonstrate a high incidence of *S. Paratyphi* A with infection rates up to 64% among all EF cases [13]. Our group has also reported high prevalence of *S. Paratyphi* A in Faisalabad, Pakistan [14].

Shigellae --- Major Cause of Bacillary Dysentery

Diarrheal diseases are the most common cause of morbidity and mortality and rank as fourth most common killer disease in the world and second as a cause of years of productive life lost due to premature mortality and morbidity [15]. Among diarrheal diseases, dysentery caused by *Shigella* is one of the most important contributors. Shigellosis is a major public health problem not only because of morbidity but also for growth retardation and malabsorption in children.

According to WHO, the annual number of *Shigella* episodes in developing countries throughout the world is 164.7 million. Out of these, 163.2 million (99%) are in developing countries (including 1.1 million mortality) and 1.5 million in industrialized countries [16]. High risk population for *Shigella* infection includes kids less than five years of age, and senior citizens. It has been found that 69% of all episodes and 61% of all deaths in shigellosis involve children under 5 years of age [17]. Improper personal hygiene and sanitation resulting in the contamination of food and drinking water are the major causes of spread of *Shigella* infection [18].

Shiga Toxin Producing *E. coli* (STEC)

Escherichia coli (E. coli) are predominately found in intestinal micro flora of humans and other mammals. These commensal *E. coli* are usually harmless but certain pathotypes are implicated in diarrhea and other enteric problems and called as "diarrheagenic *E. coli*". Diarrheagenic *E. coli* have been divided into six major pathotypes which include enteropathogenic *E. coli* (EPEC), atypical enteropathogenic *E. coli* (ATEC), enterohemorrhagic *E. coli* or Shiga toxin-producing *E. coli* (EHEC/STEC), enteroaggregative *E. coli* (EAEC), enteroinvasive *E. coli* (EIEC) and enterotoxigenic *E. coli* (ETEC) [19, 20].

STEC (EHEC) are important emerging pathogens and have been associated with number of complications like bloody diarrhea, hemorrhagic colitis and potentially fatal renal disease, hemolytic uremic syndrome (HUS). Year 1983 was momentous for microbiologists as first reconnaissance of STEC O157:H7 outbreak [21] and later its association with HUS was reported [22]. Since then STEC has been detected from an increasing number of food borne outbreaks of bloody diarrhea and HUS. The most notorious STEC serotype is O157:H7. However non-O157 STEC serotypes are also emerging as notable pathogens.

There are more than 250 different *E. coli* O serotypes implicated in Shiga toxin production and 100 of them have been found in various diarrheal outbreaks in humans [23]. Various studies indicate that among non-O157:H7, serotype O111, O26, O145 and 103 are more frequently associated with STEC outbreaks and HUS [24, 25]. The incidence of STEC mostly varies according to age group. In

USA, highest incidence (0.7 cases per 100,000) was observed in children under 15 year of age. In most of the cases (63-85%), the etiological agent is transmitted through food stuff [26].

Extraintestinal Pathogenic *E. coli* (ExPEC)

E. coli isolates capable of causing disease outside the gastrointestinal tract are known as extraintestinal *E. coli*. Extraintestinal pathogenic *E. coli* (ExPEC) can invade urinary tract, cerebrospinal fluid and blood stream. ExPEC are responsible for a variety of diseases such as urinary tract infections (UTIs), neonatal meningitis, septicemia, nosocomial pneumonia, intra abdominal infections, osteomyelitis and wound infections. These are major pathogens of UTIs in normal and unobstructed urinary tracts [27, 28].

Extraintestinal strains of *E. coli* can infect every organ, all age groups and all types of hosts. Severe illness and mortality can occur in normal, healthy hosts; however, adverse outcomes become increasingly prevalent in the presence of co-incidental disease and abnormalities in host defenses [29].

Vibrios – Cause of Cholera

There is some debate regarding inclusion of cholera in enteric bacterial diseases because the causative organisms are quite different from other enteric pathogens. But we have included cholera because it is one of the most devastating diseases of enteric system. Cholera, caused by *Vibrio* is an acute diarrhoeal disease that can kill within hours if left untreated. There are estimated 3–5 million cholera cases and 100,000–120,000 deaths due to cholera every year. However, up to 80% of cases can be successfully treated with oral rehydration salts [30].

Cholera has smoldered in an endemic fashion on the Indian subcontinent for centuries. Epidemic cholera was described in 1563 by Garcia del Huerto, a Portuguese physician at Goa, India. In 1961, the "El Tor" biotype (distinguished from classic biotypes by the production of hemolysins) reemerged and produced a major epidemic in the Philippines to initiate a pandemic. Since then, this biotype has spread across Asia, the Middle East, Africa, and parts of Europe [31].

The genus *Vibrio* consists of Gram-negative straight or curved rods, motile by means of a single polar flagellum. The Family *Vibrionaceae* is distinct from Family *Enterobacteriaceae* although members of both families are described as facultatively anaerobic Gram-negative rods [32].

V. cholerae and *V. parahaemolyticus* are pathogens human. Both produce diarrhea, but in ways that are entirely different. *V. parahaemolyticus* is an invasive organism affecting primarily the colon; *V. cholerae* is noninvasive affecting the small intestine through secretion of an enterotoxin [31].

Campylobacter – Cause of Acute Diarrhea

Campylobacter like Vibrio are different from members of Enterobacteriacae. The main pathogenic species, *C. jejuni* and *C. coli* are microaerophilic, spiral-shaped bacteria that asymptomatically colonize birds, including chicken [33, 34]. But these bacteria cause serious disease in humans. *Campylobacter* species are a leading cause of acute infectious diarrhea resulting in up to 14% of cases worldwide. In a study of 100, 000 persons in Sweden, *C. jejuni* was isolated in 56% of enteritis cases [35]. Ingestion of undercooked poultry and cross-contaminated food stuffs results in a spectrum of acute diarrheal disease, ranging from watery diarrhea to dysentery and a mesenteric adenitis syndrome mimicking acute appendicitis [36].

INTRODUCTION TO ANTIMICROBIALS

What is an Antimicrobial?

An antimicrobial is a substance that kills or inhibits the growth of microorganisms such as bacteria, fungi, or protozoa. Antimicrobial drugs either kill microbes (microbiocidal) or prevent the growth of microbes (microbiostatic). Disinfectants are antimicrobial substances used on non-living objects or outside the body.

Scientifically, antibiotics are only those substances that are produced by one microorganism that kill or prevent the growth of another microorganism. However, in common usage, the term antibiotic is used to refer to almost any drug that attempts to rid our body of a bacterial infection. Antimicrobials include not just antibiotics, but synthetically formed compounds as well.

Before penicillin became a viable medical treatment in the early 1940s, no true cure for gonorrhea, throat infections, or pneumonia existed. Patients with infected wounds often had to have a wounded limb removed, or faced death from infection. Now, most of these infections can be cured easily with a short course of antimicrobials.

There are mainly two classes of antimicrobial drugs:

1. Those obtained from natural sources:

    a. β-lactam antibiotic (such as penicillins, cephalosporins)

    b. Protein synthesis inhibitors (such as aminoglycosides, macrolides, tetracyclines, chloramphenicol, polypeptides)

2. Synthetic agents:

    c. Sulphonamides, cotrimoxazole, quinolones

Antimicrobial Modes of Action

The antimicrobials act on different targets in a bacterial cell. The β-lactams which include penicillins, cephalosporins and several other groups act on cell wall by binding to and inhibiting enzymes needed for the synthesis of peptidoglycan. This creates breaches making the organism to burst in hypotonic surroundings. Quinolones inhibit DNA replication with the formation of double-stranded DNA breaks; treatment with rifamycins arrest DNA dependent RNA synthesis. Inhibitors of protein synthesis induce cell death or stop cell growth by affecting cellular energetics, ribosome binding and protein mistranslation, as tetracycline inhibit protein synthesis by binding to 30S ribosomal unit of ribosome and chloramphenicol by binding to 50S ribosomal subunit. These are reversible bindings, but aminoglycosides bind to 30S ribosomal unit irreversibly. Polymyxins behave as detergents increasing the permeability of the membranes which encase bacteria, and causing the contents of the bacterial cell to leak out. In addition, recent evidence points towards a common mechanism of cell death involving disadvantageous cell responses to drug induced stresses that are shared by all classes of bactericidal antimicrobials (Fig. **1**), which ultimately contributes to killing by these drugs [37, 38].

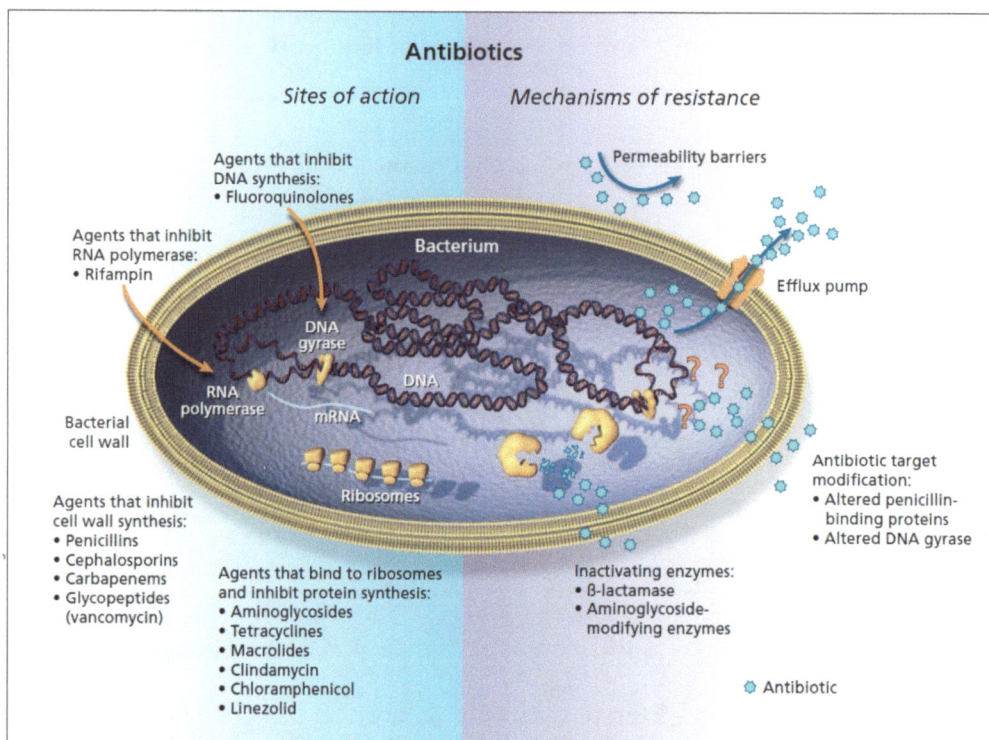

Figure 1: Sites of action and mechanisms of resistance development of various antimicrobial agents [39].

β-lactam Antibiotics

The β-lactam drugs are the most widely used antimicrobial agents exhibiting a rapid bactericidal effect and are well tolerated. All β-lactam drugs share a characteristic ring structure (the β-lactam ring) from which their name is deduced and on which antimicrobial activity of these drugs depends. These drugs target bacterial cell wall and are bactericidal. β-lactam drugs are divided into four major groups: penicillins, cephalosporins, monobactams, and carbapenems [40, 41].

Quinolones /Fluoroquinolones

Quinolones represent a group of synthetic chemotherapeutic antibacterial agents. Nalidixic acid was the first quinolone with antibacterial activity.

Fluoroquinolones are one of the several derivatives of quionolones [42]. The early quinolones such as nalidixic acid are considered first-generation quinolones; ciprofloxacin, and ofloxacin as second-generation; gatifloxacin, sparfloxacin, and temafloxacin are included in third -generation and trovafloxacin, moxifloxacin, and gemifloxacin are representatives of fourth-generation fluoroquinolones [43]. Quinolones are bactericidal and exert their antibacterial effects by inhibition of bacterial topoisomerase enzymes, namely DNA gyrase (bacterial topoisomerase II) and topoisomerase IV. These essential bacterial enzymes alter the topology of double-stranded DNA (dsDNA) within the cell.

The Antifolate Group

Trimethoprim, an antifolate is a synthetic antimicrobial agent, which interferes with folate synthesis in both Gram-negative and Gram-positive bacteria. It behaves bacteriostatically after competitive and strong binding to dihydrofolate reductase (DHFR) [44], which catalyses the formation of tetrahydrofolate from dihydrofolate. Although DHFRs from eukaryotic cells can also bind trimethoprim, the affinity of the drug to the bacterial enzymes is higher [45]. Sulfonamides are synthetic substances too, which inhibit the first step of bacterial folate synthesis pathway and work bacteriostatically [46]. Trimethoprim combined with sulfonamide have a bactericidal effect and this synergistic effect is the reason why most of the preparations on the market are a combination of trimethoprim and sulphonamides [47]. The combination is known as co-trimoxazole or trimethoprim-sulphamethoxazole.

Aminoglycosides

Aminoglycosides are among the oldest and powerful bactericidal drugs, characterized by the presence of an aminocyclitol ring linked to amino sugars in their structure. They have a broad spectrum of activity against Gram-positive and Gram-negative bacteria, mycobacteria and protozoa. These drugs act by binding irreversibly to the ribosomal acceptor (A) site and inhibiting bacterial protein synthesis. Examples of these drugs include those derived from *Streptomyces* spp. (streptomycin, neomycin and tobramycin) or *Micromonospora* spp. (gentamicin) or synthesized *in vitro* (netilmicin, amikacin, arbekacin and isepamicin) [48].

Tetracyclines

Tetracyclines were discovered in 1940s. Broad spectrum activity, low toxicity and low cost had made tetracyclines perfect therapeutic agents but their efficacy reduced by passage of time [49]. Tetracyclines act by penetrating bacterial cells by passive diffusion and binding reversibly to the ribosome, thereby preventing the attachment of aminoacyl-tRNA to the ribosomal acceptor (A) site and inhibiting bacterial protein synthesis [50].

Chloramphenicol

Chloramphenicol binds reversibly to the 50S subunit of the bacterial ribosome and inhibits peptidyl transferase reaction, which forms the peptide bonds between the amino acids, and thereby suppress bacterial protein synthesis. Chloramphenicol has broad spectrum activity and act bacteriostatically on Gram-negative and Gram-positive bacteria [51].

ANTIMICROBIAL DRUG RESISTANCE

Antimicrobial resistance is ability of the microorganisms to with stand the dose of antimicrobial that was effective in the past due to the repeated exposure of a microbe to a particular drug. Bacterial drug resistance can be attained through intrinsic properties, mutation acquired mechanisms, biochemical alterations, selective pressure or physical barriers such as biofilm formation.

Intrinsic Mechanisms

Intrinsic resistance may naturally occur as a result of the bacteria's genetic makeup. It is either due to the inaccessibility of the targets by the drug. *e.g.,* aminoglycoside resistance in strict anaerobes is due to multidrug efflux systems or drug inactivation by the bacteria. *E. coli* is intrinsically resistant to vancomycin because vancomycin is too large to pass through porin channels in outer membrane. Gram-positive bacteria, on the other hand, do not possess an outer membrane, thus are not intrinsically resistant to vancomycin. Drug resistance may also be due to naturally occurring genes found on the host's chromosome, such as, AmpC β-lactamase of Gram-negative bacteria and many efflux systems.

Mutational Resistance

Mutations in the genome results in the target site modification or reduced permeability or uptake of the drug by organism (Fig. **2**). Mutations can also cause metabolic bypass or derepression of multidrug efflux systems.

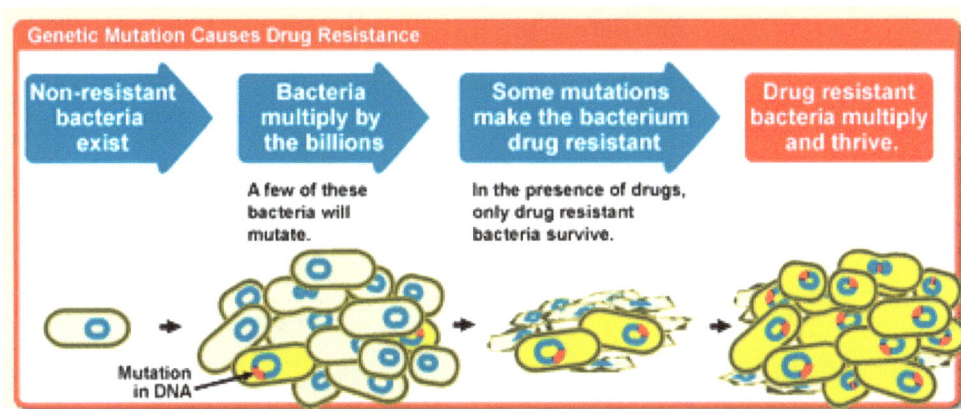

Figure 2: Genetic mutation causes drug resistance [52].

Extrachromosomal or Acquired Resistance

This type of drug resistance is disseminated by plasmids or transposones resulting in either drug inactivation, drug efflux, target site modification or metabolic bypass [53]. Among the horizontal gene transfer mechanisms, conjugation (*via* plasmids and conjugative transposons) is thought to play the most significant role in the spread of resistance genes [54]. But other mechanisms including transduction (*via* bacteriophages) and transformation (*via* incorporation of chromosomal DNA, plasmids, and transfer of DNAs from dying organisms into the chromosome) are also important [54] (Fig. **3**).

Selective Pressure

In the presence of an antimicrobial, microbes are either killed or, if they carry resistance genes, survive. These survivors will replicate, and their progeny will quickly become the dominant type throughout the microbial population.

Biochemical Mechanisms of Drug Resistance

Antimicrobial agents are rendered inactive by different mechanisms that include:

- Enzymatic drug modification and destruction

- Mutational alteration of the target protein

- Substitution and protection of drug targets

- Reduced drug accumulation due to efflux systems, porins and outer membrane proteins [56].

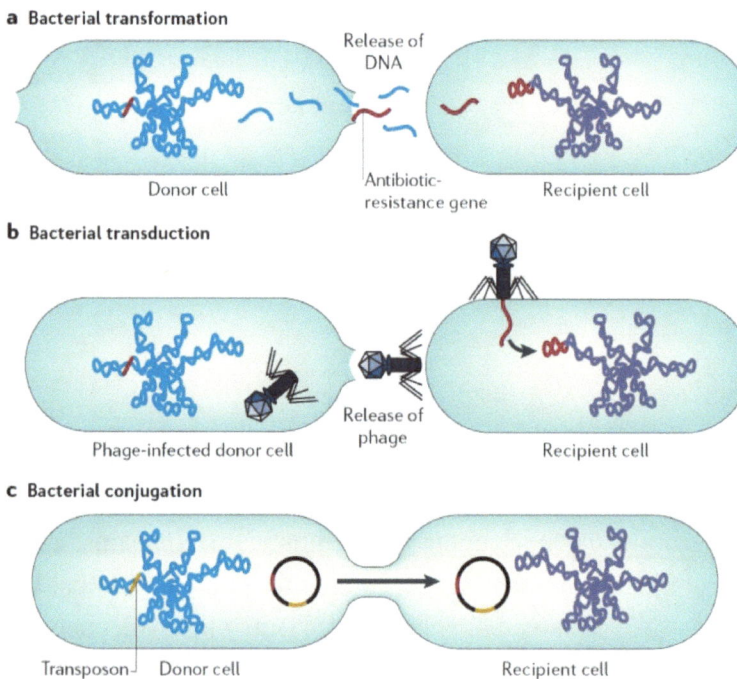

Figure 3: Horizontal gene transfer between bacteria [55].

TOOLS OF GENETIC MOBILITY

The physical movement of DNA relies on a number of molecular 'cut and paste' mechanisms that are able to control and translocate DNA fragments. Enzymes with these potentialities include recombinases, transposases, integrases and resolvases which are encoded by an assortment of selfish mobile genetic elements (Fig. **4**) insertion sequences, transposons and integrons). These genetic elements can facilitate gene deletion or capture, and accretion of genetic elements on higher order mobile elements such as conjugative plasmids [57].

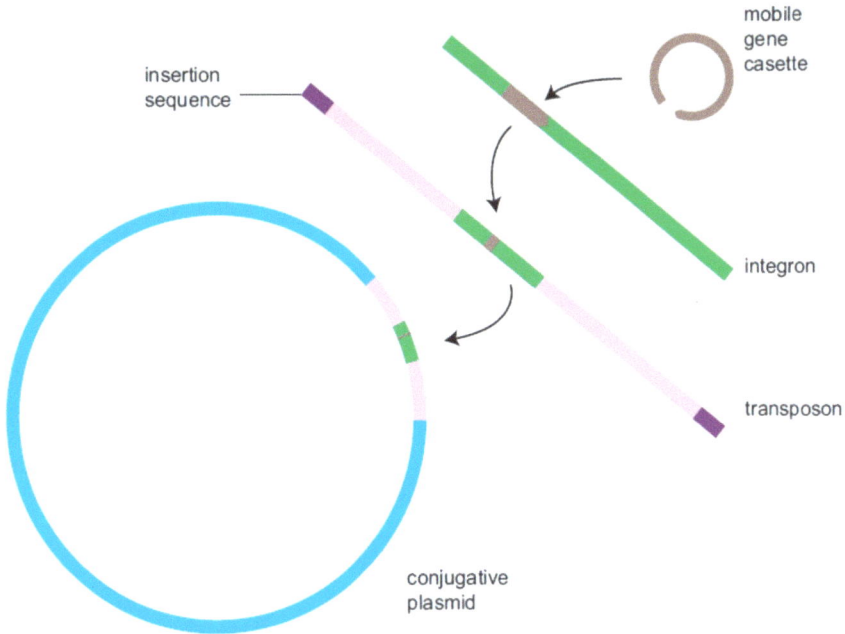

Figure 4: The schematic composition of mobile genetic elements [57].

Integrons

Integrons are "assembly platforms that incorporate exogenous open reading frames by site-specific recombination and convert them to functional genes by ensuring their correct expression". An integron includes two parts: the gene cassette and the recombination platform. The recombination platform also called the 'core' integron includes a site-specific recombinase (integrase) gene (*intI*), a recombination site (attI) and an outward- orientated promoter (Pc) that directs transcription of the captured genes (Fig. **5**). The gene cassette usually consists of one or more genes and a second type of recombination site which was originally termed the 59-base element by Hall and colleagues [58], but is now called *attC* (attachment site associated with cassettes). Although not independently mobile, integrons are widespread versatile DNA elements and can be divided into two distinct subsets: the mobile integrons and the chromosomal integrons [58, 59].

Mobile Integrons

These are primarily involved in the spread of antimicrobial resistance genes and are linked to mobile DNA elements. Five classes of mobile integrons are known to have a role in the dissemination of antibmicrobial resistance genes. Class 1, 2 and 3 integrons which are involved in multiple- antibmicrobial -resistance phenotype, belongs to 'historical' classes of mobile integrons. The other two classes of mobile integrons, class 4 and class 5, have been identified in *Vibrio* species through their involvement in the development of trimethoprim resistance [59].

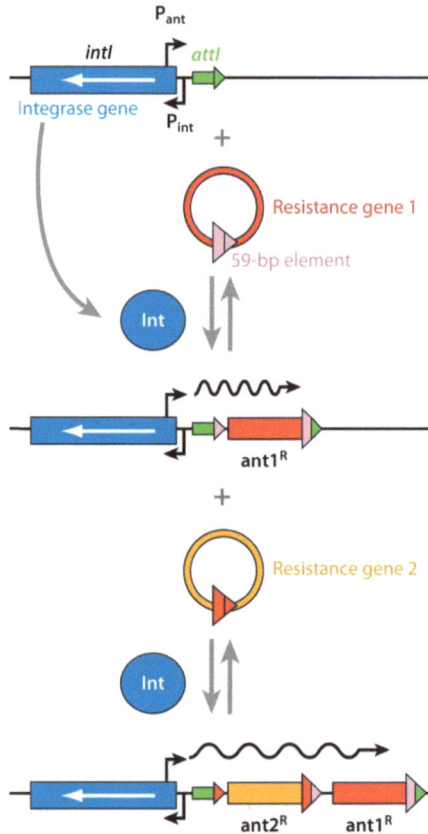

Figure 5: Mechanism of intake of resistance genes by integrons [60].

Chromosomal Integrons

These were first identified on chromosome 2 of the *V. cholerae* genome as an organization of cluster of repeated DNA sequences. The key features that define this subset include encoding of a specific integrase, VchIntIA, which is related to the integrases encoded by mobile integrons but has two characteristics that distinguish it from known mobile integrons. Firstly, large number of gene cassettes are associated with the integron and secondly, being immobile, it is located on the chromosome and not associated with mobile DNA elements [61].

Transposons

Transposons are self directed elements that code for transposases. Transposases are enzymes that bind to the ends of a transposon and catalyze its movement to another part of the genome by a cut and paste mechanism or a replicative transposition mechanism. Different types include unit, composite, conjugative and mobilizable transposons [62].

Conjugative transposons also known as integrative conjugational elements (ICEs) play a substantial role in the dispersal of antibiotic resistance genes amongst pathogenic bacteria and can only maintain themselves stably by integrating into the chromosome [63, 64]. They have an enormously broad host range (Gram-negative and Gram-positive bacteria) and confer resistance to a wide range of antimicrobials (ampicillin/penicillin, cefoxitin, chloramphenicol, erythromycin, mercuric chloride, gentamycin, kanamycin, streptomycin, tetracycline-minocycline, and vancomycin) [65].

Plasmids

Plasmids can be classified by several criteria as conjugative or mobilizable, and on the basis of incompatibility groups copy number and host range [66]. Plasmids of Gram negative bacteria are either conjugative or mobilizable depending upon the presence or absence of three different elements. These three elements are present on all conjugative plasmids and include a cis-element called *oriT* (origin of transfer), one or more *mob* (mobilizing) gene and the *tra* (transfer) gene which codes for the pilus gene anchored in the two membranes and responsible for making contact with the recipient cells [67].

Plasmids are most frequently classified into incompatibility groups according to their mode of replication and maintenance in a bacterial cell [68]. Two different plasmids are said to be compatible with each other, if they can stably coexist without selective pressure [67].

Based on the copy number per bacterial chromosome, plasmids can be arranged into four different groups: low-copy-number (1-2), medium-copy-number (5-10), high-copy-number (20-25) and very-high-copy-number (100-500). Host range is another feature for characterizing plasmids [67].

Drug Efflux Pumps

Efflux pumps are transport proteins involved in the expulsion of toxic substrates (including virtually all classes of clinically relevant drugs) into the external environment from inside of cells. These proteins are found in both Gram-positive and Gram-negative bacteria as well as in eukaryotic organisms [69]. Pumps may be specific for one substrate or may transport a range of structurally dissimilar compounds (including drugs of multiple classes); such pumps can be associated with multiple drug resistance (MDR). Drug efflux pumps are now recognized as significant contributors to both innate and acquired bacterial resistance to many of these agents because of the very broad variety of substrates they recognize [70, 71].

Pathogenicity Islands

Bacterial species can frequently exchange 5-10 kb regions of genomic DNA. These regions are generally designated as 'islands' due to their large size. These genomic islands sometimes carry virulence associated genes or drug resistance genes. Such genomic islands are referred to as pathogenicity islands (PAI or PI). PAIs are commonly found in pathogenic strains whereas in non pathogenic strains they are absent or rarely found [72]. The unstable regions are of > 30 kb size carrying bacterial virulence genes. These regions are also called plasticity zones with atypical G+C contents relative to the rest of the genome and such DNA segments are originated from a different organism through horizontal gene transfer. PAIs are associated with tRNA genes, which act as integration sites for foreign DNA. Insertion sequences or direct repeats often flank these PAIs, while transposases, origins of plasmid replication, and integrases are often found within these PAIs [73].

BIOFILMS - THE PHYSICAL BARRIERS

A biofilm is an aggregate of microorganisms in which cells adhere to each other on a surface. These adherent cells are frequently embedded within a self-produced matrix of extracellular polymeric substance (EPS). Biofilm EPS, which is also referred to as slime (although not everything described as slime is a biofilm), is a polymeric conglomeration generally composed of extracellular DNA, proteins, and polysaccharides. Biofilms may form on living or non-living surfaces and can be prevalent in natural, industrial and hospital settings [74, 75] (Fig. **6**). The microbial cells growing in a biofilm are physiologically distinct from planktonic cells of the same organism, which, by contrast, are single-cells that may float or swim in a liquid medium.

Microbes form a biofilm in response to many factors, which may include cellular recognition of specific or non-specific attachment sites on a surface, nutritional cues, or in some cases, by exposure of planktonic cells to sub-inhibitory concentrations of antibmicrobials [76, 77]. When a cell switches to the biofilm mode of growth, it undergoes a phenotypic shift in behavior in which large suites of genes are differentially regulated [78].

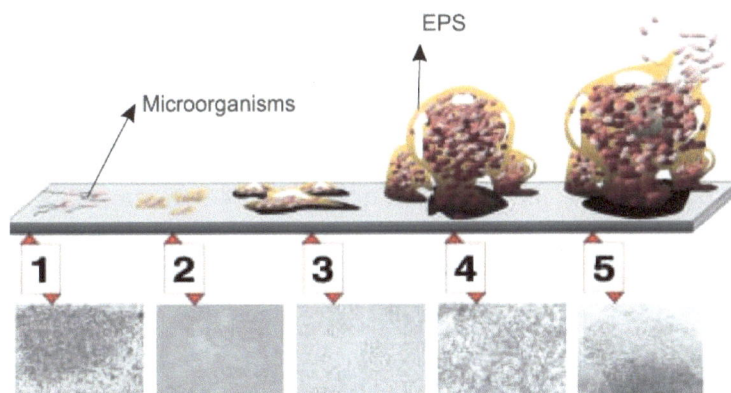

Figure 6: Steps in biofilm production [79].

BIOFILMS AND INFECTIOUS DISEASES

Biofilms are important survival mechanisms for bacterial cells. It is difficult for phagocytic cells to engulf bacteria in biofilms. Also, biofilms are much more resistant than planktonic cells to antimicrobial agents. These are highly developed colonies with bacteria at different stages of life placed in separate segments with intercommunication facilities *via* water channels. An antimicrobial has to be able to reach its target bacteria before it is inactivated during its journey through a biofilm. There are some bacteria in biofilms called "persistors" which are most difficult to kill. Flouroquinolones are effective against biofilms because of their fast penetration rate.

Biofilms have been found to be involved in a wide variety of microbial infections in the body, by one estimate 80% of all infections [80]. Infectious processes in which biofilms have been implicated include common problems such as urinary tract infections, catheter infections, middle-ear infections, formation of dental plaque, gingivitis [81], coating contact lenses [82], and less common but more lethal processes such as endocarditis, infections in cystic fibrosis, and infections of permanent indwelling devices such as joint prostheses and heart valves [83]. More recently it has been noted that bacterial biofilms may impair cutaneous wound healing and reduce topical antibacterial efficiency in healing or treating infected skin wounds [84].

SPECIFIC MECHANISMS OF DRUG RESISTANCE AGAINST MAJOR DRUG GROUPS

β-lactams

Bacteria show resistance to β-lactam antibiotics due to the hydrolysis of antibiotic by β-lactamase enzymes, the most common resistance mechanism in Gram-negative bacteria [85]. Other mechanisms of β-lactam resistance described include cellular permeability which occurs through changes in outer membrane proteins leading to a lowered permeability for the enzyme (porin deficiencies), or the export of β-lactams *via* multi-drug transporters [40]. Alterations or modification in penicillin binding proteins (PBPs) are described in some Gram-positive bacteria [86].

Quinolones /Fluoroquinolones

Mechanisms of bacterial resistance to quinolones fall into three categories: 1) alterations in target of quinolones, 2) decreased accumulation of quinolones due to impermeability of the membrane or 3) due to an over expression of efflux pump systems [91].

Quinolones target alterations occur predominately in domains near the enzyme active sites, which are known as the quinolone-resistance determining region

(QRDR). QRDR is a portion of the DNA-binding surface of the topoisomerase at which amino acid substitutions can diminish quinolone binding and subsequently cause resistance to quinolones. Quinolone resistance generally results from stepwise chromosomal point mutations mainly in the *gyrA* and *parC* genes due to amino acid substitution. Amino acid substitutions within the quinolone resistance-determining region (QRDR) mostly involve the replacement of a hydroxyl group with a bulky hydrophobic residue [92].

The initial mutations in *gyrA* result in resistance to nalidixic acid and afterwards, additional mutations lead to fluoroquinolone resistance [93]. Plasmid-mediated quinolone resistance (PMQR) is associated with low level resistance to fluoroquinolones and represent the production of Qnr proteins protecting the targets against the effects of quinolones [94].

Antifolates

So far, more than 30 different trimethoprim resistance mediating dihydrofolate reductase (*dfr*) genes have been identified [87]. These are subdivided on the basis of their structure into two major types 1 and 2 [88], which nowadays are referred to as *dfrA* and *dfrB*. A second trimethoprim resistance mechanism is to use alternative folate pathways either by usage of external supply of thymidine or by the use of other thymidylate synthases [89].

Sulfonamide resistance can result either from mutations in the chromosomal DHPS gene (*folP*), which decreases DHPS affinity for the sulfonamide inhibitors or, more frequently, from the acquisition of genes encoding alternative drug-resistant variants of the DHPS enzymes which are plasmid borne [46]. Only three genes *sul1, sul2, sul3* are currently known to code for sulfonamide-resistant DHPS [90].

Aminolycosides

Most frequently encountered aminoglycoside resistance is mediated by modifying enzymes which attach certain groups to the aminoglycoside molecule thereby destroying its antibacterial activity. These enzymes are classified as aminoglycoside N-acetyltransferases (AAC), aminoglycoside O-adenyltransferases (also named aminoglycoside nucleotidyltransferases [ANT]), and aminoglycoside O-phosphotransferases (APH) depending on their type of modification. For each of these three classes, numerous members are known which differ more or less extensively in their structure [95, 96].

In addition to aminoglycoside-modifying enzymes other mechanisms that confer resistance against these agents include decreased accumulation of the drug due to expression of efflux systems, and methylation of the 16S rRNA within the 30S subunit [97].

Tetracyclines

Bacterial resistance is mediated mainly by three mechanisms: 1) efflux of antibiotic to reduce intracellular concentration, 2) ribosome protection of the antibiotic target and 3) modification of antibiotic making it inactive [50]. The most common resistance mechanism against tetracyclines in Gram-negative bacteria is efflux of these drugs. Different classes of tetracycline specific exporters have been identified [98].

Chloramphenicol

The most common resistance mechanism to chloramphenicol in Gram-negative bacteria is the expression of a chloramphenicol acetyltransferase (CAT), which mediates O-acetylation of chloramphenicol, destroying its affinity for bacterial ribosomes and thus its ability to inhibit bacterial growth [99]. There are two separate families of CAT enzymes in bacteria, CATA and CATB [100].

DRUG RESISTANCE IN ENTERIC PATHOGENS

The global problem of antimicrobial resistance is particularly pressing in developing countries, where the infectious disease burden is high and cost constraints prevent the widespread application of newer, more expensive agents. Gastrointestinal, respiratory, sexually transmitted, and nosocomial infections are leading causes of disease and death in the developing world, and management of all these conditions has been critically compromised by the appearance and rapid spread of drug resistance. Even though surveillance of resistance in many developing countries is suboptimal, the general picture is one of accelerating rates of resistance spurred by antimicrobial misuse and shortfalls in infection control and public health. Reservoirs for resistance may be present in healthy human and animal populations. Considerable economic and health burdens emanate from bacterial resistance, and research is needed to accurately quantify the problem and propose and evaluate practicable solutions. In this section, we will discuss the emerging drug resistance against most of the previously and also currently popular antimicrobials specifically with reference to enteric pathogens.

Typhoidal Bacteria

For over 60 years, drugs have been used to treat typhoid and the first drug introduced in 1948 for this purpose was chloramphenicol. It was followed by ampicillin and co-trimoxazole. These three drugs are called the first line antityphoidal drugs. Along with emergence of resistance against these first line antimicrobials, additional resistance to streptomycin and tetracyclines has been reported in many developing countries, especially Pakistan and India. Such strains are called multidrug-resistant (MDR) [101]. To cope with this situation, fluoroquinolones became the treatment of choice along with third generation cepaholsporins and azithromycin as alternative for resistant isolates [102].

Chloramphenicol was recognized as the drug of choice to treat typhoid fever when introduced in 1948 [103]. Two years later cases of chloramphenicol resistant typhoid fever were reported [104], but chloramphenicol resistance took a long time to become established in *S. Typhi* population. In May 1972 in Kerala, India the first reported antibiotic resistant typhoid fever outbreak occurred [105], which was proved to be plasmid borne. Two other chloramphenicol resistant outbreaks were also documented in the same year in Mexico and Vietnam; both were caused by IncH plasmids carrying *S. Typhi* [106, 107]. Some recent reports show the re-emergence of sensitivity in high proportions to chloramphenicol along with ampicillin and co-trimoxazole in *S. Typhi* strains [108]. Our recent findings provide support to these observations as we have found that resistance level against chloramphenicol is midway between ampicillin and co-trimoxazole [109].

Mechanisms conferring chloramphenicol resistance described so far in *Salmonella* are enzymatic inactivation and the export of chloramphenicol by specific efflux proteins. The most common resistance mechanism to chloramphenicol in Gram-negative bacteria is the expression of a chloramphenicol acetyltransferase (CAT) which mediates O-acetylation of chloramphenicol, destroying its affinity for bacterial ribosomes and thus its ability to inhibit bacterial growth [99]. In a long duration study in Pakistan, it was unexpectedly found that the ratio of MDR isolates decreased gradually from 1995 to 2001 (Fig. **7**).

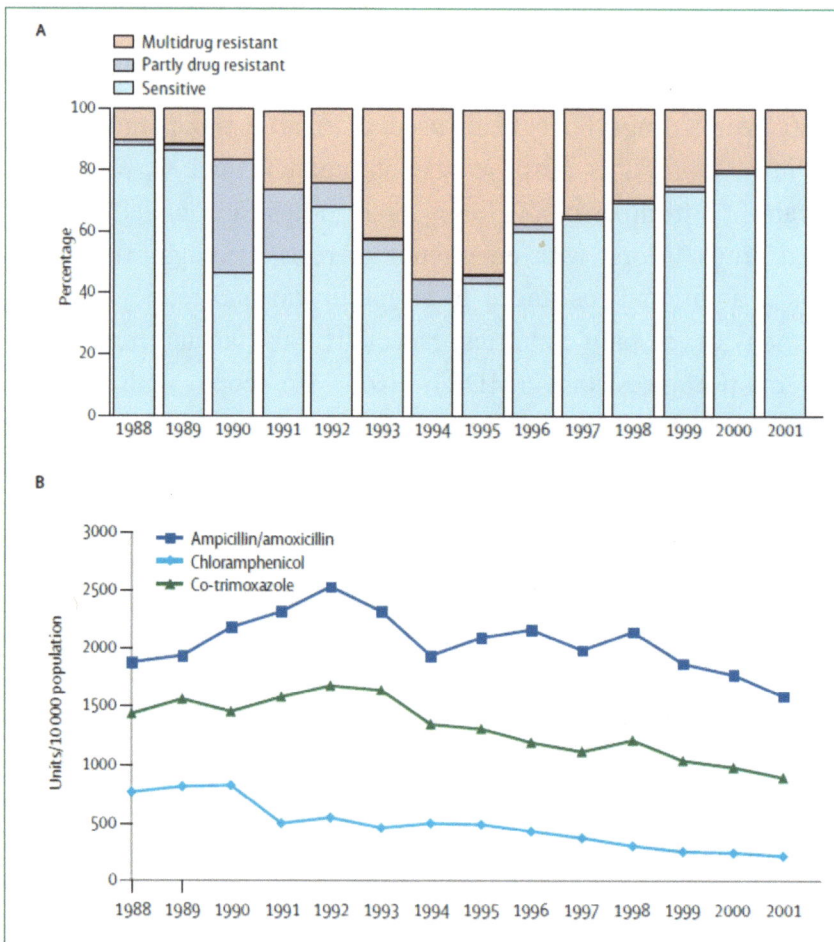

Figure 7: (A) Antimicrobial resistance patterns among *S. Typhi* isolates from children presenting at the Aga Khan University Hospital, Karachi, Pakistan (1988-2001); and (B) antimicrobial sales data for Karachi (units/10,000 population) in the same period [110].

During a large epidemic in Mexico in 1972, isolates resistant to both chloramphenicol and ampicillin were reported. However, resistance to these two drugs was transferred independently by two separate plasmids [111]. The plasmids conferring resistance to chloramphenicol were later identified as incompatibility group H where as the ampicillin resistance plasmids were of the incompatibility group I or A/C [112]. The usefulness in treatment has however decreased as a consequence of increasing resistance, mainly due to β-lactamases like TEM and SHV. The action of these enzymes can in most cases be overcome with the addition of a β-lactamase inhibitor like clavulanic acid [113].

Genes encoding for β-lactamases are known as *bla* genes and a considerable number of *bla* genes have been identified in *Salmonella* while an ESBL-producing *S. Paratyphi* A was reported in India [114]. Among the TEM-type β-lactamases class 2b includes those encoded by the genes bla_{TEM-1} and $bla_{TEM-135}$ which are broad-spectrum penicillinases. Other bla_{TEM} genes, from class 2b include bla_{TEM-3}, bla_{TEM-4}, bla_{TEM-20}, bla_{TEM-27}, bla_{TEM-52}, bla_{TEM-63}, and $bla_{TEM-131}$ and code for extended spectrum β-lactamases (ESBLs) which can also inactivate oxyiminocephalosporins and monobactams. Among SHV-and OXA type β-lactamases found in *Salmonella* are those encoded by bla_{SHV-2}, bla_{SHV-2a}, bla_{SHV-5}, bla_{SHV-9}, bla_{SHV-12} $bla_{OXA-30}$and bla_{OXA-53} [115, 116]. But fortunately we have found in an ongoing study that ESBL production is not detectable in local isolates in Pakistan at present [109].

After the development of ciprofloxacin resistance in *S. Typhi* and *S. Paratyphi* A, cephalosporins (ceftriaxone and cefixime) were one of the few choices left for the treatment of enteric fever. The first reported trial for the use of ceftriaxone to treat typhoid fever was conducted in Bangladesh in 1988 [117]. Although resistance to third generation cephalosporins in non-typhoidal *Salmonella*e had been reported as early as 1989 [118], resistance in *S. Typhi* remains rare. The first cases of reduced susceptibility or resistance to ceftriaxone were documented in Bangladesh and Kuwait in 2008 [119, 120]. A recent case of ceftriaxone resistant *S. Typhi* was reported from an Iraqi woman who returned to Germany after a month's long visit in Iraq. Interestingly, this resistance was encoded on an IncN plasmid of ~50 kb carrying the $bla_{CTX-M-15}$ and *qnrB2* genes [121]. Fortunately, in Pakistan, ceftriaxone resistance is still very rare and it is considered as one of the most effective drugs against typhoid [109].

Until the 1980s, there was no report of single isolates harboring resistance to all three first line drugs (chloramphenicol, ampicillin and cotrimoxazole). In 1980, resistance to these first line drugs was described in Bangkok [122]. Resistance to antifolates is mediated through different mechanisms including: (1) the permeability barrier and/or efflux pumps, (2) naturally insensitive target enzymes, (3) regulational changes in the target enzymes, (4) mutational or recombinational changes in the target enzymes, and (5) acquired resistance by drug-resistant target enzymes. The most common resistance mechanism to trimethoprim is the expression of a trimethoprim-resistant DHFR. This DHFR is expressed additionally to the original enzyme and the gene coding for this additional enzyme is very often located on mobile genetic elements, like plasmids, transposons or gene cassettes [45, 87, 123]. High-level trimethoprim resistance in Enterobacteriaceae is mainly due to the replacement of a trimethoprim-sensitive dihydrofolate reductase by a plasmid-, transposon- or cassette-borne trimethoprim resistant dihydrofolate reductase whose configuration escapes the action of TMP. In Pakistani isolates of *S. Typhi*, resistance against ampicillin and cotrimoxazole is very high whereas it is comparatively lower against chloramphenicol according to our findings [109]. It has been reported that this MDR resistance type found in our isolates is encoded by large plasmids belonging to H1 incompatibility group [124].

After the emergence of MDR *S. Typhi*, fluoroquinolones became the treatment of choice for typhoid treatment. However, there have been many reports of nalidixic acid resistant (NA^R) *S. Typhi* which exhibit decreased susceptibility to ciprofloxacin and show poor clinical response to fluoroquinolones [125, 126]. A major outbreak of MDR *S. Typhi* was encountered in Tajikistan in 1997, by consuming contaminated drinking water that affected nearly 9,000 individuals leading to 95 deaths. This epidemic MDR *S. Typhi* strain developed resistance to nalidixic acid and reduced susceptibility to ciprofloxacin [127] during the outbreak.

Mandal and colleagues reported a simultaneous increase in resistance levels to fluoroquinolones and a decline in the percentage of MDR in the *S. Typhi* population under fluoroquinolone treatment [128]. However, the emergence of high-level ciprofloxacin resistance in *S. Typhi* and *S. Paratyphi* A has been reported in India [129, 130]. The treatment for resistant typhoid fever now depends on third generation cephalosporins and azithromycin [131]. Ciprofloxacin, ofloxacin and gatifloxacin are still considered effective in Pakistani isolates, and very low level of resistance is observed against these florouquinolones according to our findings [109]. Ciprofloxacin is currently a suitable empirical choice in presumed enteric fever cases in Pakistan [132].

Aminoglycosides are not commonly used for the treatment of typhoid fever because they face difficulty in penetrating tissue or cells so are less likely to be effective for facultative intracellular pathogens such as *S. enterica* serovar Typhi. Even so, many authors admit the use of aminoglycoside antimicrobials such as gentamicin and amikacin in susceptibility testing in order to create a treatment regimen for MDR typhoid fever when it is urgently required. Furthermore treatment failure with ciprofloxacin and third generation cephalosporins in treating typhoid fever is also reported. In such settings, there is a need to find a cost-effective treatment regimen for typhoid fever [133-135].

S. Typhi, the causative agent of typhoid in humans, is also capable of producing biofilms which contribute to its resistance and persistence in the host. *S. Typhi* is transmitted through the fecal-oral route by contaminated water and food. Typhoid is communicable for as long as the infected person is capable of excreting bacteria in stool. These bacteria usually disappear from the stool about a week after symptoms of illness have resolved. However, a percentage of these infections can result in asymptomatic carriage of *salmonellae* possibly due to formation of biofilms as a mechanism that contributes to the development of the carrier state [136].

Bacteria in biofilms are generally considered well protected against environmental stresses, antimicrobials [137], disinfectants and the host immune system [138], and as a consequence are extremely difficult to eradicate [139]. Planktonic *Salmonella* populations are found to be sensitive to different antimicrobials as compared to biofilms. We have recently reported that most of the biofilm producing bacteria show MDR pattern of drug resistance hence delaying their clearance from the body [140]. It is reported that *S. Typhimurium* biofilms pre-formed on microplates are up to 2000-fold more resistant to ciprofloxacin as compared to planktonic cells [141]. This is of particular concern as ciprofloxacin is commonly used to treat *Salmonella* infections [142].

Antimicrobial resistance of *S. Paratyphi* A appears to be an emerging problem. The 1996 outbreak of Paratyphoid fever in India showed that the isolates were sensitive to all antimicrobials including chloramphenicol, ciprofloxacin, and ceftriaxone [143]. Two years later, there was a report from New Delhi, India, describing drug-resistant *S. Paratyphi* A. The incidence of resistance to ciprofloxacin increased to 24%, and 32% of isolates had decreased susceptibility to ciprofloxacin (minimum inhibitory concentration (MIC) >2 mg/mL), the drug of choice for enteric fever (EF) in India [144]. Reports from a north Indian tertiary care hospital showed increasing multidrug-resistant *S. Paratyphi* A strains [145]. In an outbreak of paratyphoid fever in 2001 in Nepal, 84% of the isolated strains were reported as resistant to nalidixic acid, which is considered the best predictor of clinical response to fluoroquinolones [146].

One of the largest prospective studies of EF in recent years reported a worrisome result: *S. Paratyphi* A was significantly more likely to be resistant to nalidixic acid (75.25% *vs.* 50.5%) and ofloxacin (3.6% *vs.* 0.5%) than *S. Typhi*. Moreover, MICs of other antibacterials were also higher in *S. Paratyphi* A. A high-level ciprofloxacin-resistant strain has been reported in India (MIC 8 mg/mL) and Japan (MIC 128 mg/mL) [147].

This tendency is not limited to Asian strains. A study from 10 European countries showed an increasing incidence of multidrug-resistant *S. Paratyphi* A. It rose from 9% in 1999 to 25% in 2001, and the incidence of decreased susceptibility to ciprofloxacin also increased from 6% to 18% [148].

Shigellae

Antimicrobial treatment is recommended for moderate to severe shigellosis [149] aimed at resolving the symptoms of diarrhea or reducing its duration and its transmission to close contacts. The antimicrobials are helpful in limiting the duration and shedding of bacteria [150].

Due to the misuse of antimicrobials, resistance has developed against the commonly used trimethoprim sulphamethoxazole, ampicillin and chloramphenicol.

Quinolones and fluoroquinolones are relatively effective in the treatment but resistance has started to emerge [150]. Antimicrobial resistance demands repeated reevaluation of treatment recommendation. According to a study conducted on antimicrobial resistance in *Shigella* in eight Asian countries it was reported that there was an increasing trend in multidrug resistance strains in Asia. Calculated percentage was 78%. The main cause might be the overuse of antimicrobial in these countries among humans and animals [149] and the knowledge on the epidemiology and molecular mechanisms of antimicrobial resistance is important to implement intervention strategies [151].

Resistance towards antimicrobial drugs originate either by mutations in chromosomal genes or acquisition of exogenous material carrying the resistance genes. Mutation in the chromosome may cause phenotypic mutants which enhance the likelihood of acquisition of resistance mutation. A multidrug resistance regulatory locus is widespread among *Salmonella, Shigella, E.coli, Klebsiella* and a multidrug resistance locus on chromosome in *S. flexneri 2a* strain shares homology with resistance region of *Shigella* R plasmid, NR1 [152].

Transferable drug resistance was discovered in the late 1950's and since then many types of plasmids and transposones have been discovered [153]. *Shigella* was among the first organisms shown to harbor transferable antimicrobial resistance patterns [152]. It has a tendency to acquire drug resistance frequently by mobile genetic elements, including the R plasmids, transposones, integrons and genomic islands, on the bacterial genome [154].

Antimicrobial resistance frequency among *Shigella spp.* has increased globally [155] showing great geographical variation [156] and is creating problems in the medical community. *Shigella* isolates resistant to first line drugs are present throughout the world [157]. Multidrug resistance strains occur in Europe, Africa, Asia and South America but they are more prevalent in India and China [151]. They are also increasing significantly in other parts of Asia [158]. Very little data are available describing the distribution of resistant strains of *Shigella* in Pakistan. Data of antimicrobial susceptibility of routinely used antibiotics including ampicillin, tetracycline, chloramphenicol, co-trimoxazole and nalidixic acid from different areas of Pakistan show high resistance to co-trimoxazole (87.75%), ampicillin (55.5%), nalidixic acid (39%) but interestingly chloramphenicol (11.25%), is reemerging as a useful drug [155]. Quinolones such as norfloxacin and ciprofloxacin are reported to be effective against shigellosis [151]. Our recent studies present a gloomy picture as we have found rapidly emerging resistance to ceftriaxzone (cephalosporins) in our local isolates, which is unique and alarming as it has not been reported earlier from Asia [159]. We have also found higher resistance against ciprofloxacin as compared to other Asian and Middle East countries.

According to WHO recommendations, ciprofloxacin is the drug of choice in patients with bloody diarrhea irrespective of the age [15]. These broad spectrum antimicrobials are found to be safe in the treatment of shigellosis [160]. Quinolones are contraindicated in children because they cause the bone marrow depression [157] and are considered to be an important threat to treatment of shigellosis especially in children [15]. However, some clinical studies have shown that they are safe in children and adults [160] because the risk of joint damage in children appears to be minimal with the short term courses of fluoroquinolones [161].

Drug resistance against third generation cephlosporins and flouroquinqlones is an emerging problem especially in children. In 2008, 50% of *Shigella* strains were resistant to nalidixic acid in Bangladesh and 29% in India. But no *Shigella* strain was resistant to nalidixic acid in a study conducted in Central Africa. These drugs are no longer recommended because of the risk of development of quinolones and flouroquinolones resistance and poor efficacay [161]. In India 30% of the strains were resistant to fluoroquinolones because of their extensive and indiscriminate use [160]. This trend has been increasing since 2002 [162]. The resistance to fluoroquinolones has also been found in other Asian countries. Fluoroquinolone resistant strains isolated from India were found to be susceptible to azithromycin and ceftriaxzone. In addition to this, cephalosporin resistance strains were also identified from Spain and Argentina [160]. Cephalosporin and quinolone resistance has also been reported from Pakistan [157].

A recent study has indicated that biofilm phenomenon may be present in *Shigella* species. An increase of the salts concentration enhances the ability of *Shigella* species to attach and to invade the tissue culture cells. The percentage of adherence increased to 15% and the invasion to 90% at 6% salt concentration [163].

Shiga Toxin Producing *E. coli* (STEC)

Tetracycline, sulphonamides, ampicillin and streptomycin are the major antimicrobial agents to have conferred resistance by *E.coli*. However resistance to frontline antimicrobial agents like fluoroquinolones, expanded-spectrum ß-lactams, and third-generation cephalosporins have also become a serious problem, particularly in STEC serotype O157, O26, O103, O111, O128, and O145 [164]. Initially STEC, particularly O157:H7 were found to be sensitive to many commonly used antimicrobials [165], but now both O157 and non-O157 STEC have been implicated in antimicrobial resistance. Zhao and coworkers found that 39 out of 50 (78%) STEC isolates showed resistance to at least two or more antimicrobial classes and multiple resistance to streptomycin, sulfamethoxazole, and tetracycline was frequently observed as well. Class I integrons were present in nine of the STEC strains [166]. Similar observations were made in another study of 141 STEC O157:H7 strains isolated from cattle, sheep, humans and food. Antimicrobial resistance was frequently observed against sulphisoxazole, tetracycline and streptomycin [167].

The use of antimicrobials is often mandatory in severe enteric diseases like cholera, typhoid fever and shigellosis. Unfortunately in case of STEC infection the use of antimicrobials remains controversial. Administration of antimicrobial agents for STEC infection is equivocal because of the risk pertinent with an increased release of Shiga toxin in response to various antimicrobials [168, 169].

It was reported for the first time in 1989 during an outbreak of *E. coli* O157:H7 that antimicrobial therapy of patients with diarrhea due to *E. coli* O157:H7 might be a risk factor for development of HUS [170].

In 1997, Yoh and colleagues reported that fosfomycin which was a drug of choice in Japan exacerbated the Stx1 production from *E. coli* O157:H7 *in vitro*. A seven fold increase in Stx1 release was observed in the culture exposed to fosfomycin while other antimicrobial agents like minocycline, cefazolin, gentamicin and doxycycline caused slight increase in Stx1 and had no effect on Stx2 release [171].

In another study the utility of 13 antimicrobial agents on the production and release of Stx from three different STEC O157 strains was evaluated. Culture exposed to sub-inhibitory concentration of cotrimoxazole, trimethoprim, azithromycin and gentamicin were associated with increased Stx production. It was also observed that increase in Stx production with different antimicrobials was strain specific; moreover increase in toxin production under different antimicrobials was attributed more to the strains producing Stx2 alone [172]. Bacteriostatic agents like roxithromycin, rokitamycin and clindamycin were found to suppress the release of Stx *in vitro* but not the number of viable strains, while exposure to cefdinir, fosfomycin or levofloxacin stimulated Stx release with the destruction of bacterial cells [173].

In case of Pakistani isolates, *in vitro* experimental data suggest that cefotaxime and gentamicin are safe at MIC level. However, we found that there is increase in toxin release and cytotoxicity at sub- MIC levels of ampicillin [174]. It is, therefore, suggested to avoid its use in STEC-related illness, and if proper diagnosis is not available, in all diarrhea cases. These findings are especially relevant to developing countries where, because of financial constrains, inadequate, low-dose self-treatment for insufficient period is common and usually the treatment is stopped as soon as the severity of symptoms subsides.

E. coli O157:H7 is known to produce exopolysaccharides (EPS) [175], which can provide a physical barrier to protect cells against environmental stresses. EPS is also involved in cell adhesion and biofilm formation [176]. EPS can serve as a conditioning film on inert surfaces, affect cell attachment by functioning as an adhesive or antiadhesive [177], and influence the formation of three-dimensional biofilm structures [178].

Extraintestinal *E. coli* (ExPEC)

Community acquired urinary tract infections (UTI) are highly prevalent in developing countries and are usually difficult to eradicate because the pathogenic bacteria have developed resistance to most of the drugs. UTI has been shown to be an independent risk factor for both bladder cancer and renal cell carcinoma [179]. Women are more likely to experience UTI than men. UTIs affect a large proportion of the world population and are responsible for significant morbidity and high medical costs [180, 181]. Uropathogenic *E. coli* (UPEC) cause 90% of urinary tract infections [28]. The frequent use of antimicrobials is considered the most important factor which promotes multiple drug resistance (MDR) in UPEC in both veterinary and human medicine [182].

Different pathotypes of UPEC can be identified by phylogenetic analysis. Phylogenetic studies have revealed that the UPEC are not of very diverse origins and fall into four main groups A, B1, B2, and D [183, 184]. Picard *et al.* [19] found that UPEC which correspond to phylogenetic group B2 were more susceptible to antimicrobials than those falling in A, B1 and D. Moreno and colleagues investigated that among human UPEC isolates, resistance to quinolones, fluoroquinolones and trimethoprim/sulfamethoxazole showed shift from phylogenetic group B2 towards groups A, B1 and/or D [185]. In a recent study, we have reported that among Pakistani isolates, group D isolates were highly drug resistant as compared to phylogenetic groups A, B1 and B2 which is contrary to the previous reports. This group was also found the most hemotoxic [186].

Uropathogenic strains of *Escherichia coli* (UPEC) account for 70-95% of the UTIs. Bacteria that invade the bladder cells and form biofilms may be responsible for many recurrent UTIs. Significant production of biofilm has been reported with some reports showing nearly $2/3^{rd}$ UPEC to produce biofilm [187] whereas others [188] claim biofilm production in more than 90.0%.

Vibrio

Antimicrobial treatments for one to three days shorten the course of the disease and reduce the severity of the symptoms. Patients recover without antimicrobial use if sufficient hydration is maintained [189]. Doxycycline is typically used as first line drug although some strains of *V. cholerae* have shown resistance. Testing for resistance during an outbreak can help determine appropriate future choices. Other antimicrobial proven to be effective include cotrimoxazole, erythromycin, tetracycline, chloramphenicol, and furazolidone. Fluoroquinolones, such as norfloxacin may also be used, but resistance has been reported [190].

Multiple antimicrobial resistant (MAR) *V. cholerae* with epidemic outbreaks (both classical and El Tor biotypes) have been reported in Bangladesh [191, 192].

Even though the reservations about the use of ciprofloxacin as a first line of treatment in such cases of MAR cholera have been expressed in developing countries [193], the high level resistance to nalidixic acid has led to the use of ciprofloxacin in pediatric cases. Subsequent reports of relapses and treatment failure led to the determination of MIC of ciprofloxacin, which was found to be high. This was responsible for the emergence of ciprofloxacin resistance in *V. cholerae* O1 Inaba [194]. Such resistance can be due to spontaneous mutation in *V. cholerae* or transfer of resistance from other co-inhabiting microbes, which are fluroquinolone resistant. The profiles of major MAR *V. cholerae* as documented in Kolkata and other parts of India and Bangladesh are: AFZ (Ampicilllin, Furazolidone), AFZN (Ampicillin, Furuzolidone, Neomycin), AFZ NS (Ampicillin, Furazolidone, Neomycin, Streptomycin) [195]. The antimicrobial resistance pattern of epidemic strains have changed frequently with the emergence of different *V. cholerae* O1 or O139 strains. Therefore, selection of such drug resistant clones can lead to seasonal epidemics of cholera with emergence of new clones replacing the existing clones.

V. cholerae become drug resistant by exporting drugs through efflux pumps, chromosomal mutations or developing genetic resistance *via* the exchange of conjugative plasmids, conjugative transposons, integrons or self transmissible chromosomally integrating SXT elements. *V. cholerae* use multidrug efflux pumps to export a broad range of antimicrobials, detergents and dyes that are chemically and structurally unrelated [70]. The two major groups of *V. cholerae* efflux pumps are distinguished by their energy sources: ATP hydrolysis, or the proton-motive force (PMF) of transmembrane H+ or Na+ gradients [196]. PMF pump families include MATE (multidrug and toxic compound extrusion), MFS (major facilitator superfamily), RND (resistance–nodulation–cell division) and SMR (small multidrug resistance) [70].

The spread of antimicrobial -resistant *V. cholerae* is also facilitated by horizontal gene transfer *via* self-transmissible mobile genetic elements, including SXT elements – mobile DNA elements belonging to the class of integrative conjugating elements (ICEs). Besides conferring antimicrobial resistance, SXT elements have the capacity to mobilize conjugative plasmids and genomic islands in trans [197], providing alternative mechanisms for antimicrobial resistance gene transfer.

Dissemination of antimicrobial resistance genes is also facilitated when V. cholerae cells share mobile integrons with other bacterial cells. All *V. cholerae* isolates harbor large chromosomal integrons, giving them the capacity to rapidly transfer gene cassettes containing antimicrobial resistance genes [59]. In addition, clinical and environmental *V. cholerae* can also contain mobile integrons, which are smaller (0–10 cassettes), but are embedded within mobile elements such as conjugative plasmids and transposons [59] and can disseminate horizontally.

Although virulence of Vibrios is mainly due to toxin production, the foothold to the bacteria is provided by biofilm formation. It also enables *V. cholerae* to survive in nutrient-poor conditions outside of the host [198]. Biofilm formation also increases the infectivity of *V. cholerae*, but, importantly, dispersal of biofilms is thought to occur once the bacteria colonize the host [198, 199].

Campylobacter

Campylobacter infections are among the most common causes of bacterial diarrhea in humans worldwide [200]. A recent study on illness and death due to foodborne infections in France estimated an isolation rate of 27–37/100,000 persons/year for Campylobacter infection [201]. Although the genus Campylobacter is composed of 18 described species [202], human illness is associated with thermophilic *Campylobacter*, primarily with *C. jejuni* and *C. coli* and infrequently with *C. upsaliensis, C. lari,* and *C. fetus.*

Drugs of choice for treating campylobacteriosis are erythromycin, quinolones, tetracycline, ampicillin, chloramphenicol and gentamicin. Nowadays there is a compelling evidence regarding an alarming increase in resistance of *Campylobacter* to antimicrobials administered in human treatment [203-205]. *Campylobacter* resistant strains have mainly emerged as a consequence of the use of antimicrobial agents in animal food production. Most of the strains are resistant to cloxacillin, nafcillin, oxacillin, sulfamethoxazole/trimethoprim, trimethoprim, and vancomycin [206].

In this scenario, fluoroquinolones have emerged as alternative therapy [207]. However, resistance to fluoroquinolones is also emerging [208]. Combined studies in humans and poultry have implicated the use of fluoroquinolones in poultry in the emergence of drug resistance [209].

It has been suggested that *C. jejuni* maintains itself in the environment by forming a biofilm [210]. *C. jejuni* has been found in preformed biofilms of other bacterial species [211]. *C. jejuni* in monoculture can attach to surfaces and form a biofilm, and can form a pellicle at both 37° and 30°C. It also forms a biofilm growing unattached and this aggregate biofilm has increased resistance to environmental stress. This may be relevant to the survival of the organism in the environment and in the epidemiology of *C. jejuni* infection [212].

ACKNOWLEDGEMENTS

The authors are grateful for support provided by following:

1. National Institute for Biotechnology and Genetic Engineering (NIBGE), Faisalabad.

2. Higher Education Commission (HEC), Pakistan.

3. Pakistan Science Foundation (PSF).

CONFLICT OF INTEREST

The author(s) has confirmed that there is no conflict of interest.

REFERENCES

[1]   Foxman B, Barlow R, D'Arcy H, Gillespie B, Sobel JD. Urinary Tract Infection:: Self-Reported Incidence and Associated Costs. Ann Epidemiol 2000; 10(8): 509-15.
[2]   Paterson DL. Resistance in gram-negative bacteria: Enterobacteriaceae. Am J Infect Control 2006; 34(5): S20-S28.
[3]   Aleksic S, Heinzerling F, Bockemuhl J. Human infection caused by Salmonellae of subspecies II to VI in Germany, 1977-1992. Zentralbl Bakteriol 1996; 283(3): 391-8.
[4]   Crump JA, Luby SP, Mintz ED. The global burden of typhoid fever. Bull. World Health Organ. 2004; 82(5): 346-53.
[5]   Brooks WA, Hossain A, Goswami D, *et al.* Bacteremic typhoid fever in children in an urban slum, Bangladesh. Emerg Infect Dis 2005; 11(2): 326-9.
[6]   Malla S, Kansakar P, Serichantalergs O, Rahman M, Basnet S. Epidemiology of typhoid and paratyphoid fever in Kathmandu: two years study and trends of antimicrobial resistance. JNMA J Nepal Med Assoc 2005; 44(157): 18-22.

[7]   French GL, King SD, Louis PS. Salmonella serotypes, Salmonella typhi phage types, and anti-microbial resistance at the University Hospital of the West Indies, Jamaica. J Hyg (Lond) 1977; 79(1): 5-16.

[8]   Sanborn WR, Vieu JF, Komalarini S, *et al.* Salmonellosis in Indonesia: phage type distribution of Salmonella typhi. J Hyg (Lond) 1979; 82(1): 143-53.

[9]   Jegathesan M. Phage types of Salmonella typhi isolated in Malaysia over the 10-year period 1970-1979. J Hyg 1983; 90: 91-97.

[10]   Arya SC. Salmonella typhi Vi antigen-negative isolates in India and prophylactic typhoid immunization. Natl Med J India 2000; 13(4): 220.

[11]   Baker S, Sarwar Y, Aziz H, *et al.* Detection of Vi-negative Salmonella enterica serovar typhi in the peripheral blood of patients with typhoid fever in the Faisalabad region of Pakistan. J Clin Microbiol 2005; 43(9): 4418-25.

[12]   Maskey AP, Basnyat B, Thwaites GE, Campbell JI, Farrar JJ, Zimmerman MD. Emerging trends in enteric fever in Nepal: 9124 cases confirmed by blood culture 1993-2003. Trans R Soc Trop Med Hyg 2008; 102(1): 91-5.

[13]   Khan FY, Kamha AA, Alomary IY. Fulminant hepatic failure caused by Salmonella paratyphi A infection. World J Gastroenterol 2006; 12(32): 5253-5.

[14]   Ali A, Haque A, Sarwar Y, Mohsin M, Bashir S, Tariq A. Multiplex PCR for differential diagnosis of emerging typhoidal pathogens directly from blood samples. Epidemiol Infect 2009; 137(1): 102-7.

[15]   Talukder KA, Khajanchi BK, Islam MA, *et al.* The emerging strains of Shigella dysenteriae type 2 in Bangladesh are clonal. Epidemiol Infect 2006; 134(6): 1249-56.

[16]   Sharma A, Singh SK, Bajpai D. Phenotypic and genotypic characterization of Shigella spp. with reference to its virulence genes and antibiogram analysis from river Narmada. Microbiol Res 2009.

[17]   Dutta S, Rajendran K, Roy S, *et al.* Shifting serotypes, plasmid profile analysis and antimicrobial resistance pattern of shigellae strains isolated from Kolkata, India during 1995-2000. Epidemiol Infect 2002; 129(2): 235-43.

[18]   Faruque SM, Khan R, Kamruzzaman M, *et al.* Isolation of Shigella dysenteriae type 1 and S. flexneri strains from surface waters in Bangladesh: comparative molecular analysis of environmental Shigella isolates *vs.* clinical strains. Appl Environ Microbiol 2002; 68(8): 3908-13.

[19]   Kaper JB, Nataro JP, Mobley HL. Pathogenic *Escherichia coli.* Nat Rev Microbiol 2004; 2(2): 123-40.

[20]   Muller D, Greune L, Heusipp G, *et al.* Identification of unconventional intestinal pathogenic Escherichia coli isolates expressing intermediate virulence factor profiles by using a novel single-step multiplex PCR. Appl Environ Microbiol 2007; 73(10): 3380-90.

[21]   Riley LW, Remis RS, Helgerson SD, *et al.* Hemorrhagic colitis associated with a rare Escherichia coli serotype. N Engl J Med 1983; 308(12): 681-5.

[22]   Karmali MA, Steele BT, Petric M, Lim C. Sporadic cases of haemolytic-uraemic syndrome associated with faecal cytotoxin and cytotoxin-producing Escherichia coli in stools. Lancet 1983; 1(8325): 619-20.

[23]   Johnson KE, Thorpe CM, Sears CL. The emerging clinical importance of non-O157 Shiga toxin-producing Escherichia coli. Clin Infect Dis 2006; 43(12): 1587-95.

[24] Elliott EJ, Robins-Browne RM, O'Loughlin EV, *et al.* Nationwide study of haemolytic uraemic syndrome: clinical, microbiological, and epidemiological features. Arch Dis Child 2001; 85(2): 125-31.

[25] Gerber A, Karch H, Allerberger F, Verweyen HM, Zimmerhackl LB. Clinical course and the role of shiga toxin-producing Escherichia coli infection in the hemolytic-uremic syndrome in pediatric patients, 1997-2000, in Germany and Austria: a prospective study. J Infect Dis 2002; 186(4): 493-500.

[26] WHO. Enterohaemorrhagic Escherichia coli (EHEC), Fact Sheet No. 125 (May 2005 (revised), http://www.who.int/mediacentre/factsheets/fs125/en/. 2005.

[27] Eisenstein BI, Jones GW. The spectrum of infections and pathogenic mechanisms of Escherichia coli. Adv Intern Med 1988; 33: 231-52.

[28] Johnson JR, Russo TA. Extraintestinal pathogenic Escherichia coli: "the other bad E coli". J Lab Clin Med 2002; 139(3): 155-62.

[29] Russo TA, Johnson JR. Proposal for a new inclusive designation for extraintestinal pathogenic isolates of Escherichia coli: ExPEC. J Infect Dis 2000; 181(5): 1753-4.

[30] WHO. Fact sheet N^0 107 August 2011. 2011.

[31] Todar K. http://textbookofbacteriology.net/cholera_1.html. 2005.

[32] Kandler O, Weiss N, Sneath P, Mair N, Sharpe M, Holt J. Bergey's manual of systematic bacteriology. 1986; 2.

[33] Howard SL, Jagannathan A, Soo EC, *et al.* Campylobacter jejuni glycosylation island important in cell charge, legionaminic acid biosynthesis, and colonization of chickens. Infect Immun 2009; 77(6): 2544-56.

[34] Kalra V, Chaudhry R, Dua T, Dhawan B, Sahu JK, Mridula B. Association of Campylobacter jejuni infection with childhood Guillain-Barre syndrome: a case-control study. J Child Neurol 2009; 24(6): 664-8.

[35] Ternhag A, Torner A, Svensson A, Ekdahl K, Giesecke J. Short- and long-term effects of bacterial gastrointestinal infections. Emerg Infect Dis 2008; 14(1): 143-8.

[36] Kaida K, Ariga T, Yu RK. Antiganglioside antibodies and their pathophysiological effects on Guillain-Barre syndrome and related disorders--a review. Glycobiology 2009; 19(7): 676-92.

[37] Finberg RW, Moellering RC, Tally FP, *et al.* The importance of bactericidal drugs: future directions in infectious disease. Clin Infect Dis 2004; 39(9): 1314-20.

[38] Kohanski MA, Dwyer DJ, Collins JJ. How antibiotics kill bacteria: from targets to networks. Nat Rev Microbiol 2010; 8(6): 423-35.

[39] Mulvey MR, Simor AE. Antimicrobial resistance in hospitals: How concerned should we be? Canadian Medical Association Journal 2009; 180(4): 408-15.

[40] Poole K. Resistance to beta-lactam antibiotics. Cell Mol Life Sci 2004; 61(17): 2200-23.

[41] Samaha-Kfoury JN, Araj GF. Recent developments in beta lactamases and extended spectrum beta lactamases. Bmj 2003; 327(7425): 1209-13.

[42] Walker RC. The fluoroquinolones. In: Mayo Clinic Proceedings; 1999: Elsevier; 1999. p. 1030-37.

[43] Andriole VT. Eds. The quinolones: Academic Press 2000.

[44] Hitchings GH. Mechanism of Action of Trimethoprim-Sulfamethoxazole. J Infect Dis 1973; 128(1 Suppl 3): S433-S36.

[45] Huovinen P. Trimethoprim resistance. Antimicrob Agents Chemother 1987; 31(10): 1451-6.

[46]   Skold O. Sulfonamide resistance: mechanisms and trends. Drug Resist Update 2000; 3(3): 155-60.

[47]   Bushby S, Hitchings G. Trimethoprim, a sulphonamide potentiator. Brit J Pharmacol Chemother 1968; 33(1): 72.

[48]   Durante-Mangoni E, Grammatikos A, Utili R, Falagas ME. Do we still need the aminoglycosides? Int J Antimicrob Ag 2009; 33(3): 201-05.

[49]   Standiford HC. Tetracycline and chloramphenicol, New York: Churchill Livingstone 1990.

[50]   Schnappinger D, Hillen W. Tetracyclines: antibiotic action, uptake, and resistance mechanisms. Arch Microbiol 1996; 165(6): 359-69.

[51]   Schlunzen F, Zarivach R, Harms J, *et al.* Structural basis for the interaction of antibiotics with the peptidyl transferase centre in eubacteria. Nature 2001; 413(6858): 814-21.

[52]   National Institute of Health (NIH). National Institute of Allergy and Infectious Diseases. Available   from: http://www.niaid.nih.gov/topics/antimicrobialResistance/Understanding/Pages/mutation.as px.

[53]   Hinnebusch BJ, Rosso ML, Schwan TG, Carniel E. High-frequency conjugative transfer of antibiotic resistance genes to Yersinia pestis in the flea midgut. Mol Microbiol 2002; 46(2): 349-54.

[54]   Toomey N, Monaghan A, Fanning S, Bolton D. Transfer of antibiotic resistance marker genes between lactic acid bacteria in model rumen and plant environments. Appl Environ Microbiol 2009; 75(10): 3146-52.

[55]   Furuya EY, Lowy FD. Antimicrobial-resistant bacteria in the community setting. Nat Rev Microbiol 2006; 4(1): 36-45.

[56]   Hawkey PM. The origins and molecular basis of antibiotic resistance. BMJ 1998; 317(7159): 657-60.

[57]   Frost LS, Leplae R, Summers AO, Toussaint A. Mobile genetic elements: the agents of open source evolution. Nature Rev Microbiol 2005; 3(9): 722-32.

[58]   Hall RM, Stokes HW. Integrons: novel DNA elements which capture genes by site-specific recombination. Genetica 1993; 90(2-3): 115-32.

[59]   Mazel D. Integrons: agents of bacterial evolution. Nature Rev Microbiol 2006; 4(8): 608-20.

[60]   Nikaido H. Multidrug resistance in bacteria. Annu Rev Biochem 2009; 78: 119.

[61]   Heidelberg JF, Eisen JA, Nelson WC, *et al.* DNA sequence of both chromosomes of the cholera pathogen Vibrio cholerae. Nature 2000; 406(6795): 477-83.

[62]   Roberts AP, Chandler M, Courvalin P, *et al.* Revised nomenclature for transposable genetic elements. Plasmid 2008; 60(3): 167-73.

[63]   Burrus V, Pavlovic G, Decaris B, Guedon G. Conjugative transposons: the tip of the iceberg. Mol Microbiol 2002; 46(3): 601-10.

[64]   Norman A, Hansen LH, Sorensen SJ. Conjugative plasmids: vessels of the communal gene pool. Philos T Roy Soc B 2009; 364(1527): 2275-89.

[65]   Abbani M, Iwahara M, Clubb RT. The structure of the excisionase (Xis) protein from conjugative transposon Tn916 provides insights into the regulation of heterobivalent tyrosine recombinases. J Mol Biol 2005; 347(1): 11-25.

[66]   Lipps G. Plasmids: Current research and future trends. Caister Academic Press. 2008.

[67]   Schumann W. Escherichia coli Cloning and Expression Vectors. Plasmids: current research and future trends 2008: 1.

[68]   Rychlik I, Gregorova D, Hradecka H. Distribution and function of plasmids in *Salmonella enterica*. Vet Microbiol 2006; 112(1): 1-10.

[69]   Van Bambeke F, Balzi E, Tulkens PM. Antibiotic efflux pumps. Biochem Pharm 2000; 60(4): 457-70.

[70]   Paulsen IT, Brown MH, Skurray RA. Proton-dependent multidrug efflux systems. Microbiol Rev 1996; 60(4): 575-608.

[71]   Lomovskaya O, Warren MS, Lee A, *et al.* Identification and characterization of inhibitors of multidrug resistance efflux pumps in Pseudomonas aeruginosa: novel agents for combination therapy. Antimicrob Agents Chemother 2001; 45(1): 105-16.

[72]   Hacker J, Kaper JB. Pathogenicity islands and the evolution of microbes. Annu Rev Microbiol 2000; 54: 641-79.

[73]   Ochman H, Lawrence JG, Groisman EA. Lateral gene transfer and the nature of bacterial innovation. Nature 2000; 405(6784): 299-304.

[74]   Hall-Stoodley L, Costerton JW, Stoodley P. Bacterial biofilms: from the natural environment to infectious diseases. Nat Rev Microbiol 2004; 2(2): 95-108.

[75]   Lear G, Lewis GD(editor). Microbial Biofilms: Current Research and Applications. Caister Academic Press. ISBN 978-1-904455-96-7. 2012.

[76]   Karatan E, Watnick P. Signals, regulatory networks, and materials that build and break bacterial biofilms. Microbiol Mol Biol Rev 2009; 73(2): 310-47.

[77]   Hoffman LR, D'Argenio DA, MacCoss MJ, Zhang Z, Jones RA, Miller SI. Aminoglycoside antibiotics induce bacterial biofilm formation. Nature 2005; 436(7054): 1171-5.

[78]   An D, Parsek MR. The promise and peril of transcriptional profiling in biofilm communities. Curr Opin Microbiol 2007; 10(3): 292-6.

[79]   Monroe D. Looking for chinks in the armor of bacterial biofilms. PLoS Biology 2007; 5(11): e307.

[80]   RMF. "Research on microbial biofilms (PA-03-047)". NIH, National Heart, Lung, and Blood Institute 2002-12-20. http://grants.nih.gov/grants/guide/pa-files/PA-03-047.html. 2002.

[81]   Rogers AH. Molecular Oral Microbiology. Caister Academic Press. pp. 65-108. ISBN 978-1-904455-24-0. http://www.horizonpress.com/oral2. 2008.

[82]   Imamura Y, Chandra J, Mukherjee PK, *et al.* Fusarium and Candida albicans biofilms on soft contact lenses: model development, influence of lens type, and susceptibility to lens care solutions. Antimicrob Agents Chemother 2008; 52(1): 171-82.

[83]   Parsek MR, Singh PK. Bacterial biofilms: an emerging link to disease pathogenesis. Annu Rev Microbiol 2003; 57: 677-701.

[84]   Davis SC, Ricotti C, Cazzaniga A, Welsh E, Eaglstein WH, Mertz PM. Microscopic and physiologic evidence for biofilm-associated wound colonization *in vivo*. Wound Repair Regen 2008; 16(1): 23-9.

[85]   Jacoby G, Bush K. ß-Lactam resistance in the 21st century. In White, D.G., Alekshun, M.N. & Mcdermott, P.F. (Eds.) Frontiers in antimicrobial resistance: a tribute to Stuart B. Levy. Washington DC, ASM Press. 2005.

[86]   Georgopapadakou N. Penicillin-binding proteins and bacterial resistance to beta-lactams. Antimicrob Agents Chemother 1993; 37(10): 2045.

[87]   Skold O. Resistance to trimethoprim and sulfonamides. Vet Res 2001; 32(3-4): 261-73.

[88]   Pattishall KH, Acar J, Burchall JJ, Goldstein F, Harvey RJ. Two distinct types of trimethoprim-resistant dihydrofolate reductase specified by R-plasmids of different compatibility groups. J Biol Chem 1977; 252(7): 2319.

[89]   Myllykallio H, Leduc D, Filee J, Liebl U. Life without dihydrofolate reductase FolA. Trends Microbiol 2003; 11(5): 220-23.

[90]   Perreten V, Boerlin P. A new sulfonamide resistance gene (sul3) in Escherichia coli is widespread in the pig population of Switzerland. Antimicrob Agents Chemother 2003; 47(3): 1169-72.

[91]   Hooper DC. Mechanisms of fluoroquinolone resistance. Drug Resist Updates 1999; 2(1): 38-55.

[92]   Byarugaba DK. Eds. Mechanisms of antimicrobial resistance, New York, Springer. 2009.

[93]   Hopkins KL, Davies RH, Threlfall EJ. Mechanisms of quinolone resistance in *Escherichia coli* and *Salmonella*: Recent developments. Int J Antimicrob Ag 2005; 25(5): 358-73.

[94]   Robicsek A, Jacoby GA, Hooper DC. The worldwide emergence of plasmid-mediated quinolone resistance. Lancet 2006; 6(10): 629-40.

[95]   Shaw K, Rather P, Hare R, Miller G. Molecular genetics of aminoglycoside resistance genes and familial relationships of the aminoglycoside-modifying enzymes. Microbiological Reviews 1993; 57(1): 138.

[96]   Magnet S, Blanchard JS. Molecular insights into aminoglycoside action and resistance. Chem Rev 2005; 105(2): 477-98.

[97]   Galimand M, Courvalin P, Lambert T. Plasmid-mediated high-level resistance to aminoglycosides in Enterobacteriaceae due to 16S rRNA methylation. Antimicrob Agents Chemother 2003; 47(8): 2565-71.

[98]   Guillaume G, Ledent V, Moens W, Collard JM. Phylogeny of efflux-mediated tetracycline resistance genes and related proteins revisited. Microb Drug Resist 2004; 10(1): 11-26.

[99]   Shaw WV. Chloramphenicol Acetyltransferase: Enzymology and Molecular Biolog. Cr Rev Bioch Mol 1983; 14(1): 1-46.

[100]   Schwarz S, Kehrenberg C, Doublet B, Cloeckaert A. Molecular basis of bacterial resistance to chloramphenicol and florfenicol. FEMS Microbiol Rev 2004; 28(5): 519-42.

[101]   Rowe B, Ward LR, Threlfall EJ. Multidrug-resistant Salmonella typhi: a worldwide epidemic. Clin Infect Dis 1997; 24(1 Suppl 1): S106-S09.

[102]   Caumes E, Ehya N, Nguyen J, Bricaire F. Typhoid and Paratyphoid Fever: A 10 Year Retrospective Study of 41 Cases in a Parisian Hospital. J Travel Med 2001; 8(6): 293-97.

[103]   Woodward TE, Smadel JE, Ley Jr HL, Green R, Mankikar D. Preliminary report on the beneficial effect of chloromycetin in the treatment of typhoid fever. Ann Intern Med 1948; 29(1): 131-34.

[104]   Colquhoun J, Weetch R. Resistance to chloramphenicol developing during treatment of typhoid fever. Lancet 1950: 621-3.

[105]   Paniker C, Vimala K. Transferable chloramphenicol resistance in Salmonella typhi. Nature 1972; 239: 109-10.

[106]   Gangarosa EJ, Bennett JV, Wyatt C, *et al.* An epidemic-associated episome? J Infect Dis 1972; 126(2): 215-18.

[107]   Butler T, Arnold K, Linh NN, Pollack M. Chloramphenicol-resistant typhoid fever in Vietnam associated with R factor. Lancet 1973; 302(7836): 983-85.

[108]   Kumar Y, Sharma A, Mani KR. Re-emergence of susceptibility to conventionally used drugs among strains of Salmonella Typhi in central west India. J Infect Dev Ctries 2011; 5: 227-30.

[109] Afzal A, Sarwar Y, Ali A, Haque A. Current status of fuoroquinolone and cephalosporin resistance in Salmonella enterica serovar Typhi isolates from Faisalabad, Pakistan. Pak J Med Sci 2012; 28(4): In press.

[110] Okeke IN, Laxminarayan R, Bhutta ZA, *et al.* Antimicrobial resistance in developing countries. Part I: recent trends and current status. Lancet 2005; 5(8): 481-93.

[111] Olarte J, Galindo E. Salmonella typhi resistant to chloramphenicol, ampicillin, and other antimicrobial agents: strains isolated during an extensive typhoid fever epidemic in Mexico. Antimicrob Agents Chemother 1973; 4(6): 597-601.

[112] Datta N, Olarte J. R factors in strains of Salmonella typhi and Shigella dysenteriae 1 isolated during epidemics in Mexico: classification by compatibility. Antimicrob Agents Chemother 1974; 5(3): 310-17.

[113] Wong CS. Beta lactamase inhibitors. Clin Microbiol Newsl 1988; 10(23): 177-80.

[114] Pokharel BM, Koirala J, Dahal RK, Mishra SK, Khadga PK, Tuladhar NR. Multidrug-resistant and extended-spectrum beta-lactamase (ESBL)-producing Salmonella enterica (serotypes Typhi and Paratyphi A) from blood isolates in Nepal: surveillance of resistance and a search for newer alternatives. Int J Infect Dis 2006; 10(6): 434-8.

[115] Michael GB, Butaye P, Cloeckaert A, Schwarz S. Genes and mutations conferring antimicrobial resistance in Salmonella: an update. Microbes Infect 2006; 8(7): 1898-914.

[116] Mulvey MR, Boyd DA, Baker L, *et al.* Characterization of a Salmonella enterica serovar Agona strain harbouring a class 1 integron containing novel OXA-type beta-lactamase (blaOXA-53) and 6'-N-aminoglycoside acetyltransferase genes [aac(6')-I30]. J Antimicrob Chemother 2004; 54(2): 354-9.

[117] Islam A, Butler T, Nath SK, *et al.* Randomized treatment of patients with typhoid fever by using ceftriaxone or chloramphenicol. J Infect Dis 1988; 158(4): 742-47.

[118] Garbarg-Chenon A, Vu TH, Labia R, *et al.* Characterization of a plasmid coding for resistance to broad-spectrum cephalosporins in Salmonella typhimurium. Drug Exp Clin Res 1989; 15(4): 145.

[119] Pontali E, Feasi M, Usiglio D, Mori M, Cassola G. Imported typhoid fever with hepatitis from Bangladesh: a case of delayed response to ceftriaxone? J Travel Med 2008; 15(5): 366-8.

[120] Rotimi VO, Jamal W, Pal T, Sovenned A, Albert MJ. Emergence of CTX-M-15 type extended-spectrum beta-lactamase-producing Salmonella spp. in Kuwait and the United Arab Emirates. J Med Microbiol 2008; 57(Pt 7): 881-6.

[121] Pfeifer Y, Matten J, Rabsch W. Salmonella enterica serovar Typhi with CTX-M β-lactamase, Germany. Emerg Infect Dis 2009; 15(9): 1533.

[122] Vongsthongsri U, Tharavanij S. Susceptibility of Salmonella typhi to chloramphenicol, ampicillin and cotrimoxazole. Southeast Asian J Trop Med Public Health 1980; 11: 256-61.

[123] Huovinen P, Sundstrom L, Swedberg G, Skold O. Trimethoprim and sulfonamide resistance. Antimicrob Agents Chemother 1995; 39(2): 279-89.

[124] Shanahan PM, Jesudason MV, Thomson CJ, Amyes SG. Molecular analysis of and identification of antibiotic resistance genes in clinical isolates of Salmonella typhi from India. J Clin Microbiol 1998; 36(6): 1595-600.

[125] Wain J, Hoa NT, Chinh NT, *et al.* Quinolone-resistant Salmonella typhi in Viet Nam: molecular basis of resistance and clinical response to treatment. Clin Infect Dis 1997; 25(6): 1404-10.

[126]   Parry C, Wain J, Chinh NT, Vinh H, Farrar JJ. Quinolone-resistant Salmonella typhi in Vietnam. Lancet 1998; 351(9111): 1289.

[127]   Mermin JH, Villar R, Carpenter J, *et al.* A massive epidemic of multidrug-resistant typhoid fever in Tajikistan associated with consumption of municipal water. J Infect Dis 1999; 179(6): 1416-22.

[128]   Mandal S, Mandal M, Pal N. Reduced minimum inhibitory concentration of chloramphenicol for Salmonella enterica serovar typhi. Indian J Med Sci 2004; 58(1): 16.

[129]   Capoor MR, Nair D, Walia NS, *et al.* Molecular analysis of high-level ciprofloxacin resistance in Salmonella enterica serovar Typhi and *S. Paratyphi* A: need to expand the QRDR region? Epidemiol Infect 2009; 137(6): 871-8.

[130]   Mohanty S, Gaind R, Paglietti B, Paul P, Rubino S, Deb M. Bacteraemia with pleural effusions complicating typhoid fever caused by high-level ciprofloxacin-resistant Salmonella enterica serotype Typhi. Ann Trop Paediatr 2010; 30(3): 233-40.

[131]   Rahman M. Treatment of enteric fever. ORION 2009; 32(3).

[132]   Abdullah FE, Haider F, Faima K, Irfan S, Iqbal MS. Enteric Fever in Karachi: Current antibiotic susceptibility of *Salmonellae* isolates. J Coll Physicians Surg Pak 2012; 22(3): 147-50.

[133]   Daga MK, Sarin K, Sarkar R. A study of culture positive multidrug resistant enteric fever--changing pattern and emerging resistance to ciprofloxacin. J Assoc Physicians India 1994; 42(8): 599-600.

[134]   Prabha Adhikari MR, Baliga S. Ciprofloxacin-resistant typhoid with incomplete response to cefotaxime. J Assoc Physicians India 2002; 50: 428-9.

[135]   Mandal S, Mandal MD, Pal NK. *In vitro* activity of gentamicin and amikacin against Salmonella enterica serovar Typhi: a search for a treatment regimen for typhoid fever. East Mediterr Health J 2009; 15(2): 264-8.

[136]   Reeve KE. Salmonella binding to and biofilm formation on cholesterol/gallstone surfaces in the chronic carrier state. Undergraduate Honors Thesis. School of Allied Medical Professions: The Ohio State University 2010.

[137]   Hoiby N, Bjarnsholt T, Givskov M, Molin S, Ciofu O. Antibiotic resistance of bacterial biofilms. Int J Antimicrob Agents 2010; 35(4): 322-32.

[138]   Jensen PO, Givskov M, Bjarnsholt T, Moser C. The immune system *vs.* Pseudomonas aeruginosa biofilms. FEMS Immunol Med Microbiol 2010; 59(3): 292-305.

[139]   Burmolle M, Thomsen TR, Fazli M, *et al.* Biofilms in chronic infections - a matter of opportunity - monospecies biofilms in multispecies infections. FEMS Immunol Med Microbiol 2010; 59(3): 324-36.

[140]   Raza A, Sarwar Y, Ali A, Jamil A, Haque A. Effect of biofilm formation on the excretion of Salmonella enterica serovar Typhi in feces. Int J Infect Dis 2011; 15(11): e747-52.

[141]   Tabak M, Scher K, Chikindas ML, Yaron S. The synergistic activity of triclosan and ciprofloxacin on biofilms of Salmonella Typhimurium. FEMS Microbiol Lett 2009; 301(1): 69-76.

[142]   Parry CM, Threlfall EJ. Antimicrobial resistance in typhoidal and nontyphoidal salmonellae. Curr Opin Infect Dis 2008; 21(5): 531-8.

[143]   Sood S, Kapil A, Dash N, Das BK, Goel V, Seth P. Paratyphoid fever in India: An emerging problem. Emerg Infect Dis 1999; 5(3): 483-4.

[144]   Chandel DS, Chaudhry R, Dhawan B, Pandey A, Dey AB. Drug-resistant Salmonella enterica serotype paratyphi A in India. Emerg Infect Dis 2000; 6(4): 420-1.

[145] Mohanty S, Renuka K, Sood S, Das BK, Kapil A. Antibiogram pattern and seasonality of Salmonella serotypes in a North Indian tertiary care hospital. Epidemiol Infect 2006; 134(5): 961-6.

[146] Woods CW, Murdoch DR, Zimmerman MD, *et al.* Emergence of Salmonella enterica serotype Paratyphi A as a major cause of enteric fever in Kathmandu, Nepal. Trans R Soc Trop Med Hyg 2006; 100(11): 1063-7.

[147] Adachi T, Sagara H, Hirose K, Watanabe H. Fluoroquinolone-resistant Salmonella Paratyphi A. Emerg Infect Dis 2005; 11(1): 172-4.

[148] Threlfall EJ, Fisher IS, Berghold C, *et al.* Trends in antimicrobial drug resistance in Salmonella enterica serotypes Typhi and Paratyphi A isolated in Europe, 1999-2001. Int J Antimicrob Agents 2003; 22(5): 487-91.

[149] Kuo CY, Su LH, Perera J, *et al.* Antimicrobial susceptibility of Shigella isolates in eight Asian countries, 2001-2004. J Microbiol Immunol Infect 2008; 41(2): 107-11.

[150] Ahmed SF, Riddle MS, Wierzba TF, *et al.* Epidemiology and genetic characterization of Shigella flexneri strains isolated from three paediatric populations in Egypt (2000-2004). Epidemiol Infect 2006; 134(6): 1237-48.

[151] Peirano G, Agerso Y, Aarestrup FM, dos Prazeres Rodrigues D. Occurrence of integrons and resistance genes among sulphonamide-resistant Shigella spp. from Brazil. J Antimicrob Chemother 2005; 55(3): 301-5.

[152] Hens DK, Niyogi SK, Kumar R. Epidemic strain Shigella dysenteriae Type 1 Dt66 encodes several drug resistances by chromosome. Arch Med Res 2005; 36(4): 399-403.

[153] Sunde M, Norstrom M. The prevalence of, associations between and conjugal transfer of antibiotic resistance genes in Escherichia coli isolated from Norwegian meat and meat products. J Antimicrob Chemother 2006; 58(4): 741-7.

[154] Pan JC, Ye R, Meng DM, Zhang W, Wang HQ, Liu KZ. Molecular characteristics of class 1 and class 2 integrons and their relationships to antibiotic resistance in clinical isolates of Shigella sonnei and Shigella flexneri. J Antimicrob Chemother 2006; 58(2): 288-96.

[155] Zafar A, Sabir N, Bhutta ZA. Frequency of isolation of shigella serogroups/serotypes and their antimicrobial susceptibility pattern in children from slum areas in Karachi. J Pak Med Assoc 2005; 55(5): 184-8.

[156] Phantouamath B, Sithivong N, Insisiengmay S, *et al.* Pathogenicity of Shigella in healthy carriers: a study in Vientiane, Lao People's Democratic Republic. Jpn J Infect Dis 2005; 58(4): 232-4.

[157] Sabir N, Zafar A. Cephalosporin resistant Shigella flexneri from a clinical isolate--a rare finding. J Pak Med Assoc 2005; 55(12): 560-1.

[158] Haukka K, Siitonen A. Emerging resistance to newer antimicrobial agents among Shigella isolated from Finnish foreign travellers. Epidemiol Infect 2008; 136(4): 476-82.

[159] Tariq A, Haque A, Ali A, Habeeb MA, Salman M, Sarwar Y. Molecular profiling of antimicrobial resistance and integron association of MDR clinical isolates of Shigella species from Faisalabad, Pakistan. Can J Microbiol 2012; In Press.

[160] Pazhani GP, Niyogi SK, Singh AK, *et al.* Molecular characterization of multidrug-resistant Shigella species isolated from epidemic and endemic cases of shigellosis in India. J Med Microbiol 2008; 57(Pt 7): 856-63.

[161] Bercion R, Njuimo SP, Boudjeka PM, Manirakiza A. Distribution and antibiotic susceptibility of Shigella isolates in Bangui, Central African Republic. Trop Med Int Health 2008; 13(4): 468-71.

[162] Pazhani GP, Ramamurthy T, Mitra U, Bhattacharya SK, Niyogi SK. Species diversity and antimicrobial resistance of Shigella spp. isolated between 2001 and 2004 from hospitalized children with diarrhoea in Kolkata (Calcutta), India. Epidemiol Infect 2005; 133(6): 1089-95.

[163] Ellafi A, Abdallah FB, Lagha R, Harbi B, Bakhrouf A. Biofilm production, adherence and morphological alterations of Shigella spp. under salt conditions. Ann Microbiol 2011; 61: 741-47.

[164] Schroeder CM, Meng J, Zhao S, *et al.* Antimicrobial resistance of Escherichia coli O26, O103, O111, O128, and O145 from animals and humans. Emerg Infect Dis 2002; 8(12): 1409-14.

[165] Bopp CA, Greene KD, Downes FP, Sowers EG, Wells JG, Wachsmuth IK. Unusual verotoxin-producing Escherichia coli associated with hemorrhagic colitis. J Clin Microbiol 1987; 25(8): 1486-9.

[166] Zhao S, White DG, Ge B, *et al.* Identification and Characterization of Integron-Mediated Antibiotic Resistance among Shiga Toxin-Producing *Escherichia coli* Isolates. Appl Environ Microbiol 2001; 67: 1558-64.

[167] Mora A, Blanco JE, Blanco M, *et al.* Antimicrobial resistance of Shiga toxin (verotoxin)-producing Escherichia coli O157:H7 and non-O157 strains isolated from humans, cattle, sheep and food in Spain. Res Microbiol 2005; 156(7): 793-806.

[168] Wong CS, Jelacic S, Habeeb RL, Watkins SL, Tarr PI. The risk of the hemolytic-uremic syndrome after antibiotic treatment of Escherichia coli O157:H7 infections. N Engl J Med 2000; 342(26): 1930-6.

[169] Panos GZ, Betsi GI, Falagas ME. Systematic review: are antibiotics detrimental or beneficial for the treatment of patients with Escherichia coli O157:H7 infection? Aliment Pharmacol Ther 2006; 24(5): 731-42.

[170] Pavia AT, Nichols CR, Green DP, *et al.* Hemolytic-uremic syndrome during an outbreak of Escherichia coli O157:H7 infections in institutions for mentally retarded persons: clinical and epidemiologic observations. J Pediatr 1990; 116(4): 544-51.

[171] Yoh M, Frimpong EK, Honda T. Effect of antimicrobial agents, especially fosfomycin, on the production and release of Vero toxin by enterohaemorrhagic Escherichia coli O157:H7. FEMS Immunol Med Microbiol 1997; 19(1): 57-64.

[172] Grif K, Dierich MP, Karch H, Allerberger F. Strain-specific differences in the amount of Shiga toxin released from enterohemorrhagic Escherichia coli O157 following exposure to subinhibitory concentrations of antimicrobial agents. Eur J Clin Microbiol Infect Dis 1998; 17(11): 761-6.

[173] Murakami J, Kishi K, Hirai K, Hiramatsu K, Yamasaki T, Nasu M. Macrolides and clindamycin suppress the release of Shiga-like toxins from Escherichia coli O157:H7 *in vitro*. Int J Antimicrob Agents 2000; 15(2): 103-9.

[174] Mohsin M, Haque A, Ali A, *et al.* Effects of ampicillin, gentamicin, and cefotaxime on the release of Shiga toxins from Shiga toxin-producing Escherichia coli isolated during a diarrhea episode in Faisalabad, Pakistan. Foodborne Pathog Dis 2010; 7(1): 85-90.

[175] Mao Y, Doyle MP, Chen J. Insertion mutagenesis of wca reduces acid and heat tolerance of enterohemorrhagic Escherichia coli O157:H7. J Bacteriol 2001; 183(12): 3811-5.

[176] Frank JF. Microbial attachment to food and food contact surfaces. Adv Food Nutr Res 2000; 43: 319-70.

[177] Ofek I, Doyle RJ. Bacterial adhesion to cells and tissues. Chapman and Hall, New York. 1994.

[178] Danese PN, Pratt LA, Kolter R. Exopolysaccharide production is required for development of Escherichia coli K-12 biofilm architecture. J Bacteriol 2000; 182(12): 3593-6.

[179] Parker AS, Cerhan JR, Lynch CF, Leibovich BC, Cantor KP. History of urinary tract infection and risk of renal cell carcinoma. Am J Epidemiol 2004; 159(1): 42-8.

[180] Carvalho RH, Gontijo FPP. Epidemiologically relevant antimicrobial resistance phenotypes in pathogens isolated from critically ill patients in a Brazilian Universitary Hospital. Braz J Microbiol 2008; 39(4): 623-30.

[181] Foxman B. Epidemiology of urinary tract infections: incidence, morbidity, and economic costs. Dis Mon 2003; 49(2): 53-70.

[182] Miles TD, McLaughlin W, Brown PD. Antimicrobial resistance of Escherichia coli isolates from broiler chickens and humans. BMC Vet Res 2006; 2: 7.

[183] Herzer PJ, Inouye S, Inouye M, Whittam TS. Phylogenetic distribution of branched RNA-linked multicopy single-stranded DNA among natural isolates of Escherichia coli. J Bacteriol 1990; 172(11): 6175-81.

[184] Selander RK, Caugant DA, Whittam TS. Genetic structure and variation in natural population of Escherichia coli. In: Escherichia coli and Salmonella typhimurium: cellular and molecular biology. ASM Press, Washingon, D.C. 1987: 1625-48.

[185] Moreno E, Prats G, Sabate M, Perez T, Johnson JR, Andreu A. Quinolone, fluoroquinolone and trimethoprim/sulfamethoxazole resistance in relation to virulence determinants and phylogenetic background among uropathogenic Escherichia coli. J Antimicrob Chemother 2006; 57(2): 204-11.

[186] Bashir S, Sarwar Y, Ali A, et al. Multiple drug resistance patterns in various phylogenetic groups of uropathogenic E. coli isolated from Faisalabad region of Pakistan. Braz J Microbiol 2011; 42: 1278-83.

[187] Sharma M, Yadav S, Chaudhary U. Biofilm production in uropathogenic Escherichia coli. Indian J Pathol Microbiol 2009; 52(2): 294.

[188] Suman E, Jose J, Varghese S, Kotian MS. Study of biofilm production in Escherichia coli causing urinary tract infection. Indian J Med Microbiol 2007; 25(3): 305-6.

[189] Sack DA, Sack RB, Chaignat CL. Getting serious about cholera. N Engl J Med 2006; 355(7): 649-51.

[190] Krishna BV, Patil AB, Chandrasekhar MR. Fluoroquinolone-resistant Vibrio cholerae isolated during a cholera outbreak in India. Trans R Soc Trop Med Hyg 2006; 100(3): 224-6.

[191] Faruque SM, Islam MJ, Ahmad QS, et al. An Improved Technique for Isolation of Environmental Vibrio cholerae with Epidemic Potential: Monitoring the Emergence of a Multiple-Antibiotic Resistant Epidemic Strain in Bangladesh. J Infect Dis 2006; 193(7): 1029-36.

[192] Siddique AK, Zaman K, Majumder Y, et al. Simultaneous outbreaks of contrasting drug resistant classic and el Tor Vibrio cholerae 01 in Bangladesh. Lancet 1989; 2(8659): 396.

[193] Khan WA, Begum M, Salam MA, Bardhan PK, Islam MR, Mahalanabis D. Comparative trial of five antimicrobial compounds in the treatment of cholera in adults. Trans R Soc Trop Med Hyg 1995; 89(1): 103-6.

[194] Das S, Goyal R, Ramachandran V, Gupta S. Fluoroquinolone resistance in Vibrio cholerae O1: emergence of El Tor Inaba. Ann Trop Paediatr 2005; 25(3): 211-12.

[195]   Faruque SM, Saha MN, Bag PK, *et al.* Genomic diversity among Vibrio cholerae O139 strains isolated in Bangladesh and India between 1992 and 1998. FEMS microbiology letters 2000; 184(2): 279-84.

[196]   Putman M, Van Veen HW, Konings WN. Molecular properties of bacterial multidrug transporters. Microbiol Mol Biol R 2000; 64(4): 672-93.

[197]   Daccord A, Ceccarelli D, Burrus V. Integrating conjugative elements of the SXT/R391 family trigger the excision and drive the mobilization of a new class of Vibrio genomic islands. Mol Microbiol 2010; 78(3): 576-88.

[198]   Zhu J, Mekalanos JJ. Quorum sensing-dependent biofilms enhance colonization in Vibrio cholerae. Developmental cell 2003; 5(4): 647-56.

[199]   Faruque SM, Biswas K, Udden SMN, *et al.* Transmissibility of cholera: *in vivo*-formed biofilms and their relationship to infectivity and persistence in the environment. P Natl Acad Sci 2006; 103(16): 6350-55.

[200]   Acheson D, Allos BM. Campylobacter jejuni infections: update on emerging issues and trends. Clin Infect Dis 2001; 32(8): 1201-06.

[201]   Vaillant V, Valk HD, Baron E, *et al.* Foodborne infections in France. Foodborne Pathog Dis 2005; 2(3): 221-32.

[202]   Vandamme P. Taxonomy of the family Campylobacteraceae. In: Nachamkin I, Blaser MJ, editors. Campylobacter. Washington (DC): ASM Press. pp. 3-26. 2000.

[203]   Aquino M, Pacheco A, Ferreira M, Tibana A. Frequency of isolation and identification of thermophilic campylobacters from animals in Brazil. Vet J 2002; 164(2): 159.

[204]   Avrain L, Humbert F, L'Hospitalier R, Sanders P, Vernozy-Rozand C, Kempf I. Antimicrobial resistance in Campylobacter from broilers: association with production type and antimicrobial use. Vet Microbiol 2003; 96(3): 267-76.

[205]   Butzler JP. Campylobacter, from obscurity to celebrity. Clin Microbiol Infec 2004; 10(10): 868-76.

[206]   Savasan S. Emergence of quinolone resistance among chicken isolates of Campylobacter in Turkey. Turk J Vet Anim Sci 2004; 28: 391-97.

[207]   Nachamkin I, Blaser MJ. Campylobacter. 2nd ed. ASM Press. Washington. 2000.

[208]   Lucey B, Cryan B, O'Halloran F, Wall P, Buckley T, Fanning S. Trends in antimicrobial susceptibility among isolates of Campylobacter species in Ireland and the emergence of resistance to ciprofloxacin. Veterinary record 2002; 151(11): 317-20.

[209]   Gupta A, Nelson JM, Barrett TJ, *et al.* Antimicrobial resistance among Campylobacter strains, United States, 1997-2001. Emerg Infect Dis 2004; 10(6): 1102-09.

[210]   Buswell CM, Herlihy YM, Lawrence LM, *et al.* Extended survival and persistence of Campylobacter spp. in water and aquatic biofilms and their detection by immunofluorescent-antibody and -rRNA staining. Appl Environ Microbiol 1998; 64: 733-41.

[211]   Keevil C. Rapid detection of biofilms and adherent pathogens using scanning confocal laser microscopy and episcopic differential interference contrast microscopy. Water Sci Technol 2003; 47(5): 105-16.

[212]   Joshua GWP, Guthrie-Irons C, Karlyshev A, Wren B. Biofilm formation in Campylobacter jejuni. Microbiology 2006; 152(2): 387-96.

Send Orders for Reprints to reprints@benthamscience.net

CHAPTER 8

The Potential of Bacteriophage Lysins in the Treatment of Gram-Positive Bacteria Including Multidrug Resistant Bacteria

Raymond Schuch[1], Vincent A. Fischetti[2], Hoonmo L. Koo[3] and David B. Huang[1,*]

¹ContraFect Corporation, Yonkers, New York, USA; ²Laboratory of Bacterial Pathogenesis, The Rockefeller University, USA; ³Baylor College of Medicine, USA

Abstract: Almost two million Americans per year develop hospital-acquired infections, resulting in 99,000 deaths, the vast majority of which are due to antibiotic-resistant pathogens. Because of the pressing public concern over the emergence and global spread of MDR Gram-positive bacteria, as well as the serious and life threatening nature of these diseases and the limitations of current available antibiotics, global efforts are now focused on the development of novel and alternative antibacterial. One very promising new class of antimicrobial agents includes members of a widespread family of bacteriophage-encoded, bacterial cell wall-hydrolytic enzymes, or lysins. Lysins represent a new class of antibacterial agents against Gram-positive bacteria including multidrug-resistant bacteria, with a mechanism of action distinct from antibiotics. In this review, we will describe the nature of phage lysins, how they are distinguished from antibiotics, and how they may be applied to human medicine. Finally, we will discuss the hurdles to developing a new antimicrobial class and bringing it from the lab bench to the market.

Keywords: Bacteriophage, Gram-positive bacteria, lysins, Multidrug Resistant Bacteria, multidrug-resistant.

INTRODUCTION

In the last 20 years, multidrug resistance to antibiotics has emerged among pathogenic Gram-positive bacteria. The multidrug-resistant (MDR) Gram-positive bacteria that pose the most significant public health challenges include methicillin-resistant *Staphylococcus aureus* (MRSA), vancomycin-intermediate *S. aureus* (VISA), linezolid-resistant *S. aureus* (LRSA), multi drug-resistant *Streptococcus pneumoniae*, and vancomycin-resistant *Enterococcus faecalis* and *E. faecium* (VRE) [1]. Other Gram-positive bacteria that pose significant public

Address correspondence to David B. Huang: ContraFect Corporation, 28 Wells Avenue, Third Floor, Yonkers, New York, 10701, USA; Tel: 914-207-2320; Fax: 914-207-2399 E-mail: dhuang@contrafect.com

health challenges include pathogenic streptococci (including Group A, B and G streptococci), *Listeria monocytogenes, Bacillus cereus*, and clostridial species.

In extreme cases, MDR Gram-positive bacteria have been identified that are either resistant to all licensed antibiotics or are susceptible only to very toxic antibiotics. In the face of such rapidly emerging and increasingly widespread MDR bacteria, the World Health Organization (WHO) has now warned of the potential for a return to the pre-antibiotics era for some Gram-positive pathogens [2]. There is no question that part of the problem is a significant gap between the burden of infections with MDR Gram-positive pathogens and the development of new antibiotics for these bacteria.

Almost two million Americans per year develop hospital-acquired infections (HAIs), resulting in 99,000 deaths, the vast majority of which are due to antibiotic-resistant pathogens [3]. Two common HAIs alone (sepsis and pneumonia) killed nearly 50,000 Americans and cost the U.S. health care system more than $8 billion in 2006 [4]. MRSA alone killed more Americans (approximately 19,000) than emphysema, HIV/AIDS, Parkinson's disease, and homicide combined [3].

Based on cost-of-infection studies, antibiotic-resistant pathogens cost the U.S. health care system $21 to $34 billion dollars annually and result in more than 8 million additional hospital days [5]. There are many reasons to support a conclusion that these figures correspond to an underestimate of the human and economic burden of infections due to antibiotic-resistant bacteria. For example, many of these cost-of-infection studies include a limited range of bacterial types, do not include outpatient infections, and do not take into account special patient care such as intensive care.

The increasing number of patients with serious and life-threatening MDR bacterial infections is frightening because these are infections against which we have few effective antibiotics available. Historically, from the 1940s up to the 1970s, the pharmaceutical industry provided a steady flow of new antibiotics, including several with new mechanisms of action that circumvented the problems caused by bacterial resistance to earlier agents [6]. Since then, only a few systemically-administered antibiotics (quinupristin-dalfopristin, linezolid, daptomycin, tigecycline, ceftaroline, telavancin), including two from new classes (oxazolidinones and lipopeptides) have been marketed to treat infections caused by MDR Gram-positive bacteria [7]. However, all of these available treatments are associated with limitations that include toxicity (Table **1**), resistance, commensal flora changes, and reduced activity on non-dividing bacteria and bacteria contained within biofilms.

Table 1: Marketed Intravenous Antibiotics for Gram-Positive Multidrug-Resistant Bacteria

	Vanco-mycin	Linezolid	Daptomycin	Quinupristin -Dalfopristin	Tigecycline	Ceftaroline	Telavancin
Drug class	Glyco-peptide	Oxazolidinone	Cyclic lipopeptide	Streptogramin	Glycyl-cycline	Cephalosporin	Lipo-glycopeptide
Indications	Bacteremia, IE, cSSTI, NP	cSSTI, NP	Bacteremia, rIE	cSSTI	cSSTI, cIAI	cSSTI	cSSTI
Route of administration	IV	IV, po	IV	IV	IV	IV	IV
Most common toxicity	Nephro-toxicity, red man syndrome	Myleo-suppresion, thrombo-cytopenia, lactic acidosis, optic/ peripheral neuropathy, serotonin syndrome	Elevated creatinine kinase, rhabdo-myolysis, peripheral neuropathies, nephropathy, hepatotoxicity	Venous irritation, arthralgia, myalgia, hyper-bilirubinemia	Nausea, vomiting	Diarrhea, nausea, rash	Potential terato-genicity, metallic taste, nausea, vomiting, headache, dizziness, rash, thrombocyto penia, QTc prolongation

Abbreviations: IV, intravenous; po, oral; cSSTI, complicated skin and soft-tissue infection; cIAI, complicated intraabdominal infection; NP, nosocomial pneumonia; IE, infective endocarditis; rIE, right-sided infective endocarditis.

Bacteria typically develop antibiotic resistance within only a few years after the start of clinical use [8]. Many of the antibiotics listed in Table **1** are broad spectrum and indiscriminately target both pathogenic and commensal bacteria. Broad-spectrum antibiotic killing in the human gut, for example, can kill beneficial bacteria and increase the risk of serious secondary conditions such as *Clostridium difficile*-associated diarrhea [9]. Furthermore, antibiotics generally do not have activity against non-dividing, stationary phase bacteria [10]. Antibiotics also have reduced activity against bacteria with biofilm matrices [11], which serve as physical or chemical diffusion barriers to antimicrobial penetration. Biofilms can form on medical implants and in infected tissues and cause difficult to treat conditions such as catheter related blood stream infections, ventilator associated pneumonia, endocarditis, septic arthritis and osteomyelitis [12, 13].

Because of the pressing public concern over the emergence and global spread of MDR Gram-positive bacteria, as well as the serious and life threatening nature of these diseases and the limitations of current available antibiotics, global efforts are now focused on the development of novel and alternative antibacterial tools. One very promising new class of antimicrobial agents includes members of a widespread family of bacteriophage-encoded, bacterial cell wall-hydrolytic enzymes, or lysins [14]. A large body of work has emerged over the past decade describing the identification and characterization of lysins and the application of lysin technology to control pathogens on human mucosal surfaces and in systemic infections [15]. In this review, we will describe the nature of phage lysins, how they are distinguished from antibiotics, and how they may be applied to human medicine. Finally, we will discuss the hurdles to developing a new antimicrobial class and bringing it from the lab bench to the market.

The Peptidoglycan Structure of Gram-Positive Bacteria

The peptidoglycan, or murein, layer of Gram-positive bacteria is a dense crystal lattice structure of peptide and glycan moieties that encases an organism, conferring strength and rigidity to the cell and protection from osmotic lysis [16]. The general scaffold of eubacterial peptidoglycan consists of a repeating *N*-acetylglucosamine (NAG)-*N*-acetylmuramic (NAM) disaccharide with a pentapeptide attached to the D-lactyl moiety of each NAM (Fig. **1**). The pentapeptide stem participates in interglycan cross-linking reactions to generate the final polymer. The number of residues that comprise mature, cross-linked peptidoglycan may vary among different organisms (particularly among Gram-positive organisms). For example, *Staphylococcus aureus* has a common Gram-positive structure, consisting of alternating β-1,4 linked NAG and NAM residues substituted with a pentapeptide consisting of L-alanine, D-isoglutamine, L-lysine, D-alanine, and D-alanine, that is modified in certain staphylococci during maturation by hydrolysis of the peptide bond between the terminal D-alanines and formation of a pentaglycine bridge between the lysine ε-amino group of one chain and the remaining D-alanyl carboxyl group of another chain [17]. The extreme mechanical strength of *S. aureus* cell walls ultimately depends on the high degree of cross-linking *via* pentaglycine bridges between adjacent stem peptides. Whether cross-linked by an amino bridge (as in staphylococci and other Gram-positive organisms) or by a direct peptide bond between the L-lysine (or meso-diaminopimelic acid for Gram-negative bacteria and some Gram-positive genera) at position three of one chain and the D-alanine at position 4 of another chain, the final peptidoglycan structure is ultimately cross-linked in three dimensions.

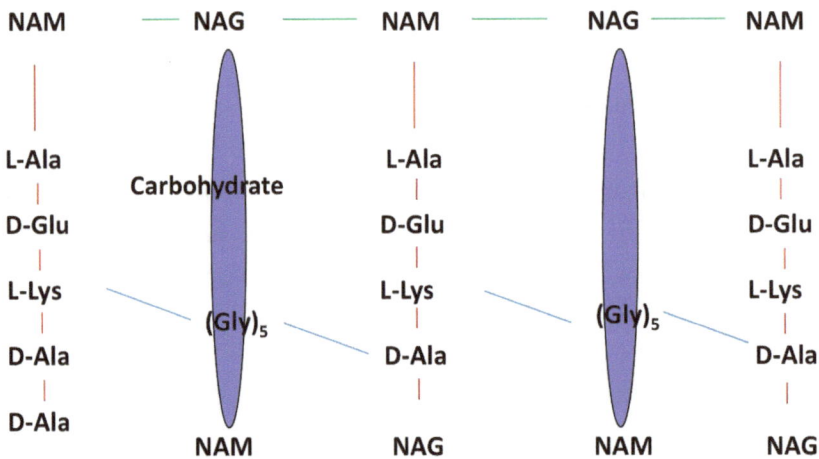

Figure 1: Peptidoglycan of Gram-positive bacteria.

Considering both the importance of peptidoglycan to bacterial growth and viability and the fact that peptidoglycan faces the extracellular environment, it is not surprising that many major antibiotic classes, including penicillins, cephalosporins, and glycopeptides, target the murein polymer [18, 19]. While most of these antibiotics either inactivate membrane-bound enzymes required for peptidoglycan biosynthesis or prevent the cross-linking of nascent, immature peptidoglycan, bacteriophage lysins are unique in that they target the mature and crosslinked peptidoglycan that comprise a majority of the cell wall.

Differences Between Lysins and Antibiotics

Lysins are distinct from antibiotics and, indeed, are now often referred to as "enzybiotics" [20]. Lysins are proteins, typically 25-40 kDa in size, encoded by the genomes of several major bacteriophage families including *Siphoviridae*, *Myoviridae*, *Tectiviridae*, and *Podoviridiae*. Antibiotics, on the other hand, are small molecules, usually <2 kDa, and may be based on products of bacterial secondary metabolism. In theory, lysins have advantages over antibiotics in certain significant areas including: 1) effectiveness against multidrug-resistant pathogenic bacteria; 2) a unique and very rapid bactericidal mechanism; 3) specificity for particular Gram-positive bacterial species and, consequently, minimal disturbance of the host microflora; 4) the low frequency of lysin resistance; and 5) low toxicity (Table **2**) [14, 21-23].

Table 2:   Differences Between Antibiotics and Lysins

Feature	Antibiotics	Lysins
Size	300-1600 Daltons	25,000-40,000 Daltons
Mechanism of action	Various	Cell wall hydrolysis
Onset of action	Slow; requires cell division	Immediate; does not require cell division
Spectrum	Broad	Narrow
Biofilm destruction	No	Yes
Potential for resistance	High	Low

As a group, lysins are distinguished by a very distinct mechanism of action. Lysins exert their effects by rapidly cleaving chemical bonds in cell wall peptidoglycan that are required for structural stability. The high internal osmotic pressure of the bacterial cytoplasm (approximately 3-5 atmospheres) is normally counterbalanced by densely cross-linked peptidoglycan. When the peptidoglycan structure is compromised, as a result of lysin-mediated hydrolysis for example, the cytoplasm will extrude and the bacterium undergoes hypotonic lysis (Fig. **3**). The ability of purified, recombinant lysins (when added exogenously to bacteria) to rapidly hydrolyze the surface peptidoglycan structure and allow extrusion of the cytoplasmic membrane and lysis (with concomitant loss of viability) is called "lysis from without", and is the hallmark feature of bacteriophage lysins [14, 24].

Loessner *et al.* [25] have shown that lysins bind with an constant (Ka = 3-6 x10^8) similar to affinity-matured antibodies, which suggests that lysins are "one-use" enzymes that do not disengage from the cell wall after binding and cleavage.

Multiple different studies confirm that binding, cleavage and lysis basically occurs upon contact. For example, nanogram quantities of PlyC lysin reduced a culture of 10^7 *Streptococcus pyogenes* cells by >6 logs of viability only seconds after enzyme addition [20]. Similarly, application of 2 units of PlyG (about 2 µg), a lysin specific for the anthrax agent *Bacillus anthracis*, mediated a >8 log decrease in viability within 3 minutes [24]. Similar findings have also been reported using lysins against *Enterococcus faecalis* and *E. faecium* [26], *Bacillus cereus* [27], *Streptococcus agalactiae* [28] and *Staphylococcus aureus* [29]. No known biological compounds, except chemical antiseptics like bleach and iodine, kill bacteria this quickly.

Peptidoglycan-Cleaving Agents in Development

The development and spread of bacterial resistance to traditional antibiotics has spurred research into new antimicrobials with novel modes of action and different cellular targets compared to antibiotics. Lysins are certainly unique in this respect, and are now the subject of increasingly intensive research efforts. Three main types of lysins are currently under development including the most common, phage-encoded endolysins, and two variations thereof including a bacterial enzyme called lysostaphin and a series of lysin-like domains encoded in bacteriophage virions.

Lysostaphin is not produced by a phage but secreted by a single strain of *Staphylococcus simulans* biovar *staphylolyticus* for the purpose of killing staphylococcal competitors in its ecological niche. It is a 27 kDa glycyl-glycine endopeptidase that cleaves the pentaglycine cross-bridge found in *S. aureus* (and the related organisms *S. epidermidis, S. haemolyticus,* and *S. saprophyticus*) leading to rapid bacterial lysis. In its secreted and mature form, lysostaphin consists of two functional domains: an N-terminal catalytic domain, which is a member of the M23 family of zinc metalloendopeptidases, and a C-terminal cell wall binding domain, which is a member of the SH3b family [30]. Lysostaphin binding to and cleavage of *S. aureus* peptidoglycan occurs very rapidly, driving >3 log reductions in viability of over 30 different *S. aureus* isolates treated for just 30 minutes [31]. The potential of lysostaphin as an antimicrobial is further highlighted by its potent synergistic effect in combination with any of a broad array of different antibiotics [32-34], and by its rapid clearance of *S. aureus* biofilms on polystyrene surfaces, catheter lines, and surgical meshes [35-38]. Furthermore, lysostaphin is efficacious *in vivo* against antibiotic-resistant *S. aureus* (*e.g.*, MRSA, VISA, and VRSA) in animal model systems for renal abscess lesions, keratitis, aortic valve endocarditis, eczema, sepsis, infected catheters, and nasal carriage [34, 36, 39, 40].

A major drawback for lysostaphin concerns the development of resistance to its activity. Exposure of *S. aureus* to low levels of lysostaphin (*i.e.*, below the minimum inhibitory concentration, or MIC) during *in vitro* growth or in the milieu of infected rabbits, led to the appearance of lysostaphin-resistant mutants at frequencies between 5.3×10^{-1} and 1.0×10^{-7} [41, 42]. Mechanisms of lysostaphin resistance vary, but the most common involves mutations at either the *femA* or *femB* locus which result in monoglycine or triglycine peptidoglycan cross-bridges that do not bind lysostaphin [43]. While it is unclear if lysostaphin resistance in *S. aureus* would ultimately result in treatment failures, these findings certainly favor the use of lysostaphin with antibiotics like β-lactams to reduce or eliminate the possibility of resistance.

Lysin-like murein hydrolytic domains can be found in the surface components of mature bacteriophages. These virion-associated, lysin-like activities are in the tail fibers, base plates, and head proteins of many phages, and serve to degrade peptidoglycan at early stages of infection to facilitate injection of phage DNA into the host. At least one such murein hydrolytic activity has now been described with activity directed toward *S. aureus* [44].

Bacteriophage Lysins

Bacteriophages (or more simply, phages), which are part of the normal human microbiome, are viruses that infect bacteria. It is estimated that there are a staggering 10^{31} phages in the biosphere, representing over 10^6 distinct species (of which only ~5,000 have been described) [45-47]. Furthermore, phage-mediated lysis is estimated to turn over roughly 40% of the bacterial population on earth every two days. Based on the ubiquity of phages in the environment and their success as obligate parasites of bacteria, it is certainly not surprising that phages and their encoded products have long been considered as sources of antimicrobial agents.

The lifecycle of a bacteriophage centers upon its interaction with and subversion of bacterial cell function. Indeed, every major bacterial activity (from transcription and translation to replication, DNA repair, and cell division) can be targeted by phage-encoded proteins that serve to disable or redirect host functions toward reproduction of phage particles. Once the new phages have been assembled, another set of phage-encoded proteins is expressed to then enable bacterial lysis and release of progeny into the environment. The double-stranded DNA phages mediate this bacterial lysis using a two-component system, whereby a phage-encoded holin makes a pore in the cytoplasmic membrane through which a phage-encoded lysin translocates into the peptidoglycan matrix [48]. Lysins are essentially cell wall hydrolases, which target the integrity of the cell wall by cleaving one of four major bonds in peptidoglycan. Once cleaved, bacterial peptidoglycan can no longer withstand the internal osmotic pressure imposed by the cytoplasm, and the organism rapidly ruptures.

Scientists have been aware of the lytic activity of phage for nearly a century. Indeed, whole phage therapy, relying on the lytic lifecycle of infective viruses to kill disease-causing bacteria, predates the discovery of penicillin by Fleming in 1928. It has only been over the last 12 years that lysins (the actual bacteriocidal component of phages) have received a great deal of interest as an antimicrobial agent [21]. Lysin therapy is essentially based on the ability of purified recombinant enzyme to elicit a very rapid "lysis from without" of target bacteria. While phages must use holin and lysin systems to trigger host lysis "from within", only the lysin is needed to drive lysis when applied from the outside. The ability of exogenous lysin to rapidly kill bacteria is limited to organisms with surface exposed peptidoglycan (*i.e.*, Gram-positive bacteria); with Gram-negative bacteria, the outer membrane protects subjacent peptidoglycan from the activity of exogenously applied lysins.

Lysins of phages infecting Gram-positive bacteria are generally between 25-40 kDa in size and display a modular organization of at least two distinct functional domains [14] (Fig. **2**). Generally, the N-terminal domain specifies one of four different enzymatic mechanisms based on cleavage sites in target peptidoglycan. Included among the catalytic (or CAT) domains are N-acetyl-β-D-glucosaminidases (glycosidases) and N-acetlymuramidases (lysozymes) which hydrolyze different glycosidic bonds in the glycan strand, endopeptidases which cleave peptide moieties, and N-acetylmuramoyl-L-alanine amidases which cleave the amide bond connecting the glycan moiety and the peptide moiety [49]. The C-terminal domain of most lysins specify binding to peptidoglycan ligands (Lu *et al.*, 2005) or secondary cell wall polymers like teichoic acids and neutral polysaccharides [24]. The cell wall binding domain (or CBD) often targets ligands that are restricted to species or even strain levels, thus imposing a level of specificity to the binding (and thus, activity) of lysins. While the CAT and CBD domains of lysins typically act in concert to mediate rapid and specific bacterial lysis, the CBD is dispensable in rare cases for both high-level lytic activity and specificity either *in vitro* [50, 51] or using *in vivo* models of infection [28, 52].

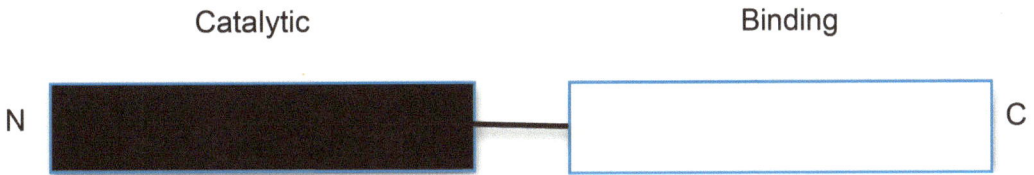

Figure 2: Two domain structure of phage lytic enzymes range from 25 to 40 kDa in size. The N-terminal catalytic domain cleaves one of the five major peptidoglycan bonds and may be a: N-acetylmuramidase, glucosaminidase, endopeptidase, γ-D-glutaminyl-L-lysine endopeptidase, or N-acetylmuramoyl-L-alanine amidase. The C-terminal binding domain binds with high affinity to cell wall-associated structures essential for bacterial growth and viability.

Lysin Efficacy

Generally, lysins only kill the species (or subspecies) of bacteria from which they were produced. For example, lysins produced from streptococcal phage kill certain streptococci, and lysins produced by pneumococcal phage kill pneumococci [53, 54]. In particular, a lysin from the C1 streptococcal phage (PlyC) kills group C streptococci, as well as groups A and E streptococci, the bovine pathogen *S. uberis* and the horse pathogen, *S. equi*, with no effect on streptococci normally found in the oral cavity of humans and other gram-positive

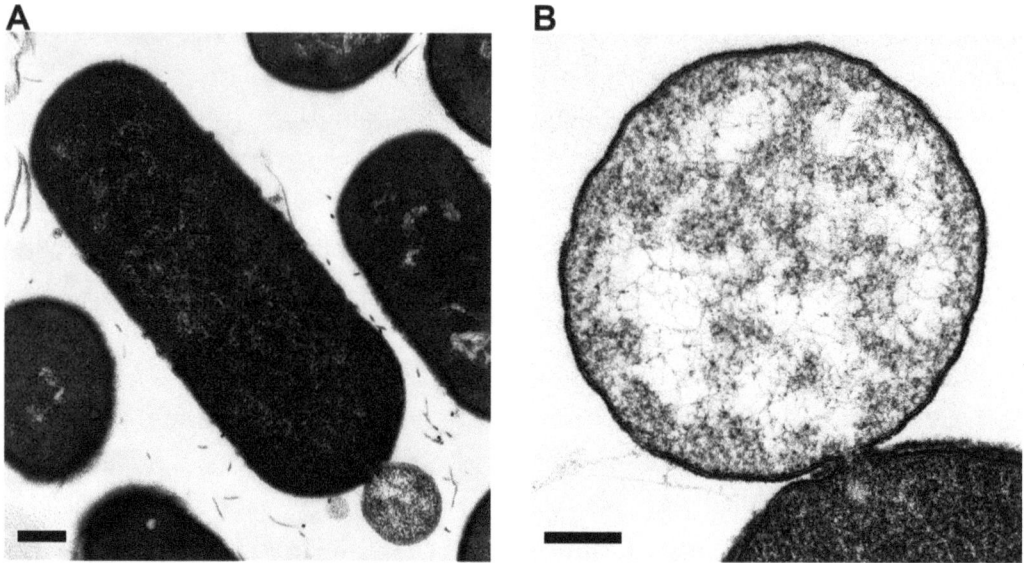

Figure 3: Transmission electron micrograph of *Bacillus anthracis* treated with the PlyG lysin. Images are captured at 5 (**A**) and 15 (**B**) minutes after lysin addition. In both images note the cytoplasm streaming out of the organism. Scale bars represent 200 nm and 100 nm in A and B, respectively.

bacteria. Similar results are seen with a pneumococcal-specific lysin, however in this case, the enzyme was also tested against strains of penicillin-resistant pneumococci and the killing efficiency was the same as 3 above [53, 55]. In contrast to antibiotics, which are usually broad spectrum and kill many different bacteria found in the human body (some of which are beneficial) lysins may be identified which kill only the disease organism with little to no effect on the normal human bacterial flora. One of the most specific lysins reported is the lysin for *B. anthracis* (PlyG); this enzyme only kills *B. anthracis* and rare but unique *B. cereus* strains [56]. Another highly specific lysin is a chimeric lysin for staphylococci called ClyS [57]. Because this enzyme is an endopeptidase that cleaves the peptidoglycan cross-bridge, and only staphylococci have polyglycine in their cross bridge, this enzyme was shown to have lytic activity on all staphylococci and no other species of bacteria tested [57]. In some cases however, phage enzymes may be identified with broad lytic activity. For example, an enterococcal phage lysin PlyV12 has recently been reported to not only kill enterococci but also a number of other gram-positive pathogens such as *S. pyogenes*, group B streptococci and *Staphylococcus aureus*, making it one of the broadest acting lysins identified [58]. However, its activity for these other pathogens was somewhat lower than for enterococci. Thus, the specificity of lysis may be either the result of the cell wall substrate bound by the lysin's binding domain or the unique cross-bridge for that bacterial species.

The Promise of Hybrid Lysins

The CAT and CBD domains of lysins are functionally distinct, meaning they can be swapped to create chimeric enzymes with altered catalytic activities and binding specificities. For example, Garcia *et al.* [59] swapped catalytic domains among *S. pneumoniae* phage lysins, resulting in novel enzymes with redirected cleavage specificities in the pneumococcal peptidoglycan. The ability to shuffle lysin domains brings the potential to create designer lysins with optimized functional properties related to solubility, thermostability, binding specificity and catalytic efficiency. For example, Daniel *et al.* [29] and Fernandes *et al.* [60] constructed chimeric lysins to solve the solubility problems associated with many anti-staphylococcal lysins. Here, Daniel *et al.* [29] fused a CAT domain of the *S. aureus* phage Twort lysin phage to the CBD of the *S. aureus* phage phiNM3 lysin, creating a highly soluble hybrid called ClyS with potent Staphylococcus-specific activity. Interestingly, ClyS was ultimately shown to be superior to mupirocin (a standard-of-care antibiotic) at eradicating staphylococcal skin infections in mice [61]. Fernandes *et al.* [60] similarly replaced the CAT of a poorly soluble anti-staphylococcal lysin Lys87 with the highly soluble CAT domains of two *Enterococcus faecalis* lyins Lys168 and Lys170, generating soluble and active anti-staphylococcal agents. Still other chimeric molecules are described that demonstrate improved catalytic efficiency. Schmelcher *et al.* [62] combined the CAT and CBD domains of *Listeria* phage lysins PlyPSA and Ply118 to not only demonstrate swapped binding properties among listerial serotypes, but also to generate a CAT(Ply118)-CBD(PlyPSA) fusion with a 3-fold increase in activity compared to either parental lysin. By combining multiple distinct CBDs recognizing different *Listeria* serotypes, Schmelcher *et al.* [62] ultimately created tandem CBD constructs able to recognize a broad range of *Listeria* strains. Based on these findings, one could envision a large toolbox of lysin domains to be mixed and matched in every combination to ultimately optimize potent antibacterial agents directed against any Gram-positive pathogen of interest.

Pharmacokinetics of Lysins

Very little data is published regarding the pharmacokinetics of lysins (*e.g.*, area under the curve, half-life, volume of distribution, clearance, kinetics, protein binding and other characteristics). A study showed that the half-life of lysins delivered systemically to animals is about 15-20 minutes [63]. If the half-life of animals is similar in humans, systemic administration of lysins will need to be modified to extend their half-life, or they may need to be delivered frequently (*e.g.*, tid or qid) or by constant IV infusion. However, modifications to the lysins may not prove to be a useful approach, because in a recent study it was shown that the addition of polyethylene glycol molecules of different size and placed at different locations in the Cpl-1 lysin resulted in the inactivation of the lytic activity of the molecule [64].

Absence of Bacterial Resistance to Lysins

While resistance mechanisms specific for lysostaphin have been reported, no such resistance to the activity of bacteriophage-encoded lysins has been described. Exposure of both pneumococci and staphylococci to low (*i.e.*, sub-MIC) concentrations of lysin does not lead to the recovery of resistant mutants, even after 40 daily passages on agar plates [53, 61]. Similarly, a lysin-sensitive *B. cereus* strain was serially passaged in liquid culture daily over a 10 day period (in the presence of a range of lysin concentrations) with no apparent resistance [24]. Even after mutagenesis of a lysin-sensitive *B. cereus* strain with a powerful DNA alkylating agent (and a concomitant 1,000 to 10,000-fold increase in streptomycin and novobiocin antibiotic resistance), no lysin-resistant derivatives were found. While these findings suggest that spontaneous mutations allowing lysin resistance will be rare, they cannot account for the acquisition of resistance by horizontal gene transfer in the manner described for high-level vancomycin resistance in staphylococci.

It is surprising that no bacterial modification has yet been identified that confers bacteriophage lysin resistance. Indeed, even the expression of thick polysaccharide capsules by vancomycin intermediate *S. aureus,* pneumococci, streptococci and *Bacillus anthracis* or the formation of dense biofilms by staphylococci or pneumococci cannot block lysin activity [20, 24, 28, 36, 53, 65,

66]. One theory for the apparent lack of resistance postulates that lysins evolved over eons to bind and cleave targets in host peptidoglycan that are essential for bacterial viability [14]. Thus, lysin-resistant mutants cannot be isolated because they cannot grow. This is supported by findings that lysin cell wall binding ligands in pneumococci (specifically, choline), streptococci (polyrhamnose), and *B. anthracis* (neutral polysaccharide) are required for viability [14, 24, 67].

These findings suggest that, unlike with antibiotic therapy, lysins will be less prone to the evolution of resistance and, thus, the loss of clinical efficacy. It thus stands to reason that if lysins are now added to antibiotic treatment regimens, even the prospect of antibiotic resistance will also be diminished. Combinations can also be envisioned that would "resensitize" MDR Gram-positive organisms to the antibiotics against which they have become resistant. Finally, the fact that lysins are so specific in their antibacterial activity, their use cannot drive the evolution of resistance in reservoirs of commensal bacteria (a problem associated with broad-spectrum antibiotics).

Synergy Between Lysins and Antibiotics

Three different methods are standardly used to determine synergy *in vitro*: 1) time kill assays in liquid culture, 2) disk diffusion on the surface of agar plates and 3) checkerboard broth microdilution analysis (also called isobolograms). Using each of these methods, a clear synergistic effect has been observed between lysins and multiple distinct antibiotics. For example, a combination of pneumococcal lysin Cpl-1 and gentamicin was increasingly synergistic in killing pneumococci strains with decreasing penicillin MICs, while Cpl-1 and penicillin showed synergy against a penicillin-resistant strain [68]. Synergistic lethal effects were also observed between lysin MV-L and glycopeptide antibiotics against vancomycin-intermediate *S. aureus* [69] and between the ClyS lysin and oxacillin against MRSA cells [29]. The right combination of lysin and antibiotic can, therefore, not only treat antibiotic resistant bacteria but also resensitize bacteria to antibiotics against which they have evolved resistance.

Synergistic effects have also been observed using *in vivo* models of infection and treatment. Daniel *et al.* [29] used combinations of the ClyS lysin and oxacillin, at doses of each agent that were not protective individually, to demonstrate a potent synergistic effect against MRSA septic death in a mouse model. Interestingly, even a combination of two anti-pneumococcal lysins with different cleavage specificities (*i.e.*, the Pal amidase and Cpl-1 lysozyme) was more effective in a murine septicemia model than each of the single enzymes alone [70]. The use of lysin combinations or lysin/antibiotic combinations may provide a duel benefit by reducing the emergence of enzyme- or antibiotic-resistant mutants and enabling the use of lower drug concentrations (thus lowering costs and reducing dose-related toxicity).

Biofilms

Biofilms are sessile bacterial communities that form on environmental surfaces ranging from healthy teeth, to central venous catheters and surgical meshes, to prosthetic devices and diseased bone and heart tissue. In biofilms, bacteria are protected within dense exopolysaccharide frameworks that result in 1,000- to 10,000-fold increases in resistance to antibiotic activity, compared to their planktonic, or free-living forms [71, 72]. The antibiotic resistance may be attributable to a combination of factors including the exopolysaccharide acting as a physical barrier to chemical assaults, the presence of bacteria in a highly resistant "persistor" state (defined by a slow metabolism and the absence of division), high local accumulations of antibiotic-detoxifying enzymes, and the efficient removal of antibiotics *via* biofilm channels normally used to exchange nutrients and waste products with the external environment.

A majority of work describing lysin activity against biofilms has focused on lysostaphin. Both Walencka *et al.* [73] and Wu *et al.* [35] demonstrated the effectiveness of lysostaphin in the eradication of dense biofilms formed by either *S. aureus* or *S. epidermidis* on artificial surfaces, including polystyrene dishes and catheters. Furthermore, Kun *et al.* [36] also showed that lysostaphin can be administered to mice *via* a jugular vein catheter to either prevent biofilm formation on the luminal surface of the catheter or to destroy mature biofilms already present on the luminal surface. In a second study of *in vivo* biofilms, Entenza *et al.* [74] administered the anti-pneumococcal Cpl-1 lysin as an infusion to reduce bacterial titers on heart valve vegetations (*i.e.*, biofilms) by >4 log

CFU/g in 2 hours. Biofilm clearance from cardiac valves was associated with the disappearance of pneumococci from circulation (a reduction of 10^5 CFU/ml of blood) and animal rescue in the rat endocarditis model. *In vitro*, the Pal, Cpl-1 and Cpl-7 phage lysins were all highly effective at either removing pneumococcal biofilms or killing pneumococci within mature biofilms on polystyrene surfaces [66].

In the case of severe staphylococcal bacteremia, antibiotics may be administered with apparent success as they eliminate most bacteria in the bloodstream, only to have a recurrence of infection within a few days. This failure likely results from residual staphylococci in a biofilm at a secondary site (bone, for example, in the case of osteomyelitis) which was resistant to the antibiotic and which served to reseed the infection. Thus a treatment that addresses both the acute cause of illness (planktonic bacteria in the blood) and chronic cause (biofilm bacteria in the bone) would be of great benefit. It is possible that antibiotic/lysin combinations may serve this purpose.

In Vitro Lysin Activity

The bacteriolytic activity of phage lysins have been characterized using a variety of *in vitro* assays. Crude assays of lysin activity are based on the simple clearance of bacterial lawns on agar petri dishes, whereby purified lysin is dropped onto the surface of a freshly-plated bacterial lawn; after a subsequent growth phase, the appearance of a clearing zone in the dense bacterial lawn denotes bacteriolytic activity of that lysin. The lysis of bacteria on agar plates actually forms the basis of all genetic screens used to identify bacteriophage lysins [24, 75]. More sophisticated techniques exist for examining lysin activity and involve: 1) turbidometric methods to follow decreases in the optical density of lysin-treated cultures, 2) luminescence-based methods to measure the ATP released from lysin-treated bacteria for quantification of the extent of bacteriolysis, and 3) variations of time-kill methods, whereby known quantities of bacteria and lysin are combined in liquid growth media, and bacterial viability is followed over time. Using these methods, the *in vitro* efficacy of a large number of phage lysins has been examined against a range of pathogenic and multidrug resistant Gram-positive bacteria, including *Enterococcus faecalis* and *faecium, Staphylococcus aureus* and *epidermidis, Streptococcus pneumoniae, pyogenes, agalactiae,* and *dysgalactiae, Bacillus cereus, anthracis* and *thuringiensis, Listeria monocytogenes,* and *Clostridium perfringens.* In every case examined, the *in vitro* efficacy of a particular lysin has been confirmed using *in vivo* models of animal infection.

In Vivo Lysin Activity

The majority of *in vivo* animal studies have focused on the anti-pneumococcal lysins Cpl-1 and Pal which have now been evaluated in mouse models for *S. pneumoniae* bacteremia, pneumonia, meningitis, endocarditis, otitis media and nasopharyngeal carriage. In the model for mucosal colonization, a dose-dependent effect was observed for the ability of the Pal amidase to eradicate pneumococci from the mouse nares within 5 hours of treatment [53, 63]. In a mouse model of pneumococcal bacteremia, Cpl-1 was administered every 12 hours after a 24 hour infection and 100% of the mice survived an otherwise fatal pneumonia and showed rapid recovery. Here, Cpl-1 dramatically reduced pulmonary bacterial counts, and prevented bacteremia [76]. In a study of experimental pneumococcal meningitis using infant Wistar rats, a single intracisternal injection (20 mg/kg) of Cpl-1 reduced pneumococci in the cerebrospinal fluid (CSF) by 3 logs within 30 minutes; here, intraperitoneal injection of Cpl-1 (200 mg/kg) also yielded a 2 log decrease in pneumococci found within the CSF [77]. In still another striking study, the authors used the Cpl-1 lysin to reduce intranasal pneumococci and ultimately prevent the development of acute otitis media (AOM) following infection with the influenza virus. In the AOM model, intranasal application of Cpl-1 (two-1 mg treatments, 4 hours apart) was 100% effective at preventing ear infections, while 80% of buffer-treated mice ultimately developed the secondary infection (AOM) [78]. Taken together, the pneumococcal lysins are effective at both eliminating pneumococci from mucosal surfaces and from infections of deeper tissues.

The efficacy of lysin-based therapy targeting a range of human pathogens has now been described. At least two lysins have been used in mouse models of anthrax. Intraperitoneal injection with 50 μg of PlyG, a *B. anthracis* phage lysin, rescued 68.4% of animals infected with an otherwise 100% lethal dose of bacilli.

A second lysin active against *B. anthracis* was reported by Yoong *et al.* [79]. This lysin was called PlyPH (so named because of its activity across a large pH range, from 4 to 10.5) and was found to protect 40% mice in the *i.p.* anthrax model compared to 100% death in control mice. Still another lysin, called PlyGBS [28], has been described with activity against group *B streptococci* (*Streptococcus agalactiae*) in either the vagina or oropharynx of mice. Mice treated intravaginally with 10 ug of PlyGBS demonstrated a 3-log drop in colonizing streptococci within 2 hours, while those treated either orally or intranasally demonstrated a 2-log drop in the viability of streptococci colonizing the upper respiratory tract. Many other examples of lysins used to either decolonize animals or to treat active infections have been reported, targeting pathogens that include group A streptococci [20], *Enterococcus faecalis* and *Enterococcus faecium* [26], and *Staphylococcus aureus* [29, 61]. Together, these findings present a large body of work that supports the use of phage lysins to treat a range of infections with Gram-positive bacterial pathogens.

Alternate Uses of Lysins

Numerous inventions have been patented regarding the production and application of lysins. The often highly pathogen-specific nature of lysin binding and lysin activity, suggest a possible use in the area of diagnostics. Indeed, fusions of lysin CBDs to magnetic beads are under development as a means to separate Listeria from contaminated food [80]. Alternatively, an assay based on bacteriolysis has been described whereby a targeted bacterium is killed by a lysin, and specific endogenous antigens are released and rapidly detected with antibodies specific for those antigens. Here, the lysin of *Listeria* phage phi LM4 is used to kill and detect *Listeria* in contaminated food samples. If *Listeria* is present, an intracellular biochemical such as ATP will be released as a result of bacteriolysis and can be detected by a luciferase reaction. By monitoring the photon release or by spectrophotometric methods, the biochemical components are detected and associated with target bacteria. A similar ATP-based method to detect *Bacillus anthracis* spores has also been described [24]. Lysins may also be used for DNA extraction by lysis of bacterial cells [81]. For example, a recombinant lysostaphin expressed in *E. coli* is sold commercially by Sigma-Aldrich and may be used for staphylococcal genetic studies, including DNA isolation, formation of protoplasts, and differentiation of staphylococcal strains [82].

Disinfection of medical tools and devices represents another practical application of lysins. Lysins targeting clinically relevant bacteria can serve to sterilize medical tools by preventing and reducing bacterial colonization on these surfaces and thus destroying the bacterial biofilms, which are responsible for infections associated with devices such as catheters, heart valves and joint replacements.

Identifying and Isolating New Lysins

Several ways have been developed to identify new lysins. The simplest is to identify a phage, shotgun clone its DNA and identify lytic activity by overlaying the plated clones with the phage-sensitive bacterium. This methodology allows for identification of a lysin for the organism or species that the phage infects. A more general way of isolating lysins in order to understand the diversity in this class of enzymes in the environment is through functional metagenomic analysis [83]. This technique uses random environmental phage populations (from water, soil, or feces) processed for metagenomic analysis. The twist here is to add an amplification step and an expression step to express and produce the products of the isolated lysin genes. This approach has the potential of identifying novel lysins en masse with powerful biotechnological value. Another approach, which combines the general and the specific approach mentioned above, is to exploit the lysogens in a host genome. This approach, termed multigenomics, identifies the lysin genes in the lysogens within many strains of the same species. In this case the DNA from tens to hundreds of strains of the same species is processed as in the metagenomic analysis, except here the enzymes are from the variety of lysogens in the single species [84].

The Use of Lysin Preparations for Food Decontamination

The specificity of bacteriophages renders them ideal candidates for applications designed to increase food safety during the production process as phages can be used for biocontrol of bacteria without interfering with the natural microflora or the cultures in fermented products. An advantage of phage decontamination is that phage compositions are non-toxic and do not alter color, texture or taste of food to be decontaminated [85]. The U.S. Food and Drug Administration (FDA) approved the use of a bacteriophage-based additive (a cocktail of six bacteriophages to be sprayed on ready-to-eat meat) for the control of *Listeria monocytogenes* contamination [86]. A number of animal studies have revealed the potential of bacteriophage for the control of various foodborne pathogens within the animal gastrointestinal tract and to subsequently decrease the likelihood of foodborne outbreaks. From a biopreservative perspective, phage have a number of key properties, including relative stability during storage, an ability to self-replicate, and a nontoxic nature which make lysins a potential use in the biocontrol agent for food safety applications [23].

Mastitis, an inflammatory reaction of the mammary gland that is usually caused by microbial infection, is recognized as the most costly disease in dairy cattle due to rejected milk, degraded milk quality, early culling of cows, drug costs, veterinary expenses and increase labor costs for farmers [87]. In the U.S., annual losses are estimated at two billion dollars. This is approximately 10% of the total value of farm milk sales, and about two-thirds of these losses are due to reduced milk production in subclinical infected cows. *S. aureus and S. uberis* are the major Gram-positive bacteria responsible for mastitis [88]. Intramammary infections caused by *S. aureus* are usually chronic and subclinical in nature. Many drugs have been used for mastitis treatment, but several factors, including the sponge-like construction of the utter, the ability of *S. aureus* to survive inside neutrophils and to induce formation of microabscesses, and its resistance to the antibiotics used for treatment, result in infections that are difficult to manage therapeutically. Lysins have the potential to be efficacious in the treatment of bovine mastitis caused by Gram-positive bacteria including staphylococci. Several excellent studies now describe the development of multiple distinct bacteriophage lysins for the treatment of both streptococcus and staphylococcus in infected milk and in a mammary gland environment [89].

Clinical Potential for the Lysins

In general, lysins kill the species of bacteria from which they were produced. Unlike antibiotics, which are usually broad spectrum and kill many different bacteria found in the human body, some of which are beneficial, lysins typically kill only the bacteria from which they were produced. Therefore, lysins have little to no effect on the normal human bacteria flora because of their narrow spectrum of activity. There are some lysins, however, that may be identified with broad lytic activity [90].

Lysins specifically kill Gram-positive bacteria on mucous membranes without affecting the surrounding normal flora, thus reducing a significant pathogen reservoir in the population. The majority of human infections caused by bacteria begin at a mucous membrane sites such as the upper and lower respiratory, intestinal, urogenital and ocular regions [91, 92]. The human mucous membranes are the reservoir for many pathogenic bacteria, such as pneumococci, enterococci, staphlylococci, streptococci, found in the environment. Many of these pathogenic bacteria are resistant to antibiotics. With the exception of polysporin and mupirocin ointments, which are the most widely used topically, there are no antibiotics that are designed to control colonizing pathogenic bacteria on mucous membranes. Because of the increase in MDR Gram-positive bacteria, antibiotics are not indicated to control the carrier state of pathogenic bacteria. However, lysins may prevent infection by safely and specifically destroying pathogenic Gram-positive bacteria on mucous membranes, thereby minimizing systemic absorption. This may be especially useful in hospitals and nursing homes. Animal studies have shown that enzymes specific for *S. pneumoniae* and *S. pyogenes* may be used nasally and orally to control these organisms in the community as well as in nursing homes and hospital to prevent or markedly reduce the serious infections that these bacteria can cause. Importantly, lysins have the clinical potential to treat serious life threatening infections, which include bacteremia, infective endocarditis, catheter related blood stream infections, hospital acquired and community acquired pneumonia, meningitis, localized or diffuse skin infections such as necrotizing fasciitis, and hospital acquired diarrheal illnesses.

Regulatory Challenges to the Development of Phages and Lysins

Nonhuman applications of phages and lysins are already implemented in the United States. These applications include use in the food safety, agriculture, animal husbandry and veterinary medicine (*e.g.*, mastitis among cattle), waste-water treatment, and environmental remediation [23, 85, 93-97]. The initial approved uses of phages in humans have been for topical administration or for elimination of colonization with multi-drug resistant Gram-positive bacteria [98, 99]. A few companies, including French companies Le LAboratoire du Bacteriophage and L'Oreal, the German company Antipiol and the German Bacteriophage Society, and American companies Eli-Lilly, Swan-Myers of Abbot Laboratories, Squibb and Sons (now belonging to Bristol-Myers, Squibb Company) and the Parke & Davis Company (now a part of Pfizer) manufactured and/or marketed a variety of different phage preparations to treat various infections including abscesses, suppurating wounds, vaginitis, and infections of the upper respiratory tract [100]. Unlike topical administration of lysins, the systemic administration of lysins for the treatment of deep infections will likely be met with significant regulatory obstacles in the United States since there are no published US Food and Drug Administration (FDA) guidelines about the evaluation of the safety of lysin therapy. However, the FDA has proposed a new FDA approval mechanism, called "Special Population Limited Medical Use (SPLMU) Drug" which would provide an important new approval pathway option for Sponsors interested in and able to develop drugs to treat patients with the most serious infections where few or no therapeutic options exist [5]. The initial systemic use of lysin therapy would likely be limited, in the majority of instances, to circumstances in which the pathogen is known, as adjunct or combination therapy and, preferably, its *in vitro* susceptibility to the available lysin(s) is also known. Because of the specificity of lysins to specific Gram-positive bacteria, adjunct or combination therapy of antibiotic plus lysin would likely initially be used as step-down therapy after an empiric broad spectrum antibiotic therapy was initiated while culture and susceptibility was pending. Initial indications for lysin will likely be MDR Gram-positive bacteria especially those that are difficult to eradicate, such as those growing in biofilm. Although the Eastern Europe/former Soviet Union experience seems to strongly support the notion that phage therapy is in general a safe and effective treatment of bacterial infections, there is much to be done in the western countries to bring it to the market. It is highly expected that as more safe therapies are developed, more evidences regarding the safety and efficacy of lysins therapy are unveiled by pre-clinic studies, in combination with the rapid increase of pressure from multi-resistant bacterial pathogenic strains, lysin therapy will be accepted by the public and the medical authorities of the western world in the future.

Potential Safety Issues for Lysins

There are no known published toxicities, including allergic hypersensitivity, to lysins. Nonetheless, there are potential safety concerns with the use of lysis such as 1) neutralization of lysins by the host immune system leading to failure of lysin therapy; and 2) rapid cell lysis of bacteria resulting in the release of large amount of membrane-bound endotoxins in the hosts.

A potential concern about lysin therapy is the generation of neutralizing antibodies during a treatment course that could hinder such a treatment or future applications of the enzyme. Unlike antibiotics, which are generally non-immunogenic small molecules, enzymes are proteins that stimulate an immune response when delivered mucosally or systemically, which could neutralize lysin activity. To examine this, rabbit hyperimmune serum has been raised against lysins specific for *B. anthracis, S. pyogenes, S. pneumoniae* and *S. aureus* and shown, in each case, to only slow, but not block, the lysin activity *in vitro* [63, 69]. Similarly, *in vivo*, Jado *et al.* [76] showed that antibodies generated in mice during treatment with either of two therapeutic pneumococcal lysins, did not affect the efficacy of treatment. These results may be partially explained by the fact that the binding affinity of a lysin for it cell wall substrate is higher than that of IgG antibodies for the lysin. Thus, antibodies will not be neutralizing. Collectively, these results are encouraging since it suggests that such enzymes can be used repeatedly to treat Gram-positive bacteria in life-threatening illnesses such as bacteremia.

Because lysins kill Gram-positive bacteria on contact, release of bacterial debris including immunostimulatory peptidoglycan and toxins could cause an increase in cytokine production and a host inflammatory response. To examine this, a mouse model of pneumonia was used in which trans-nasal infection with pneumococci causes an increase of proinflammatory cytokines and chemokines within 36 hours [101]. In this model, treatment with Cpl-1 at 24 hour post-infection actually resulted in a profound decrease of IL-1b, IL-6, and the chemokines KC, MIP-1a, MCP-1, and G-CSF in the lung. Decreases of IL-1, KC, MIP-1a, MCP-1, G-CSF, and IFNγ concentrations were similarly observed in the plasma. Pulmonary and systemic IL-10 synthesis was only found in septic animals 60h after infection and was completely prevented by Cpl-1 treatment. These findings suggest that lysin treatment will not necessarily trigger cytokine releases during the treatment course.

CONCLUSIONS

Lysins represent a new class of antibacterial agents, against Gram-positive bacteria including multidrug-resistant bacteria, with a mechanism of action distinct from antibiotics. *In vitro* and *in vivo* studies have shown the lysins are bacteriolytic on contact, providing a fast kill not seen with conventional antibiotics. Lysins have narrow spectrum activity and may not affect commensal flora. Compared to conventional antibiotic therapy, to date, resistance to lysins is very low. Lysins are amenable to protein engineering which may be used to expand their spectrum of antimicrobial activity, reduce their size, increase their half-life, improve activity, and generate large libraries of anti-infective agents specific for any pathogen of interest. Given the pressing need for new anti-infectives to combat infections caused by Gram-positive organisms, including multi-drug resistant pathogens, and to overcome the problem of resistance to conventional antibiotics, lysins are a welcome addition to the treatment options for the management of life-threatening infections caused by Gram-positive bacteria.

ACKNOWLEDGEMENTS

There are no acknowledgments to be made to this manuscript.

CONFLICT OF INTEREST

The authors confirm that this chapter content has no conflict of interest.

AUTHOR DISCLOSURE STATEMENT

VAF is a shareholder in ContraFect Corporation. VAF and HLK were not paid for their contributions to this manuscript. RS and DBH are employees and share holders of ContraFect Corporation.

REFERENCES

[1]   Rivera AM, Boucher HW. Current concepts in antimicrobial therapy against select gram-positive organisms: methicillin-resistant *Staphylococcus aureus*, penicillin-resistant pneumococci, and vancomycin-resistant enterococci. Mayo Clin Proc 2011; 86(12): 1230-43.

[2]   WHO Global Strategy for containment of antimicrobial resistance. World Health Organization 2001.

[3]   Klevens RM, Morrison MA, Nadle J, *et al.* Invasive methicillin-resistant *Staphylococcus aureus* infections in the United States. JAMA 2007; 298(15): 1763-71.

[4]   Eber MR, Laxminarayan R, Perencevich EN, Malani A. Clinical and economic outcomes attributable to health care-associated sepsis and pneumonia. Arch Intern Med 2010; 170(4): 347-53.

[5]   IDSA, Promoting Anti-Infective Development and Antimicrobial Stewardship through the U.S. Food and Drug Administration, Prescription Drug User Fee Act (PDUFA) Reauthorization. Before the House Committee on Energy and Commerce's Subcommittee on Health 2012.

[6]   Projan SJ, Youngman PJ. Antimicrobials: new solutions badly needed. Current Opinion in Microbiology 2002; 5(5): 463-5.

[7]   Antibiotic Resistance 2012 [homepage on the Internet]. The Antibiotic Development Pipeline and Strategies to Combat Antibiotic Resistance. [cited 14th January 2013]. Available from: ReportLinker.com

[8]   Pechere JC. Patients' interviews and misuse of antibiotics. Clin Infect Dis 2001; 33 Suppl 3: S170-3.

[9]   Noren T. *Clostridium difficile* and the disease it causes. Methods Mol Biol 2010; 646: 9-35.

[10]   Allison KR, Brynildsen MP, Collins JJ. Heterogeneous bacterial persisters and engineering approaches to eliminate them. Curr Opin Microbiol 2011; 14(5): 593-8.

[11]   Arciola CR, Campoccia D, Speziale P, Montanaro L, Costerton JW. Biofilm formation in Staphylococcus implant infections. A review of molecular mechanisms and implications for biofilm-resistant materials. Biomaterials 2012; 33(26): 5967-82.

[12]   Hall-Stoodley L, Stoodley P. Evolving concepts in biofilm infections. Cell Microbiol, 2009; 11(7): 1034-43.

[13]   Burmolle M, Thomsen TR, Fazil M, Dige I, Christensen L, Homoe P, *et al.* Biofilms in chronic infections - a matter of opportunity - monospecies biofilms in multispecies infections. FEMS Immunol Med Microbiol 2010; 59(3): 324-36.

[14]   Fischetti VA, Nelson D, Schuch R. Reinventing phage therapy: are the parts greater than the sum? Nat Biotechnol 2006; 24(12): 1508-11.

[15]   Nelson DC, Schmelcher M, Rodriguez-Rubio L, Klumpp J, Pritchard DG, Dong S, *et al.* Endolysins as antimicrobials. Adv Virus Res 2012; 83: 299-365.

[16]   Silhavy TJ, Kahne D, Walker S. The bacterial cell envelope. Cold Spring Harb Perspect Biol 2010; 2(5): a000414.

[17]   Gally D. Archibald AR. Cell wall assembly in *Staphylococcus aureus*: proposed absence of secondary crosslinking reactions. J Gen Microbiol 1993; 139(8): 1907-13.

[18]   Gautam A, Vyas R, Tewari R. Peptidoglycan biosynthesis machinery: a rich source of drug targets. Crit Rev Biotechnol 2011; 31(4): 295-336.

[19]   Bugg TD, Braddick D, Dowson CG, Roper DI. Bacterial cell wall assembly: still an attractive antibacterial target. Trends Biotechnol 2011; 29(4): 167-73.

[20]   Nelson D, Loomis L, Fischetti VA. Prevention and elimination of upper respiratory colonization of mice by group A streptococci by using a bacteriophage lytic enzyme. Proc Natl Acad Sci U S A 2001; 98(7): 4107-12.

[21]   Fischetti VA. Bacteriophage endolysins: a novel anti-infective to control Gram-positive pathogens. Int J Med Microbiol 2010; 300(6): 357-62.

[22]   O'Mahony J, Fenton M, Henry M, Sleator RD, Coffey A. Lysins to kill - a tale of viral weapons of mass destruction. Bioeng Bugs 2011; 2(6): 306-8.

[23]   Coffey B, Mills S, Coffey A, McAuliffe O, Ross RP. Phage and their lysins as biocontrol agents for food safety applications. Annu Rev Food Sci Technol 2010; 1: 449-68.

[24]   Schuch R, Nelson D, Fischetti VA. A bacteriolytic agent that detects and kills Bacillus anthracis. Nature 2002; 418(6900): 884-9.

[25]   Loessner MJ, Kramer K, Ebel F, Scherer S. C-terminal domains of Listeria monocytogenes bacteriophage murein hydrolases determine specific recognition and high-affinity binding to bacterial cell wall carbohydrates. Mol Microbiol 2002; 44(2): 335-49.

[26]   Yoong P, Schuch R, Nelson D, Fischetti VA. Identification of a broadly active phage lytic enzyme with lethal activity against antibiotic-resistant *Enterococcus faecalis* and *Enterococcus faecium*. Journal of Bacteriology 2004; 186(14): 4808-12.

[27]   Porter CJ, Schuch R, Pelzek AJ, Buyckle AM, McGowan S, Wilce MC, *et al.* The 1.6 A Crystal Structure of the Catalytic Domain of PlyB, a Bacteriophage Lysin Active Against *Bacillus anthracis.* J Mol Biol 2007; 366(2): 540-50.

[28]   Cheng Q, Nelson D, Zhu S, Fischetti VA. Removal of Group B Sttreptococci Colonizing the Vagina and Oropharynx of Mice with a Bacteriophage Lytic Enzyme. Antimicrobial Agents and Chemotherapy 2005; 49(1): 111-117.

[29]   Daniel A, Euler C, Collin M, Chahales P, Goerelick KJ, Fischetti VA. Synergism between a novel chimeric lysin and oxacillin protects against infection by methicillin-resistant *Staphylococcus aureus.* Antimicrob Agents Chemother 2010; 54(4): 1603-12.

[30]   Baba T, Schneewind O. Target cell specificity of a bacteriocin molecule: a C-terminal signal directs lysostaphin to the cell wall of *Staphylococcus aureus.* EMBO J 1996; 15(18): 4789-97.

[31]   Kusuma CM, Kokai-Kun JF. Comparison of four methods for determining lysostaphin susceptibility of various strains of *Staphylococcus aureus.* Antimicrob Agents Chemother 2005; 49(8): 3256-63.

[32]   Kiri N, Archer G, Climo MW. Combinations of lysostaphin with beta-lactams are synergistic against oxacillin-resistant *Staphylococcus epidermidis.* Antimicrob Agents Chemother 2002; 46(6): 2017-20.

[33]   Desbois AP, Coote PJ. Bactericidal synergy of lysostaphin in combination with antimicrobial peptides. Eur J Clin Microbiol Infect Dis 2011; 30(8): 1015-21.

[34]   Kumar JK. Lysostaphin: an antistaphylococcal agent. Appl Microbiol Biotechnol 2008; 80(4): 555-61.

[35]   Wu JA, Kusuma C, Mond JJ, Kokai-Kun JF. Lysostaphin disrupts *Staphylococcus aureus* and Staphylococcus epidermidis biofilms on artificial surfaces. Antimicrob Agents Chemother 2003; 47(11): 3407-14.

[36]   Kokai-Kun JF, Chanturiya T, Mond JJ. Lysostaphin eradicates established *Staphylococcus aureus* biofilms in jugular vein catheterized mice. J Antimicrob Chemother 2009; 64(1): 94-100.

[37]   Belyansky I, Tsirline VB, Montero PN, Satishkumar R, MArtin TR, Lincourt AE, *et al.* Lysostaphin-coated mesh prevents staphylococcal infection and significantly improves survival in a contaminated surgical field. Am Surg 2011; 77(8): 1025-31.

[38]   Belyansky I, Tsirline VB, MArtin TR, Klima DA, Heath KLincourt AE, *et al.* The addition of lysostaphin dramatically improves survival, protects porcine biomesh from infection, and improves graft tensile shear strength. J Surg Res 2011; 171(2): 409-15.

[39]   Kokai-Kun JF, Chanturiya T, Mond JJ. Lysostaphin as a treatment for systemic *Staphylococcus aureus* infection in a mouse model. J Antimicrob Chemother 2007; 60(5): 1051-9.

[40]   Sabala I, Jonsson IM, Tarkowski A, Bochtler M. Anti-staphylococcal activities of lysostaphin and LytM catalytic domain. BMC Microbiol 2012; 12(1): 97.

[41]   Climo MW, Ehlert K, Archer GL. Mechanism and suppression of lysostaphin resistance in oxacillin-resistant *Staphylococcus aureus*. Antimicrob Agents Chemother 2001; 45(5): 1431-7.

[42]   Kusuma C, Jadanova A, Chanturiya T, Kokai-Kun JF. Lysostaphin-resistant variants of *Staphylococcus aureus* demonstrate reduced fitness *in vitro* and *in vivo*. Antimicrob Agents Chemother 2007; 51(2): 475-82.

[43]   Stranden AM, Ehlert K, Labischinski H, Berger-Bachi B. Cell wall monoglycine cross-bridges and methicillin hypersusceptibility in a femAB null mutant of methicillin-resistant *Staphylococcus aureus*. J Bacteriol 1997; 179(1): 9-16.

[44]   Rodriguez L, Martinez B, Zhou Y, Rodriguez A, Donovan DM, Garcia P. Lytic activity of the virion-associated peptidoglycan hydrolase HydH5 of Staphylococcus aureus bacteriophage vB_SauS-phiIPLA88. BMC Microbiol 2011; 11: 138.

[45]   Ackermann HW. Frequency of morphological phage descriptions in the year 2000. Brief review. Arch Virol 2001; 146(5): 843-57.

[46]   Brussow H, Hendrix RW. Phage genomics: small is beautiful. Cell 2002; 108(1): 13-6.

[47]   Thurber RV. Current insights into phage biodiversity and biogeography. Curr Opin Microbiol 2009; 12(5): 582-7.

[48]   Wang IN, Smith DL, Young R. Holins: the protein clocks of bacteriophage infections. Annu Rev Microbiol 2000; 54: 799-825.

[49]   Diaz E, Lopez R, Garcia JL. Chimeric phage-bacterial enzymes: a clue to the modular evolution of genes. Proc Natl Acad Sci U S A 1990; 87(20): 8125-9.

[50]   Becker SC, Dong S, Baker JR, Foster-Frey J, Pritchard DG, Donovan DM. LysK CHAP endopeptidase domain is required for lysis of live staphylococcal cells. FEMS Microbiol Lett 2009; 294(1): 52-60.

[51]   Low LY, Yang C, Perego M, Osterman A, Liddington R. Role of net charge on catalytic domain and influence of cell wall binding domain on bactericidal activity, specificity, and host range of phage lysins. J Biol Chem 2011; 286(39): 34391-403.

[52]   Fenton M, Casey PG, Hill C, Gahan CG, Ross RP, McAuliffe O, *et al.* The truncated phage lysin CHAP(k) eliminates *Staphylococcus aureus* in the nares of mice. Bioeng Bugs 2010; 1(6): 404-7.

[53]   Loeffler JM, Nelson D, Fischetti VA. Rapid killing of *Streptococcus pneumoniae* with a bacteriophage cell wall hydrolase. Science 2001; 294(5549): 2170-2.

[54]   Nelson D, Loomis L, Fischetti VA. Prevention and elimination of upper respiratory colonization of mice by group A streptococci by using a bacteriophage lytic enzyme. Proc Natl Acad Sci USA 2001; 98(7): 4107-4112.

[55]   Djurkovic S, Loeffler JM, Fischetti VA. Synergistic killing of *Streptococcus pneumoniae* with the bacteriophage lytic enzyme Cpl-1 and penicillin or gentamicin depends on the level of penicillin resistance. Antimicrob Agents Chemother 2005; 49: 1225-1228.

[56]   Schuch R, Nelson D, Fischetti VA. A bacteriolytic agent that detects and kills *Bacillus anthracis*. Science 2002; 418: 884-889.

[57]   Daniel A, Euler C, Collin M, Chahales P, Gorelick KJ, Fishcetti VA. Synergism between a novel chimeric lysin and oxacillin protects against infection by methicillin-resistant *Staphylococcus aureus*. Antimicrob Agents Chemother 2010; 54(4): 1603-1612.

[58]   Yoong P, Schuch R, Nelson D, Fischetti VA. Identification of a broadly active phage lytic enzyme with lethal activity against antibiotic-resistant *Enterococcus faecalis* and *Enterococcus faecium*. Journal of Bacteriology 2004; 186: 4808-4812.

[59]   Garcia P, Garcia JL, Garcia E, Sanchez-Puelles JM, Lopez R. Modular organization of the lytic enzymes of *Streptococcus pneumoniae* and its bacteriophages. Gene 1990; 86(1): 81-8.

[60]   Fernandes S, Proenca D, Cantante C, Silva FA, Leandro C, Lourenceo S, *et al*. Novel Chimerical Endolysins with Broad Antimicrobial Activity Against Methicillin-Resistant *Staphylococcus aureus*. Microb Drug Resist 2012; 18(3): 333-43.

[61]   Pastagia M, Euler C, Chahales P, Fuentes-Duculan K, Krueger JG, Fischetti VA. A novel chimeric lysin shows superiority to mupirocin for skin decolonization of methicillin-resistant and -sensitive *Staphylococcus aureus strains*. Antimicrob Agents Chemother 2011; 55(2): 738-44.

[62]   Schmelcher M, Tchang VS, Loessner MJ. Domain shuffling and module engineering of Listeria phage endolysins for enhanced lytic activity and binding affinity. Microb Biotechnol 2011; 4(5): 651-62.

[63]   Loeffler JM, Djurkovic S, Fischetti VA. Phage lytic enzyme Cpl-1 as a novel antimicrobial for pneumococcal bacteremia. Infect Immun 2003; 71(11): 6199-204.

[64]   Resch G, Moreillon P, Fischetti VAF. PEGylating a bacteriophage endolysin inhibits its bactericidal activity. AMB Express 2011; 1: 29.

[65]   Aguinaga A, Frances ML, Del Pozo JL, Alonso M, Serrera A, Lasa I, *et al*. Lysostaphin and clarithromycin: a promising combination for the eradication of *Staphylococcus aureus* biofilms. Int J Antimicrob Agents 2011; 37(6): 585-7.

[66]   Domenech M, Garcia E, Moscoso M. *In vitro* destruction of *Streptococcus pneumoniae* biofilms with bacterial and phage peptidoglycan hydrolases. Antimicrob Agents Chemother 2011; 55(9): 4144-8.

[67]   Yother J, Leopold K, White J, Fischer W. Generation and properties of a *Streptococcus pneumoniae* mutant which does not require choline or analogs for growth. J Bacteriol 1998; 180(8): 2093-101.

[68]   Djurkovic S, Loeffler JM, Fischetti VAF. Synergistic killing of *Streptococcus pneumoniae* with the bacteriophage lytic enzyme Cpl-1 and penicillin or gentamicin depends on the level of penicillin resistance. Antimicrob Agents Chemother 2005; 49(3): 1225-8.

[69]   Rashel M, Uchiyama J, Ujihara T, Uehara Y, Kuramoto S, Sugihara S, *et al*. Efficient elimination of multidrug-resistant *Staphylococcus aureus* by cloned lysin derived from bacteriophage phi MR11. J Infect Dis 2007; 196(8): 1237-47.

[70]   Loeffler JM, Fischetti VAF. Synergistic lethal effect of a combination of phage lytic enzymes with different activities on penicillin-sensitive and -resistant *Streptococcus pneumoniae* strains. Antimicrob Agents Chemother 2003; 47(1): 375-7.

[71]   Stewart PS. Mechanisms of antibiotic resistance in bacterial biofilms. Int J Med Microbiol 2002; 292(2): 107-13.

[72]   Parsek MR, Singh PK. Bacterial biofilms: an emerging link to disease pathogenesis. Annu Rev Microbiol 2003; 57: 677-701.

[73]   Walencka E, Sadowska B, Rozalska S, Hryniewicz W, Rozalska B. Lysostaphin as a potential therapeutic agent for staphylococcal biofilm eradication. Pol J Microbiol 2005; 54(3): 191-200.

[74]   Entenza JM, Loeffler JM, Grandgirard D, Fischetti VA, Moreillon P. Therapeutic effects of bacteriophage Cpl-1 lysin against *Streptococcus pneumoniae* endocarditis in rats. Antimicrob Agents Chemother 2005; 49(11): 4789-92.

[75]   Schuch R, Fischetti VA, Nelson DC. A genetic screen to identify bacteriophage lysins. Methods Mol Biol 2009; 502: 307-19.

[76]   Jado I, Lopez R, Garcia E, Fenoll A, Casal J, Garcia P. Phage lytic enzymes as therapy for antibiotic-resistant *Streptococcus pneumoniae* infection in a murine sepsis model. J Antimicrob Chemother 2003; 52(6): 967-73.

[77]   Grandgirard D, Loeffler JM, Fischetti VA, LEib SL. Phage lytic enzyme Cpl-1 for antibacterial therapy in experimental pneumococcal meningitis. J Infect Dis 2008; 197(11): 1519-22.

[78]   McCullers JA, Karlstrom A, IVerson AR, Loeffler JM, Fischetti VA. Novel strategy to prevent otitis media caused by colonizing *Streptococcus pneumoniae*. PLoS Pathog 2007; 3(3): e28.

[79]   Yoong P, Schuch R, Nelson D, Fischetti VA. PlyPH, a bacteriolytic enzyme with a broad pH range of activity and lytic action against *Bacillus anthracis*. J Bacteriol 2006; 188(7): 2711-4.

[80]   Schmelcher M, Shabarova T, Eugster MR, Eichenseher F, Tchang VS, Banz M, *et al.* Rapid multiplex detection and differentiation of listeria cells by use of fluorescent phage endolysin cell wall binding domains. Appl Environ Microbiol 2010; 76(17): 5745-56.

[81]   Salazar O, Asenjo JA. Enzymatic lysis of microbial cells. Biotechnol Lett 2007; 29(7): 985-94.

[82]   Klesius PH, Schuhardt VT. Use of lysostaphin in the isolation of highly polymerized deoxyribonucleic acid and in the taxonomy of aerobic Micrococcaceae. J Bacteriol 1968; 95(3): 739-43.

[83]   Schmitz JE, Daniel A, Collin M, Schuch R, Fischetti VA. Rapid DNA library construction for functional genomic and metagenomic screening. Appl Environ Microbiol 2008; 74(5): 1649-1652.

[84]   Schmitz JE, Ossiprandi MC, Rumah KR, Fischetti VA. Lytic enzyme discovery through multigenomic sequence analysis in *Clostridium perfringens*. Appl Microbiol Biotechnol 2011; 89(6): 1783-95.

[85]   Hagens S, Loessner MJ. Application of bacteriophages for detection and control of foodborne pathogens. Appl Microbiol Biotechnol 2007; 76(3): 513-9.

[86]   Lang LH. FDA approves use of bacteriophages to be added to meat and poultry products. Gastroenterology 2006; 131(5): 1370.

[87]   Halasa T, Huilps K, Osteras O, Hogeveen H. Economic effects of bovine mastitis and mastitis management: a review. Vet Q 2007; 29(1): 18-31.

[88]   Zadoks RN, Middleton JR, McDougall S, Katholm J, Schukken YH. Molecular epidemiology of mastitis pathogens of dairy cattle and comparative relevance to humans. J Mammary Gland Biol Neoplasia 2011; 16(4): 357-72.

[89]   Schmelcher M, Powell AM, Becker SC, CAmp MJ, Donovan DM. Chimeric phage lysins act synergistically with lysostaphin to kill mastitis-causing *Staphylococcus aureus* in murine mammary glands. Appl Environ Microbiol 2012; 78(7): 2297-305.

[90]   O'Flaherty S, Coffey A, MEaney W, Fitzgerald GF, Ross RP. The recombinant phage lysin LysK has a broad spectrum of lytic activity against clinically relevant staphylococci, including methicillin-resistant *Staphylococcus aureus*. J Bacteriol 2005; 187(20): 7161-4.

[91]   von Eiff, C, Becker K, Machka K, Stammer H, Peters G. Nasal carriage as a source of *Staphylococcus aureus* bacteremia. Study Group. N Engl J Med 2001; 344(1): 11-6.

[92]   de Lencastre H, Kristinsson KG, Brito-Avo A, Sances IS, Sa-Leao R, Saldanha J, *et al.*, Carriage of respiratory tract pathogens and molecular epidemiology of *Streptococcus pneumoniae* colonization in healthy children attending day-care centers in Lisbon, Portugal. Microbiol Drug Resist 1999; 5(1): 19-29.

[93]   Hoopes JT, Stark CJ, Kim HA, Sussman DJ, Donovan DM, Nelson DC. Use of a bacteriophage lysin, PlyC, as an enzyme disinfectant against *Streptococcus equi*. Appl Environ Microbiol 2009; 75(5): 1388-94.

[94]   Shapiro OH, Kushmaro A. Bacteriophage ecology in environmental biotechnology processes. Curr Opin Biotechnol 2011; 22(3): 449-55.

[95]   Withey S, Cartmell E, Avery LM, Stephenson T. Bacteriophages--potential for application in wastewater treatment processes. Sci Total Environ 2005; 339(1-3): 1-18.

[96]   Smartt AE, Xu T, Jegier P, Carswell JJ, Blount SA, Sayler GS, *et al*. Pathogen detection using engineered bacteriophages. Anal Bioanal Chem 2012; 402(10): 3127-46.

[97]   Monk AB, Rees CD, Barrow P, Hagens S, Harper DR. Bacteriophage applications: where are we now? Lett Appl Microbiol 2010; 51(4): 363-9.

[98]   Kutateladze, Adamia R. Bacteriophages as potential new therapeutics to replace or supplement antibiotics. Trends Biotechnol 2010; 28(12): 591-5.

[99]   Burrowes B, Harper DR, Andereson J, McConville M, Enright MC. Bacteriophage therapy: potential uses in the control of antibiotic-resistant pathogens. Expert Rev Anti Infect Ther 2011; 9(9): 775-85.

[100]   Kutateladze M, Adamia R. Phage therapy experience at the Eliava Institute. Med Mal Infect 2008; 38(8): 426-30.

[101]   Witzenrath M, Schmeck B, Doehn JM, Tschernig T, Zahlten J, Leoffler JM, *et al*. Systemic use of the endolysin Cpl-1 rescues mice with fatal pneumococcal pneumonia. Crit Care Med, 2009; 37(2): 642-9.

Index

A

Acetylation 6, 282

Acinetobacter baumannii 8, 54, 237-8

Activity 5, 16, 34-8, 40-4, 46-7, 59, 62, 64-5, 85, 161-2, 241-2, 248-50, 315-17, 319-20, 325

  bacteriolytic 323

  biological 76

  broad spectrum 39-40, 42-3, 46, 272

Adjuvants 167-8, 203-4, 213, 217

African green monkey kidney cells 93, 110, 116, 123

Alfalfa 206, 209, 214-17, 219

Alkylation 6

AMCA 116, 123

Amidation 6

Amino acid (AA) 6-7, 10, 13-14, 24, 73, 78, 123, 125, 127, 170-2, 175, 179, 244-5, 272, 281

Aminoglycosides 32, 34, 36-7, 71, 239, 269, 271, 281, 287

Ampicillin 34-5, 240, 250, 283, 286, 288, 290-2, 294-5

AMPs 5-7, 12

Anti-endotoxic 9-10

Anti-infectious drugs 6

Anti-infective agents 158-9, 161, 163, 165, 167, 169, 171, 173, 175, 177, 179, 181, 183, 185, 187

Anti-LTB antibodies 212

Anti-microbial and anti-infective drugs 4

Antibacterial 3-7, 32-7, 35, 41-2, 44-5, 48, 54-5, 65, 67-8, 71, 75, 80-2, 129, 238-42, 246, 249, 251-2, 256, 256, 268, 271, 280-81, 308-15, 321-3, 321-2, 321, 327-31

Antibacterial and antiviral therapeutics 53

Antibodies 64, 108, 165, 167, 169, 174, 207, 212, 216-18, 223-4, 253, 325, 330

Antimicrobial 5, 11, 32-7, 39, 41-2, 49, 237-8, 247, 251, 254, 256, 263, 268-9, 272-3, 275, 277, 279, 282, 287-9, 291-5, 295, 314, 316

R

V

www.ingramcontent.com/pod-product-compliance
Lightning Source LLC
Chambersburg PA
CBHW050803220326
41598CB00006B/109